Lectures on Quantum Groups

by

Pavel Etingof and Olivier Schiffmann

International Press
www.intlpress.com

Lectures on Quantum Groups, Second Edition
by Pavel Etingof and Olivier Schiffmann

ISBN 978-1-57146-207-7

Paperback reissue 2010.
Previously published in 2002 under ISBN 1-57146-094-2 (clothbound).

Printed in the United States of America.

to our parents

Contents

Introduction

Quantum groups is a new exciting area of mathematics, which originated from mathematical physics (field theory, statistical mechanics), and developed greatly over the last 15 years. It is connected with many other, old and new, parts of mathematics, and remains an area of active, fruitful research today.

This book arose from a graduate course on quantum groups given by the first author at Harvard in the Spring of 1997, when it was written down in an extended and improved form by the second author.

The purpose of this book is to give an elementary introduction to the aspect of the theory of quantum groups which has to do with the notion of quantization. It is written for a general mathematical audience: we tried to do everything from scratch, assuming only the basic algebra and geometry.

The first seven lectures are devoted to the theory of quasiclassical objects which are relevant in the theory of quantum groups: Poisson manifolds (algebras), Poisson-Lie groups, Lie bialgebras, the classical Yang-Baxter equation and its solutions (classical r-matrices). The material here is largely standard. At the end of this part we consider in detail the classification of classical r-matrices for simple Lie algebras, given by Belavin and Drinfeld. Our exposition in Lectures 1-7 is similar to that of Chari and Pressley [CP].

In Lectures 8-12, we discuss the definition and properties of the main characters in our story – bialgebras and Hopf algebras. Here we discuss quantum R-matrices, the double construction, and the notion of quantization of Lie bialgebras. We formulate the results about existence of quantization, anticipated by Drinfeld [Dr1] and proved recently in [EK1].

In Lectures 13-14 we discuss monoidal categories. This material is standard, and contained in the book of Maclane [Mac], as well as in several textbooks on quantum groups. We give it in a form suitable for subsequent exposition. In particular, we stress the importance of non-symmetric and non-strict monoidal categories.

In Lectures 15-16 we discuss quasi-bialgebras and quasi-Hopf algebras, which are algebraic counterparts of non-strict monoidal categories, in the same sense as bialgebras and Hopf algebras are algebraic counterparts of strict monoidal categories. We consider the main properties of quasi-bialgebras, and the simplest examples of them. Then we study quasitriangular quasi-Hopf algebras, the Knizhnik-Zamilodchikov equation and the corresponding quasi-Hopf algebra, define and study equivalence by a twist, and cite Drinfeld's classification result. From this result, we deduce the Drinfeld-Kohno theorem about the monodromy of the Knizhik-Zamolodchikov equations.

In Lecture 17 we introduce Lie associators and the Grothendieck-Teichmuller group, give their main properties, and define, following Drinfeld, the free, transitive action of the Grothendieck-Teichmuller group on the space of Lie associators.

In Lecture 18 we discuss the Tannaka-Krein philosophy for tensor categories, which allows one to get a better understanding of the notion of a bialgebra and a quasitriangular bialgebra.

In Lectures 19-22 we describe the method of quantization of Lie bialgebras developed recently in [EK1, EK2]. This part is the culmination point of the book, where many methods and notions of the previous chapters come together. In this part, we prove that any Lie bialgebra can be quantized, and that this quantization is given by a universal, functorial construction.

Finally, in the Appendix we give some applications of the material of the book to number theory (counting independent values of zeta-functions). The contents of the Appendix is discussed somewhat differently in [Kass].

Now a few words about the nature of this book. It is written in the spirit of lecture notes rather than that of a serious monograph. Our goal was not to cover the maximal amount of material, nor to present it in the most complete form, but to expose a number of deep and interesting results in a reader-friendly way. In view of this, we did not discuss many important parts of the theory of quantum groups (representation theory, quantum groups at roots of unity, knot invariants, Drinfeld new realizations, relations to q-special functions, etc.), and did not give many basic references. Luckily, there exist many textbooks on quantum groups [Kass, CP, ShSt, Jos, J, Lu, Maj], where this missing information can be readily found.

Two unusual features of this book, compared to other textbooks, are extensive use of pictorial language for writing and checking algebraic relations, and over fifty problems and exercises (with solutions).
We hope that these features will facilitate active reading of the book, and make it accessible to a wide audience.

The authors would like to thank the Harvard mathematics department for hospitality and the participants of the quantum groups course in the Spring of 1997 for many useful discussions. They are very grateful to Ping Xu and

Eric Vasserot for careful reading of the manuscript and many helpful remarks. Above all, they are grateful to Tanya and Christelle for their endless patience and support.

Lecture 1

Poisson algebras and quantization

Throughout this lecture, k will be a field of characteristic zero and an associative algebra will mean an associative algebra over k with unity.

1.1 Modules over rings of power series

Let $K = k[[h]]$. We will be considering formal deformations, and we need to define a suitable category of K-modules.

1.1.1 Topologically free K-modules

First note that K carries a natural norm given by

$$\| a_n h^n + a_{n+1} h^{n+1} + \dots \| = C^{-n} \qquad (a_n \neq 0)$$

where $C > 1$ is any fixed constant (the *h-adic norm*), with respect to which it is complete. The topology defined by this norm is called the *h-adic topology*. It coincides with the topology of inverse limit on K, defined by the construction of K as $K = \varprojlim k[h]/h^n$.

Now let V be any vector space over k, and set

$$V[[h]] = \{\sum_{n \geq 0} v_n h^n \mid v_n \in V\}.$$

The *h-adic* norm and topology are defined on $V[[h]]$ in the same way as on K, and $V[[h]]$ is complete.

Definition: A topological K-module isomorphic to $V[[h]]$ for some k-vector space V is called a *topologically free* K-module.

Any morphism $f : V[[h]] \to W[[h]]$ between topologically free K-modules is continuous since $f(h^n v) = h^n f(v)$, and hence $\| f(v) \| \leq \| v \|$.

Remarks: (i)If M is a topologically free K-module, the space V can be reconstructed from M, up to isomorphism, as $V = M/hM$.
(ii)If V is finite dimensional, then $V \otimes_k K = V[[h]]$. Note that this is false if V is infinite dimensional (an element of $V \otimes_k K$ is a *finite* sum of terms of the form $v \otimes a$, $v \in V$, $a \in K$).
(iii)A morphism $\phi : V[[h]] \to W[[h]]$ is determined by its restriction to V.

1.1.2 Completion of K-modules

Let M be any K-module. The spaces $M_n = M/h^n M$ form a projective system, and we will call $\hat{M} = \varprojlim M_n$ the completion of M. The projections $M \to M_n$ induce a natural map $i : M \to \hat{M}$, which is not necessarily injective: we have $\mathrm{Ker}(i) = \cap_n h^n M$. A K-module will be called *separated* if $\cap_n h^n M = \{0\}$ (i.e if $M \hookrightarrow \hat{M}$), and *complete* if the map $i : M \to \hat{M}$ is surjective.

Exercise 1.1. Show that a torsion-free K-module is topologically free if and only if it is both complete and separated.

Exercise 1.2. If M and N are two torsion-free K-modules then

$$\widehat{M \otimes_K N} = \widehat{\hat{M} \otimes_K \hat{N}}.$$

When M, N are topologically free, we will simply denote this tensor product by $M \otimes N$. Thus when M and N are topologically free K-modules, $M \otimes N$ stands for the completed tensor product.

1.2 Poisson algebras

1.2.1 Definition

Definition: A commutative associative algebra A over k is called a *Poisson algebra* if it is equipped with a k-bilinear Lie bracket $\{,\} : A \otimes A \to A$ satisfying the *Leibniz Identity*:

$$\forall e, f, g \in A, \qquad \{ef, g\} = e\{f, g\} + \{e, g\}f.$$

By a map of Poisson algebras, we will mean a map preserving the Poisson bracket. If $(A, \{,\}_A)$ and $(B, \{,\}_B)$ are two Poisson algebras then their tensor product $A \otimes B$ is naturally a Poisson algebra, with the bracket

$$\{a \otimes b, a' \otimes b'\}_{A \otimes B} = \{a, a'\}_A \otimes bb' + aa' \otimes \{b, b'\}_B.$$

We now give a few examples.

1.2.2 Examples of Poisson algebras

Example 1.1 (The trivial Poisson structure). Any commutative associative algebra is a Poisson algebra for the bracket $\{,\} = 0$.

Example 1.2 (The symplectic plane). Let $A = k[x,p]$ be the algebra of polynomial functions on the plane. Define

$$\{f,g\} = \frac{\partial f}{\partial x}\frac{\partial g}{\partial p} - \frac{\partial f}{\partial p}\frac{\partial g}{\partial x}.$$

It is an easy exercise to verify that $(A, \{,\})$ is a Poisson algebra. This generalizes to $A = k[x_1, \dots x_n, p_1, \dots p_n]$ with

$$\{f,g\} = \sum_{i=1}^{n} (\frac{\partial f}{\partial x_i}\frac{\partial g}{\partial p_i} - \frac{\partial f}{\partial p_i}\frac{\partial g}{\partial x_i}).$$

Example 1.3 (Symplectic manifolds). We can generalize the last example to provide an important class of Poisson algebras. We let $k = \mathbb{R}$ here. Recall that a symplectic manifold is a smooth manifold M equipped with a nondegenerate closed 2-form ω. The simplest example is $\mathbb{R}^{2n}$ with coordinates $(x_i, p_i)_{i=1,\dots n}$ and with $\omega = \sum_i dx_i \wedge dp_i$. The algebra $\mathcal{C}^\infty(M)$ of smooth functions on M comes with a natural Poisson bracket which is defined in the following way: for $f \in \mathcal{C}^\infty(M)$, let V_f be the vector field defined by the condition $df(u) = \omega(u, V_f)$ for any vector field u. Now set

$$\{f,g\} = \omega(V_f, V_g).$$

It is easily seen that $(\mathcal{C}^\infty(M), \{,\})$ is a Poisson algebra (see, e.g [CG]). For example, if $M = \mathbb{R}^{2n}$, $\omega = \sum_i dx_i \wedge dp_i$, we have

$$V_f = \sum_i (\frac{\partial f}{\partial x_i}\frac{\partial}{\partial p_i} - \frac{\partial f}{\partial p_i}\frac{\partial}{\partial x_i}), \tag{1.1}$$

$$\{f,g\} = \sum_i (\frac{\partial f}{\partial x_i}\frac{\partial g}{\partial p_i} - \frac{\partial f}{\partial p_i}\frac{\partial g}{\partial x_i}). \tag{1.2}$$

Furthermore, Darboux's Theorem states that locally there is only one symplectic structure: for any point z of a symplectic manifold M we can find a system of local coordinates (x_i, p_i) in which $\omega = \sum_i dx_i \wedge dp_i$ (Darboux coordinates). In particular, the Poisson structure in this neighborhood is given by formula (1.2).

Example 1.4 (The dual of a Lie algebra). Let $\mathfrak{g}$ be a finite dimensional Lie algebra defined over k and let $\mathfrak{g}^*$ be its dual. The algebra $S\mathfrak{g}$ (polynomial functions on $\mathfrak{g}^*$) is equipped with the following Poisson bracket:

$$\{\theta, \theta'\}(f) = f([d\theta(f), d\theta'(f)]), \qquad \theta, \theta' \in S\mathfrak{g},\ f \in \mathfrak{g}^*, \tag{1.3}$$

where we identify $\mathfrak{g}^{**}$ and $\mathfrak{g}$ in the usual way. If $k = \mathbb{R}$, this Poisson bracket extends to $\mathcal{C}^\infty(\mathfrak{g}^*)$ in an obvious way.

We end this section with a problem.

Problem 1.1. Let $(A, \{,\})$ be a Poisson algebra with no zero divisors. Suppose that the transcendence degree of A over k is less than two. Show that the Poisson bracket $\{,\}$ is necessarily trivial.

1.3 Quantization of Poisson algebras

1.3.1 Deformations

Definition: A *deformation algebra* is a topologically free K-algebra, i.e it is a topologically free K-module A together with a K-bilinear (multiplication) map $A \times A \to A$ making A into an associative algebra.

Now let A_0 be an associative k-algebra. A deformation of A_0 is by definition a deformation algebra A such that $A_0 = A/hA$.

1.3.2 Quantization

Let us now restrict ourselves to the case of primary interest to us, when A_0 is commutative, but A not necessarily. In this case, A_0 inherits from A a natural Poisson structure, which is constructed as follows.

We will write $a * b$ for the multiplication in A to distinguish it from the multiplication in A_0.

Let $f_0, g_0 \in A_0$, and choose arbitrary liftings f, g to A. Since A_0 is commutative, we have $f * g - g * f \equiv 0 \pmod{h}$, and we set

$$\{f_0, g_0\} = \frac{1}{h}(f * g - g * f) \pmod{h}.$$

Notice that if $f' = f + hx$ and $g' = g + hy$ are two other liftings of f_0, g_0, then $\frac{1}{h}(f' * g' - g' * f') \equiv \frac{1}{h}(f * g - g * f) \pmod{h}$ so that $\{,\}$ is well defined. It is easy to check that $(A_0, \{,\})$ is a Poisson algebra. For instance, the Jacobi identity follows from the associativity of A.

Definition: The Poisson algebra $(A_0, \{,\})$ is called the *quasiclassical limit* of A. Conversely, A is called a *quantization* of A_0.

Notice that the notion of quantization is compatible with the notion of tensor product of Poisson algebras. In other words, if A is a quantization of A_0 and B a quantization of B_0, then $A \otimes B$ is a quantization of $A_0 \otimes B_0$.

Although quasiclassical limit and quantization are in some sense inverse to each other, there is an essential asymmetry between them. Namely, the quasiclassical limit of any deformation algebra is unique and well defined, while the question of existence of quantizations for a given Poisson algebra is a difficult problem (in particular, there exist Poisson algebras which admit no quantizations, cf [Ma]). Moreover, a quantization is in general non-unique. We shall see constructions of quantizations in several special cases.

Let us describe the notion of quantization more explicitly. Namely, identify A with $A_0[[h]]$ (as K-modules) and expand the product $f * g$ in a power series of h:

$$f * g = fg + hc_1(f,g) + h^2c_2(f,g) + \dots \qquad (f,g \in A_0) \tag{1.4}$$

where $c_i : A_0 \otimes A_0 \to A_0$. The associativity of $*$ is equivalent to the conditions

$$\sum_{i+j=n} c_j(c_i(e,f),g) = \sum_{k+l=n} c_k(e, c_l(f,g)) \qquad n = 1,2,\dots \tag{1.5}$$

where we set $c_0(f,g) = fg$. With these notations, $\{f_0, g_0\} = c_1(f,g) - c_1(g,f)$. Note that $*$ on A is completely determined by its restriction to A_0.

This shows that we can think of a deformation of A_0 as an infinite sequence of maps $c_i : A_0 \otimes A_0 \to A_0$ satisfying the associativity conditions (1.5).

Remarks: i)Observe that if $*$ is associative, A automatically has a unit. Namely, it is the unique invertible solution of the equation $x^2 = x$ (but may be different from the unit of A_0).
ii)If $c_1, c_2, \dots c_{n-1}$ identically vanish (or are symmetric), then the bilinear map $\{f_0, g_0\}_n = c_n(f,g) - c_n(g,f)$ is a Poisson structure on A_0.
iii)The obstruction to quantization of A_0 is described by the space $H^3(A_0, A_0)$- see [G] for more details.

Problem 1.2. Let A_0 be a commutative algebra without zero divisors with transcendence degree over k less than two. Show that any deformation of A_0 is commutative.

1.3.3 Examples of quantization

Example 1.5 (The trivial Poisson structure). Let $(A_0, \{,\} = 0)$ be a trivial Poisson algebra. Any deformation A of A_0 with $c_1 = 0$ is a quantization (for example, take any deformation, and set $h \mapsto h^2$). This shows that quantization is in general not unique.

Example 1.6 (Moyal-Weyl quantization). Let $A_0 = k[x,p]$ with Poisson structure as in example 1.2. Set

$$\begin{aligned} f * g &= m(e^{\frac{1}{2}h(\frac{\partial}{\partial x}\otimes\frac{\partial}{\partial p} - \frac{\partial}{\partial p}\otimes\frac{\partial}{\partial x})} f \otimes g) \\ &= \sum_{n \geq 0} \frac{h^n}{2^n n!} m((\frac{\partial}{\partial x} \otimes \frac{\partial}{\partial p} - \frac{\partial}{\partial p} \otimes \frac{\partial}{\partial x})^n f \otimes g) \end{aligned}$$

where $m : A_0 \otimes A_0 \to A_0$ is the multiplication.

Exercise 1.3. Check that this defines an associative multiplication, and that it gives a quantization of $(A_0, \{,\})$.

The generalization of this construction to $k[x_1, \dots x_n, p_1, \dots p_n]$ is given by

$$f * g = m(e^{\frac{h}{2}\sum_i (\frac{\partial}{\partial x_i} \otimes \frac{\partial}{\partial p_i} - \frac{\partial}{\partial p_i} \otimes \frac{\partial}{\partial x_i})} f \otimes g)$$

Example 1.7 (Symplectic manifolds). As we have seen, the algebra $C^\infty(M)$ of smooth functions on a symplectic manifold M is naturally endowed with a Poisson structure $\{,\}$. A quantization of the symplectic manifold M is by definition a quantization of $(C^\infty(M), \{,\})$. Furthermore, we will call a quantization *local* if the maps $c_i : C^\infty(M) \times C^\infty(M) \to C^\infty(M)$ are bidifferential operators. The existence of quantization of symplectic manifolds is settled by the following theorem, which was proved by Lecomte, DeWilde [DeLe] and later by different methods by Maeda, Omori and Yoshioka [MOY] and Fedosov [Fed].

Theorem 1.1. *Any symplectic manifold admits a local quantization.*

We now give a construction of a quantization in a special case (of importance in physics): the cotangent bundle T^*X of a smooth manifold X.

Recall the canonical symplectic structure on T^*X: let $\pi : T^*X \to X$ be the projection, and $d\pi : T_{x,p}(T^*X) \to T_xX$ be its differential. Consider the 1-form η on T^*X given by $\eta_{x,p}(v) =< p, d\pi(v) >$, and set $\omega = d\eta$. The 2-form ω is closed and nondegenerate.
Let A_0 be the algebra of smooth functions on T^*X whose restriction to fibers of π are polynomials of uniformly bounded degree. A_0 inherits a Poisson structure from $C^\infty(T^*X)$. It is easy to see that A_0 is graded, $A_0 = \bigoplus_{n\geq 0} A_0^n$, with $A_0^n = \Gamma(S^nTX)$ (polynomial functions on T^*X of degree n).
The fundamental idea in the construction of a quantization of T^*X is to use the algebra of differential operators on X. Recall Grothendieck's inductive definition of differential operators on a commutative algebra B:

$$\begin{array}{lll} D_0 & = \{m_b : B \to B,\ x \mapsto bx\} & \text{(diff. op. of deg. 0)} \\ D_1 & = \{d : B \to B \mid \forall b \in B,\ [d, m_b] \in D_0\} & \text{(diff. op. of deg. 1),} \\ \vdots & = \vdots \quad \vdots & \vdots \\ D_n & = \{d : B \to B \mid \forall b \in B,\ [d, m_b] \in D_{n-1}\} & \text{(diff. op. of deg. n).} \end{array}$$

It is important to notice that $B \simeq D_0 \subset D_1 \dots \subset D_n \subset \dots$ and that we have $D_iD_j \subset D_{i+j}$. Thus $D_B = \cup_{i\geq 0} D_i$ is a filtered algebra (however, it is in general not graded, i.e there is no splitting $D_n = D_{n-1} \oplus C_n$ such that $C_nC_m \subset C_{n+m}$). Consider the case $B = C^\infty(X)$. The above definition then coincides with the usual notion of differential operators on a smooth manifold. In particular, the maps $\sigma_i : D_i \to D_i/D_{i-1}$ are the *principal symbol* maps and we have an isomorphism $D_i/D_{i-1} \simeq \Gamma(S^iTX) = A_0^i$. This shows that the graded algebra $Gr(D)$ of the filtered algebra D is isomorphic to A_0.

Let us consider the following topologically free algebra

$$A = (\bigoplus_{n\geq 0} h^n D_n)[[h]] = \{d_0+hd_1+h^2d_2+\dots \mid i \geq ord(d_i),\ \ i-ord(d_i) \underset{i\to\infty}{\longrightarrow} \infty\}$$

(with topology defined by $\| h^n d \| = C^{ord(d)-n}$) and the following map $\phi : A \to A_0$:

$$\phi(d_0 + hd_1 + h^2 d_2 + \dots) = \sigma_0(d_0) + \sigma_1(d_1) + \sigma_2(d_2) + \dots$$

Notice that the sum on the r.h.s is finite since $ord(d_i) < i$ for almost all i, so that ϕ is well defined. Furthermore, notice that, since $Gr(D) = A_0$ is commutative, we have $[D_i, D_j] \subset D_{i+j-1}$, and so ϕ is an algebra morphism. Finally, we have $\mathrm{Ker}(\phi) = hA$, so that we can view A as a deformation of A_0.

To verify that this is indeed a quantization of the Poisson structure on T^*X, we notice that it is enough to check it on A_0^1, which generates A_0 over $C^\infty(X)$, and on which it is obvious.

The following problem can be solved using the above ideas.

Problem 1.3. Let $L \in D(X)$ be a differential operator on a connected smooth affine algebraic curve X over a field k of characteristic zero. Let $Z(L) \subset D(X)$ be the centralizer of L. Show that if $L \notin k$, $Z(L)$ is commutative (this is proved in [A]).

Generalization: X is a connected smooth affine algebraic variety of dimension m, $L_1, \dots L_m \in D_X$ are algebraically independent, pairwise commuting differential operators on X. Then $Z(L_1, \dots L_m)$ is commutative ([ML]; see alo [BEG]).

1.3.4 Loss of symmetry in quantization

Let M be a symplectic manifold. The group SDiff(M) of symplectic diffeomorphisms of M acts on $C^\infty(M)$, and preserves the Poisson bracket. We have an embedding Φ_0 : SDiff$(M) \hookrightarrow \mathrm{Aut}(A_0)$. One might ask if this can be extended to a quantization. The (negative) answer is given by the following

Theorem 1.2 (Groenwald- van Hove). *Let A be a quantization of $A_0 = C^\infty(M)$. There is no homomorphism*

$$\Phi : \mathrm{SDiff}(M) \to \mathrm{Aut}(A)$$

such that $\Phi \equiv \Phi_0 \pmod h$.

This theorem means that there is breaking of symmetry in the process of quantization.

In particular, there is no functor from the category of symplectic manifolds (with morphisms given by symplectic diffeomorphisms) to the category of associative K-algebras, which assigns to any symplectic manifold a quantization of this manifold. However, we will see at the end of these lectures that such a functor does exist in the case of *Poisson-Lie groups*, and which we will define in the next lecture.

Remark. For a discussion of the Groenwald- van Hove theorem for symplectic plane, see [GuSt].

1.4 Poisson manifolds and quantization

1.4.1 Definition

By definition, a Poisson manifold is a smooth manifold M with a Poisson structure on $C^\infty(M)$. For any function $f \in C^\infty(M)$, the map $\{f, .\} : C^\infty(M) \to C^\infty(M)$, $g \mapsto \{f, g\}$ is a derivation, so it can be written $\{f, g\} =< V_f, dg >$ for some vector field V_f. Such a vector field is called Hamiltonian. In particular, $\{f, g\}$ only depends on $df \wedge dg$, and there exists a *Poisson bivector field* $\Pi \in \Gamma(\Lambda^2 TM)$ uniquely defined by

$$\{f, g\} = df \otimes dg(\Pi). \tag{1.6}$$

Conversely, an element $\Pi \in \Gamma(\Lambda^2 TM)$ defines by (1.6) a Poisson bracket if and only if it satisfies a certain nonlinear differential equation (coming from the Jacobi identity), which can be written in local coordinates in which we set $\Pi = \sum_{i,j} \Pi_{i,j} \frac{\partial}{\partial x_i} \wedge \frac{\partial}{\partial x_j}$ as

$$\forall i, j, k, \qquad \sum_r (\Pi_{r,i} \frac{\partial \Pi_{j,k}}{\partial x_r} + \Pi_{r,j} \frac{\partial \Pi_{k,i}}{\partial x_r} + \Pi_{r,k} \frac{\partial \Pi_{i,j}}{\partial x_r}) = 0.$$

Remark: the above equation can be rephrased using the *Schouten bracket*. By definition, the Schouten bracket of $\Pi, \Pi' \in \Gamma(\Lambda^2 TM)$ is

$$[\Pi_1, \Pi_2]_s = [\Pi_{12}, \Pi'_{13}] + [\Pi_{12}, \Pi'_{23}] + [\Pi_{13}, \Pi'_{23}] \in \Gamma(\Lambda^3 TM)$$

with the usual Lie algebra structure of $\Gamma(TM)$. Thus, an element $\Pi \in \Gamma(\Lambda^2 TM)$ defines a Poisson bracket if and only if $[\Pi, \Pi]_s = 0$.

In the particular case of the natural Poisson structure on a symplectic manifold with a nondegenerate, closed 2-form $\omega \in \Gamma(\Lambda^2 T^* M)$, we get an identification $\Lambda^2 T_x^* M \stackrel{\sim}{\to} \Lambda^2 T_x M$ under which ω goes to the Poisson bivector Π. Conversely, one can check that a nondegenerate Poisson bivector on M induces a symplectic structure on M (via the identification $\Lambda^2 T_x M \stackrel{\sim}{\to} \Lambda^2 T_x^* M$).

The notions of a map of Poisson manifolds, direct product of Poisson manifolds are obvious analogs of the corresponding notions for Poisson algebras. By definition, a submanifold N of a Poisson manifold M is a Poisson submanifold if $\Pi_{|N} \in \Gamma(\Lambda^2 TN)$.

1.4.2 Symplectic leaves of a Poisson manifold

Let M be a Poisson manifold of dimension $2k$ or $2k+1$, with Poisson bivector Π. For $x \in M$, let $T_x^\Pi \subset T_x M$ be the subspace spanned by the components of Π (i.e if $\Pi = \sum_i \Pi_i^1 \otimes \Pi_i^2$ is an irreducible expression, then $T_x^\Pi = \mathrm{Span}(\Pi_i^1) = \mathrm{Span}(\Pi_i^2)$, or, alternatively, $T_x^\Pi = \mathrm{Span}(f \otimes 1(\Pi)) = \mathrm{Span}((1 \otimes f)(\Pi))$ where f runs over $T_x M^*$). We will call the dimension d_x of T_x^Π the rank of the Poisson structure at x. It is an even integer.

Now let $M_{2l} = \{x \in M \mid d_x \leq 2l\}$. We have a stratification

$$M = M_{2k} \supset M_{2k-2} \supset \dots$$

by closed subsets M_{2l} which are in general singular.

Nevertheless, it is possible to decompose M as a disjoint union of (immersed) submanifolds, each bearing a symplectic structure (the *symplectic leaves*).

To do this, let us introduce the following equivalence relation between points of M: $x \sim y$ if there exists a smooth path $\gamma : [0,1] \to M$ with $\gamma(0) = x$, $\gamma(1) = y$ and such that $\gamma'(t) \in T^\Pi_{\gamma(t)}$ and $\gamma'(t) \neq 0$ for $t \in [0,1]$. Such paths are called Hamiltonian paths.

Definition: A symplectic leaf is an equivalence class for this relation, i.e it is of the form

$$M_x = \{y \in M \text{ such that there exists a Hamiltonian path } x \mapsto y\}.$$

We have $M = \cup_x M_x$. One can show that M_x has a natural structure of an immersed submanifold of dimension d_x. Furthermore, since $T_x M_x = T_x^\Pi$, the restriction of Π to M_x is nondegenerate, and hence endows M_x with a symplectic structure (notice however that M_x is not in general a submanifold of M with the induced topology: a typical example of what can go wrong is the 2-dimensional dense winding around a 4-torus:

$$M = \mathbb{R}^4/\mathbb{Z}^4 \qquad \Pi(x,y,w,z) = \frac{\partial}{\partial x} \wedge \frac{\partial}{\partial y} + \sqrt{2}\frac{\partial}{\partial w} \wedge \frac{\partial}{\partial z}.$$

in which M_x is not locally closed).

We obtain in this way a foliation of the smooth manifolds $M_{2i} \backslash M_{2i-2}$ into symplectic leaves of rank $2i$.

Generic points: A point $x \in M$ is called generic if the rank of Π at x is locally constant. The foliation into symplectic leaves around x behaves like a direct product of a trivial Poisson manifold with a symplectic manifold; more precisely, we can find local coordinates (x_i, p_i, z_j) near x in which

$$\Pi = \sum_i \frac{\partial}{\partial x_i} \wedge \frac{\partial}{\partial p_i}.$$

and the symplectic leaves are (locally) defined by z_j =constant.

Example (the dual of a Lie algebra): We let $k = \mathbb{R}$ here. Let $\mathfrak{g}$ be a finite dimensional Lie algebra, and $\mathfrak{g}^*$ its dual space. Endow $\mathcal{C}^\infty(\mathfrak{g}^*)$ with the Poisson structure defined by (1.3). Let us now describe the symplectic leaves of $\mathfrak{g}^*$. Let us choose a basis $\theta_1, \ldots \theta_n$ of $\mathfrak{g}$. The Hamiltonian fields V_{θ_i} on $\mathfrak{g}^*$ are given by

$$V_{\theta_i}(f)(\theta) = f(ad_{\theta_i}(\theta)) = -ad^*_{\theta_i}(f)(\theta) \qquad f \in \mathfrak{g}^*, \theta \in \mathfrak{g}. \tag{1.7}$$

Introduce a connected Lie group G such that Lie$(G) = \mathfrak{g}$, and let us write Ad and Ad^* for the adjoint and coadjoint action of G on $\mathfrak{g}$ and $\mathfrak{g}^*$ respectively. Then by (1.7), the integral curve $\gamma(t)$ of V_{θ_i} going through f at $t = 0$ is nothing but

$$\gamma(t) = Ad^*(exp(-t\theta_i))(f). \tag{1.8}$$

But G is generated by the one-parameter subgroups $exp(-t\theta_i)$, and any Hamiltonian vector field is of the form $V_g = \sum_i c_i V_{\theta_i}$ where $c_i \in \mathcal{C}^\infty(\mathfrak{g}^*)$. Hence we see that the symplectic leaves of $\mathfrak{g}^*$ are precisely the coadjoint orbits. This implies that coadjoint orbits carry a natural symplectic structure, as was discovered by Kirillov, Kostant and Sourieau.

1.4.3 Quantization of Poisson manifolds

By definition, a quantization of a Poisson manifold M is a quantization of the Poisson algebra $\mathcal{C}^\infty(M)$. The question of existence of quantization has been recently settled by Kontsevich [Ko]:

Theorem 1.3 (Kontsevich). *Any Poisson manifold admits a local quantization.*

1.4.4 Example of quantization of a Poisson manifold (Geometric quantization)

Let $\mathfrak{g}$ be a finite dimensional Lie algebra, $\mathfrak{g}^*$ its dual, $A_0 = S\mathfrak{g}$ with its natural Poisson structure. We describe a general construction reminiscent of the quantization of T^*X, and then apply it to $\mathfrak{g}^*$.

Let $A_0 = (\bigoplus_{n\geq 0} A_0^n, \{,\})$ be a graded Poisson algebra:

$$\{,\} : A_0^n \times A_0^m \to A_0^{n+m-1}$$

and $\tilde{A} = \bigcup_{n\geq 0} \tilde{A}^n$ a filtered associative algebra such that $Gr(\tilde{A}) = A_0$ as commutative algebras. Suppose that

$$\sigma_{i+j-1}([\tilde{a},\tilde{b}]) = \{\sigma_i(\tilde{a}), \sigma_j(\tilde{b})\} \qquad \forall\, \tilde{a} \in \tilde{A}^i,\ \tilde{b} \in \tilde{A}^j \tag{1.9}$$

where $\sigma_i : \tilde{A}^i \to \tilde{A}^i/\tilde{A}^{i-1} \simeq A_0^i$. Define A to be the following topologically free algebra:

$$(\bigoplus_{n\geq 0} h^n \tilde{A}^n)[[h]] = \{a_0 + ha_1 + h^2 a_2 + \dots \mid i \geq deg(a_i),\ i - deg(d_i) \underset{i\to\infty}{\longrightarrow} \infty\}. \tag{1.10}$$

Then A is a quantization of A_0. Indeed, consider the map $\phi : A \to A_0$

$$\phi(a_0 + a_1 h + a_2 h^2 + \dots) = \sigma_0(a_0) + \sigma_1(a_1) + \sigma_2(a_2) + \dots$$

where $\sigma_i : \tilde{A}^i \to \tilde{A}^i/\tilde{A}^{i-1} \simeq A_0^i$ is the projection. Notice that the sum on the r.h.s is finite. Then ϕ is a morphism of algebras and $\mathrm{Ker}(\phi) = hA$, so A is a deformation of A_0. Furthermore, the Poisson bracket $\{,\}$ on A_0 induced from A coincides with $\{,\}_{A_0}$, as is easily seen from (1.9).

In our case, we have $A_0 = \bigoplus_n S^n\mathfrak{g}$, and we can take $\tilde{A} = U(\mathfrak{g})$, the universal enveloping algebra of $\mathfrak{g}$. Notice that the canonical quantization of T^*X is obtained in exactly the same way, with $A_0 = C^\infty_{pol}(T^*X)$ (smooth functions on T^*X, which are polynomial on the fibers of $\pi : T^*X \to X$) and $\tilde{A} = D_X$ (differential operators on X). In fact, polynomial functions on $\mathfrak{g}^*$ are nothing more than, say, *left-invariant* functions on T^*G which are polynomial on fibers of $T^*G \to G$, and $U(\mathfrak{g})$ is the algebra of *left-invariant* differential operators on G, so our construction here coincides with the construction of canonical quantization, restricted to left-invariant functions.

Remark: In both the canonical and the geometric quantization, the formulas for the $*$-product (i.e for the functions $c_i(f,g)$) are local: they are given by bidifferential operators acting on polynomials. Hence, when the ground field is $\mathbb{R}$ or $\mathbb{C}$, these quantizations automatically extend to smooth functions.

1.5 Rational forms of a quantization

We have considered only formal deformations of k-algebras (over $K = k[[h]]$). For some applications, and in particular to give a numerical value to the deformation parameter h, we need "rational forms", or deformations over rings of functions on some affine curves. Though we will mainly focus on formal quantization in these lectures, we give the following definition of a "rational form of a quantization":

Let Σ be a connected affine algebraic curve over k, and $0 \in \Sigma$ a smooth point. Choose a formal parameter h around 0 (a generator of the (completed) local ring $\mathcal{O}(\Sigma)_0 = \varprojlim \mathcal{O}(\Sigma)/I^n$ where $\mathcal{O}(\Sigma)$ is the ring of functions on Σ and I is the ideal of functions vanishing at 0), and denote by $i : R = \mathcal{O}(\Sigma) \hookrightarrow K$ the embedding induced by h.

Definition: A deformation defined over R of a k-algebra A_0 is an associative algebra A_R isomorphic to $A_0 \otimes_k R$ as an R-module such that $A_R/IA_R \simeq A_0$. In a similar way, a deformation A of A_0 (over K) admits a rational form if there exists a deformation A_R defined over R of A_0 such that

$$A \simeq \varprojlim{}_{n\to\infty} A_R \otimes_R (R/I^n R).$$

For example, if we pick $\Sigma = \mathbb{A}^1$ then $R = k[h]$, and a deformation A of A_0 is said to be polynomial if it admits an $\mathbb{A}^1$-rational form.

If we have a Σ-rational deformation of A_0, then to any point ξ of Σ defined over k corresponding to the ideal I_ξ (of functions vanishing at ξ) we can associate an algebra

$$A(\xi) = A_R/I_\xi A_R.$$

In other words we have a family of "deformations" of A_0 parameterized by the closed points of Σ, which are canonically identified with A_0 as a vector space. In particular, when the deformation parameter is the point 0, we have $A(0) = A_0$.

Example: It is easy to check that the quantization procedure we described in 1.4.4 is polynomial. Thus, the canonical quantization $A(T^*X)$ of T^*X and the geometric quantization $A(\mathfrak{g}^*)$ of $\mathfrak{g}^*$ are both polynomial quantizations. Notice that we have $A(T^*X)(\lambda) = D_X$ and $A(\mathfrak{g}^*)(\lambda) = U(\mathfrak{g})$ for $\lambda \neq 0$, while we have $A(T^*X)(0) = C^\infty_{pol}(T^*X)$ and $A(\mathfrak{g}^*)(0) = S\mathfrak{g}$ -there is a "jump" at the point $\lambda = 0$.

1.6 Physical meaning of quantization

In this section, we very briefly outline the basic mathematical models of classical and quantum mechanics, to explain some of the terminology.

Classical mechanics: A classical mechanical system consists of a phase space M which is a Poisson (usually symplectic) manifold, and a function $H : M \to \mathbb{R}$ (the *Hamiltonian*, or *energy* function). Functions on M are called *observables*, and there are two fundamental operations on them, the usual product and the Poisson bracket. The equations of motion are the Hamiltonian equations:

$$\frac{df}{dt} = \{f, H\}.$$

In this formulation, the law of conservation of energy follows from $\{H, H\} = 0$ For example, it is often the case that the phase space is the cotangent bundle T^*X of a smooth manifold X (a given state of the system is determined by position $x \in X$ with coordinates (x_i) and momentum $p \in T_xX$ with coordinates (p_i) defined by $p = \sum_i p_i dx_i$), and the dynamics of a point are described by Hamilton's equations for $f = (x_i, p_i)$:

$$\frac{dx_i}{dt} = \frac{\partial H}{\partial p_i}, \qquad \frac{dp_i}{dt} = -\frac{\partial H}{\partial x_i}.$$

Quantum mechanics: The preceding model is not compatible with the *indeterminacy principle* which states that it is not possible to know precisely both the position and the momentum of the particles of a system at a given time, and is one of the main principles of quantum theory. Therefore, in quantum theory, the phase space manifold M is replaced by an (infinite dimensional) Hilbert space $\mathcal{H}$, (which is morally the space of L^2 functions on a Lagrangian submanifold of M when M is symplectic), and the observables now form some algebra A of (unbounded) operators on $\mathcal{H}$ which is noncommutative. The Hamiltonian H is now an element of A, and the dynamics are now described by *Schrödinger's equations*

$$-i\hbar\frac{da}{dt} = [a, H]$$

for some observable $a \in A$. Here $\hbar$ is the Planck constant. In reality, $\hbar$ is a small but finite quantity, but physicists often treat $\hbar$ as a formal deformation parameter. Such an approach is called "perturbation theory".

We can now make sense of the statement that Quantum mechanics becomes Classical mechanics in the quasiclassical limit: the algebra A is a polynomial quantization of A_0, i.e a polynomial deformation of A_0 such that $A_0 = A/\hbar A$ with the Poisson bracket

$$\{\ ,\ \}_{A_0} = \lim_{\hbar\to 0}\frac{i[\ ,\ ]}{\hbar}$$

and Schrödinger's equations for motion become Hamilton's equations in the quasiclassical limit.

Example: Consider the following classical mechanical system: $M = T^*\mathbb{R}$ with coordinates (x, p) and usual Poisson structure. The Hamiltonian is given by

$$H(x,p) = \frac{p^2}{2} + V(x)$$

where $V(x) \in C^\infty(\mathbb{R})$ is the potential. The Hamiltonian equations for this system are

$$\frac{dx}{dt} = p, \qquad \frac{dp}{dt} = -\frac{dV}{dx}$$

which is equivalent to *Newton's equation* $\frac{d^2x}{dt^2} = -V'(x)$.

Recall the quantization $A = (\bigoplus_{n\geq 0} h^n D^n_{\mathbb{R}})[[h]]$ of the Poisson algebra $C^\infty_{pol}(T^*\mathbb{R})$. This quantization admits a polynomial rational form and defines a family of algebras $A(\hbar)$ for $\hbar \in \mathbb{R}$. As an algebra, $A(\hbar) = D^{pol}_{\mathbb{R}}$-the algebra of polynomial differential operators on $C^\infty(\mathbb{R})$-when $\hbar \neq 0$, and $A(0) = C^\infty_{pol}(T^*\mathbb{R})$. The algebra $A(\hbar)$ acts on the Hilbert space $L^2(\mathbb{R})$. The position coordinate is $\hat{x} = x$, the momentum coordinate is now expressed as $\hat{p} = -i\hbar\frac{d}{dx}$, and the quantum Hamiltonian is

$$\hat{H} = -\frac{\hbar^2}{2}\frac{d^2}{dx^2} + V(x).$$

The evolution equations for operators are now Schrödinger's equations:

$$i\hbar\frac{da}{dt} = [-\frac{\hbar^2}{2}\frac{d^2}{dx^2} + V(x), a] \qquad a \in A.$$

Lecture 2

Poisson-Lie groups

In this lecture, we define the notions of Poisson-Lie groups and Lie bialgebras, which will be one of the main topics of these notes.

2.1 Poisson-Lie groups

2.1.1 Definition

A Poisson-Lie group is a Lie group with a compatible Poisson structure. More precisely:

Definition: A Poisson manifold endowed with a structure of a Lie group is a *Poisson-Lie group* if the multiplication map

$$m : G \times G \to G$$

is a map of Poisson manifolds.

A morphism between two Poisson-Lie groups is a morphism for both the Lie group and the Poisson structures. A Lie subgroup of a Poisson-Lie group is a Poisson-Lie subgroup if it is also a Poisson submanifold.

Let us write the above definition explicitly: a Lie group endowed with a Poisson bracket is a Poisson-Lie group if and only if, for any $x_0, y_0 \in G$, and any functions $f, g \in C^\infty(G)$, we have

$$\{f, g\}(x_0 y_0) = \{f, g\}_x(xy_0)_{|x=x_0} + \{f, g\}_y(x_0 y)_{|y=y_0} \tag{2.1}$$

where, for example, $\{f, g\}_x(xy_0)_{|x=x_0} = \{f(xy_0), g(xy_0)\}(x_0)$ and $f(xy_0), g(xy_0)$ are considered as functions of x. In terms of the Poisson bivector $\Pi \in \Gamma(\Lambda^2 TG)$, the above condition can be written as

$$\Pi(xy) = (d_x(\rho_y) \otimes d_x(\rho_y))\Pi(x) + (d_y(\lambda_x) \otimes d_y(\lambda_x))\Pi(y) \tag{2.2}$$

where $\rho_y : G \to G,\ h \mapsto hy$ and $\lambda_x : G \to G,\ h \to xh$ are the right multiplication and left multiplication maps respectively (i.e the Poisson bivector $\Pi(xy)$ is the sum of the left-translate of $\Pi(y)$ by x and the right-translate of $\Pi(x)$ by y). In particular, setting $x = y = e$, we see that $\Pi(e) = 0$.

Remark: It follows from the definition that the inversion map

$$i : G \to G$$
$$g \mapsto g^{-1}$$

is an *anti*-Poisson map, i.e $\{f \circ i, g \circ i\}(x) = -\{f,g\}(x^{-1})$ or, at the Poisson bivector level, $(d_x i \otimes d_x i)\Pi(x) = -\Pi(x^{-1})$. Indeed, by (2.2) for $y = x^{-1}$ and using $\Pi(e) = 0$, we have

$$(d_x\rho_{x^{-1}} \otimes d_x\rho_{x^{-1}})\Pi(x) + (d_{x^{-1}}\lambda_x \otimes d_{x^{-1}}\lambda_x)\Pi(x^{-1}) = 0$$

where $d_x\rho_{x^{-1}} : T_xG \to T_eG, d_{x^{-1}}\lambda_x : T_x^{-1}G \to T_eG$, hence

$$(d_e\lambda_{x^{-1}} \otimes d_e\lambda_{x^{-1}})(d_x\rho_{x^{-1}} \otimes d_x\rho_{x^{-1}})\Pi(x) = -\Pi(x^{-1})$$

where $d_e\lambda_{x^{-1}} : T_eG \to T_{x^{-1}}G$, $d_x\rho_{x^{-1}} : T_xG \to T_eG$, and the result follows from $d_x i = -d_e\lambda_{x^{-1}} d_x\rho_{x^{-1}}$.

2.2 Lie bialgebras

2.2.1 Definition

The tangent space of a Lie group at the identity e has a Lie algebra structure. In the case of a Poisson-Lie group, it inherits an additional structure.

To see this, recall that by (2.2), $\Pi(e) = 0$. In other terms, $\{e\}$ is a symplectic leaf of G. Now consider the following general situation: X is a Poisson manifold, and $x_0 \in X$ is such that $\Pi(x_0) = 0$. Then the cotangent space $T^*_{x_0}X$ has a natural Lie algebra structure. The construction of this structure is the following: let $\mathcal{O}(X)_{x_0}$ be the ring of germs of smooth functions defined in a neighborhood of x_0, and denote by I its unique maximal ideal (of functions vanishing at x_0). Consider the Poisson bracket

$$\{,\} : \mathcal{O}(X)_{x_0} \otimes \mathcal{O}(X)_{x_0} \to \mathcal{O}(X)_{x_0},$$
$$f \otimes g \to df \otimes dg(\Pi).$$

Since $\Pi_{x_0} = 0$, we have $\{,\} : \mathcal{O}(X)_{x_0} \otimes \mathcal{O}(X)_{x_0} \to I$ and hence we have a Lie bracket $\{,\} : I \otimes I \to I$. Moreover, if $f \in I, g \in I^2$, then $\{f,g\} = df \otimes dg(\Pi) \in I^2$, so that I^2 is a Lie ideal of I. This induces a Lie algebra structure on the quotient $I/I^2 \simeq T^*_{x_0}X$.

In particular, if G is a Poisson-Lie group and $\mathfrak{g} = \mathrm{Lie}(G) \simeq T_eG$ then the preceding construction defines a Lie algebra structure $[\,,\,] : \Lambda^2\mathfrak{g}^* \to \mathfrak{g}^*$ on $\mathfrak{g}^*$. Taking the dual of this commutator, we obtain a map $\delta : \mathfrak{g} \to \Lambda^2\mathfrak{g}$. The Jacobi identity for $[\,,\,]$ is then equivalent to the *coJacobi identity* for δ:

$$\forall x \in \mathfrak{g} \qquad Alt(\delta \otimes Id)\delta(x) = 0,$$

where $Alt(a \otimes b \otimes c) = a \otimes b \otimes c + b \otimes c \otimes a + c \otimes a \otimes b$.

Remark: The map δ is easy to describe in terms of the Poisson bivector Π: let us use left translations λ_g to identify T_gG with $\mathfrak{g}$. This allows us to view the bivector as a map $\Pi : G \to \Lambda^2\mathfrak{g}$. Then $\delta = d\Pi : \mathfrak{g} \to \Lambda^2\mathfrak{g}$. The result is the same if we use right translations.

A vector space $\mathfrak{a}$ equipped with a linear map $\delta : \mathfrak{a} \to \Lambda^2\mathfrak{a}$ satisfying the coJacobi identity is called a *Lie coalgebra*. Thus the tangent space $T_eG = \mathfrak{g}$ of a Poisson-Lie group G is both a Lie algebra and a Lie coalgebra. Moreover these structures are not independent :

Lemma 2.1. *We have*

$$\delta([a,b]) = [\delta(a), 1\otimes b + b\otimes 1] + [a\otimes 1 + 1\otimes a, \delta(b)] \qquad a,b\in\mathfrak{g}. \tag{2.3}$$

Proof: Let us use right translations to identify T_xG with $\mathfrak{g}$, and view the Poisson bivector as a map $\tilde{\Pi} : G \to \Lambda^2\mathfrak{g}$. By (2.2), we have

$$\tilde{\Pi}(x_0y_0) = \tilde{\Pi}(x_0) + (Ad_{x_0}\otimes Ad_{x_0})\tilde{\Pi}(y_0). \tag{i}$$

Similarly,

$$\tilde{\Pi}(y_0x_0) = \tilde{\Pi}(y_0) + (Ad_{y_0}\otimes Ad_{y_0})\tilde{\Pi}(x_0). \tag{ii}$$

Now let $x_0 = e^{ta}, y_0 = e^{tb}$ for some $a,b\in\mathfrak{g}$. The difference (i)-(ii) vanishes up to second order as $t\mapsto 0$, and the t^2 term reads

$$d\tilde{\Pi}([a,b]) = [a\otimes 1 + 1\otimes a, d\tilde{\Pi}(b)] - [b\otimes 1 + 1\otimes b, d\tilde{\Pi}(a)].$$

The lemma now follows from the fact that $d\tilde{\Pi} = \delta$.

□

We will call condition (2.3) the *cocycle* condition, as it means that δ is a 1-cocycle for $\mathfrak{g}$ with coefficients in $\Lambda^2\mathfrak{g}$. Another way of formulating the argument of Lemma 2.1 is to say that $\tilde{\Pi} : G \to \Lambda^2\mathfrak{g}$ is a group 1-cocycle (by (i)), and that its derivative $d\tilde{\Pi} = \delta : \mathfrak{g} \to \Lambda^2\mathfrak{g}$ is therefore a Lie algebra 1-cocyle. We are thus led to make the following definition:

Definition: A *Lie bialgebra* $(\mathfrak{g}, [\,,\,], \delta)$ is a Lie algebra $(\mathfrak{g}, [\,,\,])$ equipped with a map $\delta : \mathfrak{g} \to \Lambda^2\mathfrak{g}$ (the *cocommutator*, or *cobracket*) satisfying the coJacobi identity and the cocyle condition.

A morphism of Lie bialgebras is a Lie algebra morphism preserving the co-commutator. A *Lie subbialgebra* of a Lie bialgebra $\mathfrak{g}$ is a Lie subalgebra $\mathfrak{h}$ such that $\delta(\mathfrak{h}) \subset \Lambda^2\mathfrak{h}$. If $\mathfrak{h} \subset \mathfrak{g}$ is a Lie ideal then the quotient Lie algebra $(\mathfrak{g}/\mathfrak{h}, [\,,\,])$ inherits of a Lie bialgebra structure from $\mathfrak{g}$ if and only if $\delta(\mathfrak{h}) \subset \mathfrak{g}\otimes\mathfrak{h} + \mathfrak{h}\otimes\mathfrak{g}$. In this case, $\mathfrak{h}$ is said to be a *Lie coideal.*

We will denote by $LBA(k)$ (resp. $LBA_f(k)$) the category of Lie bialgebras (resp. finite-dimensional Lie bialgebras) defined over the field k.

The results of this section can be summarized as follows:

Proposition 2.1. *Let G be a Poisson-Lie group. The Lie algebra $\mathfrak{g} = \mathrm{Lie}(G)$ is naturally a Lie bialgebra.*

2.2.2 Examples of Lie bialgebras

Example 2.1 (Trivial Poisson structure). Any Lie group G equipped with the trivial Poisson bracket is a Poisson-Lie group. The corresponding bialgebra is $\mathrm{Lie}(G)$ with trivial cocommutator.

Example 2.2 (Two dimensional Lie bialgebras). It is easy to see that any two-dimensional non abelian Lie algebra is isomorphic to $T_2 = \{ \begin{pmatrix} a & b \\ 0 & 0 \end{pmatrix} \}$, with basis $x = \begin{pmatrix} 1 & 0 \\ 0 & 0 \end{pmatrix}$, $y = \begin{pmatrix} 0 & 1 \\ 0 & 0 \end{pmatrix}$ and relation $[x, y] = y$. Let us classify all possible bialgebra structures on T_2. Since $\Lambda^2 T_2 = kx \wedge y$, these are given by

$$\delta(x) = \alpha x \wedge y, \qquad \delta(y) = \beta x \wedge y.$$

One can check that this indeed defines a Lie bialgebra structure $T_2^{(\alpha,\beta)}$ on T_2 for any choice of $\alpha, \beta \in k$. Moreover, the automorphisms of T_2 are given by

$$x \mapsto x + by, \qquad y \mapsto ay, \qquad a, b \in k,\ a \neq 0.$$

Using these it is easy to check that, if $\beta \neq 0$ then $T_2^{(\alpha,\beta)} \simeq T_2^{(0,\beta)}$, and if $\alpha \neq 0$, then $T_2^{(\alpha,0)} \simeq T_2^{(1,0)}$. In this way we get a one parameter family $b_2(\beta) = T_2^{(0,\beta)}$, for $\beta \neq 0$ of Lie bialgebra structures on T_2 which degenerates into $\tilde{b}_2 = T_2^{(1,0)}$, and the trivial Lie bialgebra structure.

Example 2.3 (A Lie bialgebra structure on $sl_2(\mathbb{C})$). Recall that $sl_2(\mathbb{C})$ is the Lie algebra of traceless 2×2 matrices, with basis

$$h = \begin{pmatrix} 1 & 0 \\ 0 & -1 \end{pmatrix}, \qquad e = \begin{pmatrix} 0 & 1 \\ 0 & 0 \end{pmatrix}, \qquad f = \begin{pmatrix} 0 & 0 \\ 1 & 0 \end{pmatrix},$$

with relations

$$[h, e] = 2e, \qquad [h, f] = -2f, \qquad [e, f] = h.$$

The following formulas define a Lie bialgebra structure on $sl_2(\mathbb{C})$:

$$\delta(e) = \frac{1}{2} e \wedge h, \qquad \delta(f) = \frac{1}{2} f \wedge h, \qquad \delta(h) = 0.$$

This structure is called the *standard* Lie bialgebra structure. Notice that the subalgebras $\mathfrak{b}^+ = \mathbb{C}e \oplus \mathbb{C}h$ and $\mathfrak{b}^- = \mathbb{C}f \oplus \mathbb{C}h$ are Lie subbialgebras of $sl_2(\mathbb{C})$.

2.2.3 Duality

It turns out that the notion of finite dimensional Lie bialgebra is self-dual.

Proposition 2.2. *Let* $(\mathfrak{g}, [\,,\,], \delta)$ *be a finite dimensional Lie bialgebra and let* $[\,,\,]^* : \mathfrak{g}^* \to \Lambda^2\mathfrak{g}^*$ *and* $\delta^* : \Lambda^2\mathfrak{g}^* \to \mathfrak{g}^*$ *be the dual maps to* $[\,,\,]$ *and* δ *respectively. Then* $(\mathfrak{g}^*, \delta^*, [\,,\,]^*)$ *is a Lie bialgebra.*

Proof: It is clear that $(\mathfrak{g}^*, \delta^*)$ is a Lie algebra, and that $(\mathfrak{g}^*, [\,,\,]^*)$ is a Lie coalgebra. We have to check the cocycle condition. We will use a pictorial technique, about which we will say more in Lecture 20. We attach to each combination of maps δ, $[\,,\,]$ a diagram in the following way: to the basic operations δ and $[\,,\,]$, we assign the pictures

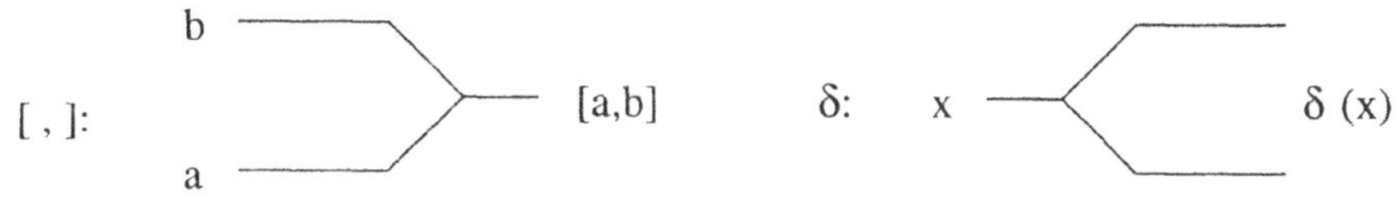

Composition of maps is obtained by adjoining diagrams from left to right. For example, $\delta([a,b])$ corresponds to the following diagram:

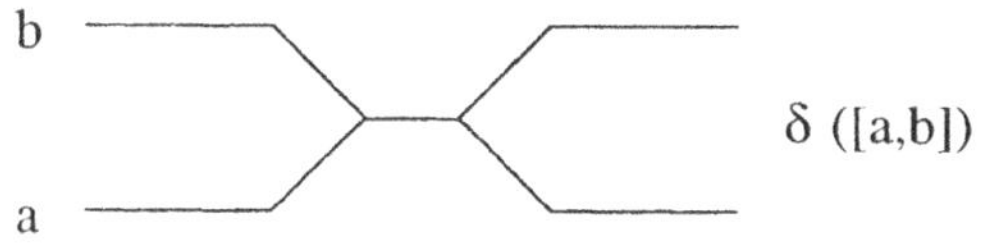

In this formulation, the operation of taking the dual is nothing but interchanging left and right. In particular, the cocycle condition can be written as:

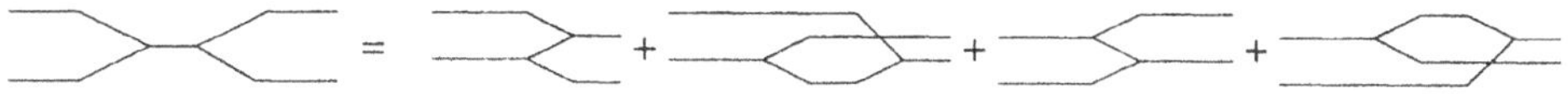

and is easily seen to be self dual.

□

2.3 Poisson-Lie theory

2.3.1 Main theorem of Poisson-Lie theory

In classical Lie group theory, the correspondence between Lie groups and Lie algebras (real or complex) is summarized in the following theorem:

Theorem 2.1 (Lie). *The functor* $F : G \mapsto Lie(G)$ *between the category of simply connected Lie groups and the category of finite dimensional Lie algebras is an equivalence of categories.*

It turns out that this equivalence extends to the case of Poisson-Lie groups and Lie bialgebras:

Theorem 2.2 (Drinfeld). : *The functor $\tilde{F} : G \to Lie(G)$ between the category of simply connected Poisson-Lie groups and the category of finite dimensional Lie bialgebras is an equivalence of categories.*

Proof: We need to show two things:

1. To any finite dimensional Lie bialgebra $\mathfrak{g}$ there corresponds a Poisson-Lie group G, unique up to isomorphism, such that $\tilde{F}(G) = \mathfrak{g}$.

2. If G_1 and G_2 are simply connected Poisson-Lie groups and $\mathfrak{g}_1 = \tilde{F}(G_1)$, $\mathfrak{g}_2 = \tilde{F}(G_2)$, then there is a one-to-one correspondence between the morphisms $G_1 \to G_2$ and the morphisms $\mathfrak{g}_1 \to \mathfrak{g}_2$.

Let us prove 1). The proof of 2) is left to the reader. Let $(\mathfrak{g}, [\,,\,], \{\,,\})$ be a Lie bialgebra. Using Lie's theorem, we reduce the problem to showing that the simply connected Lie group G such that $\mathrm{Lie}(G) = \mathfrak{g}$ admits a unique Poisson structure compatible with the bialgebra structure of $\mathfrak{g}$. This is a consequence of the fact that there is in this case a one-to-one correspondence between group 1-cocycles $\tilde{\Pi} : G \to \Lambda^2\mathfrak{g}$ and Lie algebra 1-cocycles $\delta : \mathfrak{g} \to \Lambda^2\mathfrak{g}$, but we will give a direct proof. Let us deal with uniqueness first. Suppose that G is a Poisson-Lie group such that $\tilde{F}(G) = \mathfrak{g}$, and let us again use right translations to trivialize $\Lambda^2 TG$. Viewing the Poisson bivector as a map $\tilde{\Pi} : G \to \Lambda^2\mathfrak{g}$, we have

$$\tilde{\Pi}(xy) = \tilde{\Pi}(x) + (Ad(x) \otimes Ad(x))\tilde{\Pi}(y). \tag{2.4}$$

Setting $x = e^{ta}$ for some $a \in \mathfrak{g}$ and differentiating at $t = 0$, we see that $\tilde{\Pi}$ is the unique solution of the following system of nonhomegeneous linear differential equations

$$\begin{aligned} \nabla_a \tilde{\Pi}(y) &= d_e\tilde{\Pi}(a) + [a \otimes 1 + 1 \otimes a, \tilde{\Pi}(y)] \\ &= \delta(a) + ad(a)(\tilde{\Pi}(y)) \end{aligned} \tag{2.5}$$

with initial condition $\tilde{\Pi}(e) = 0$, where we denote by $\nabla_a(f)(y)$ the Lie derivative along the *right*-invariant vector field on G corresponding to a. This implies uniqueness.

Moreover the system (2.5) is coherent, i.e we have $[\nabla_a, \nabla_b]\tilde{\Pi} = \nabla_{[b,a]}\tilde{\Pi}$:

$$\begin{aligned}
[\nabla_a, \nabla_b](\tilde{\Pi}) &= \nabla_a(\delta(b) + ad(b)\tilde{\Pi}) - \nabla_b(\delta(a) + ad(a)\tilde{\Pi}) \\
&= \nabla_a(\delta(b)) - \nabla_b(\delta(a)) + (\nabla_a ad(b) - \nabla_b ad(a))\tilde{\Pi} \\
&= (\nabla_a ad(b) - \nabla_b ad(a))\tilde{\Pi} \\
&= (ad(b)\nabla_a - ad(a)\nabla_b)\tilde{\Pi} \\
&= [ad(b), ad(a)]\tilde{\Pi} + \delta([b,a]) \\
&= \nabla_{[b,a]}\tilde{\Pi}
\end{aligned}$$

where we used the fact that $\nabla_a(\delta(b)) = 0$ (since $\delta(b)$ is a right-invariant vector field), the relation $[ad(a), \nabla_b] = 0$ and the cocycle condition.

This implies that, conversely, if H is a simply connected Lie group such that $\mathrm{Lie}(H) = \mathfrak{g}$, then the system (2.5) together with the initial condition $\tilde{\Pi}(e) = 0$ defines a unique section Π of $\Lambda^2 TH$. It is clear that this section satisfies (2.4), and thus defines the desired Poisson-Lie structure. This concludes the proof of Theorem 2.2.

□

Remark: The condition that H be simply connected is essential in the construction of the Poisson structure. It is not true that a Lie bialgebra structure on $\mathrm{Lie}(G)$ for any Lie group G lifts to a Poisson structure on G (see example 2.7 below).

Note that the system (2.5) can be explicitly solved: the solution is given by

$$\tilde{\Pi}(e^a) = \frac{1 - e^{-ad(a)}}{ad(a)}\delta(a) = \sum_{n \geq 0} \frac{(-ad(a))^n}{(n+1)!}\delta(a). \tag{2.6}$$

Notice that this is in general not algebraic.

2.3.2 Dual Poisson-Lie group

Definition: Let G be a simply connected Poisson-Lie group and let $\tilde{F}(G) = \mathfrak{g}$ be its Lie algebra with its canonical Lie bialgebra structure. The *dual Poisson-Lie group* G^* of G is the simply connected Poisson-Lie group such that

$$\tilde{F}(G^*) = \mathfrak{g}^*.$$

Remark: There is no easy geometric realization of the dual of a Poisson-Lie group. In particular, G^* and G could have very different topologies (see examples 2.7,2.9).

2.3.3 Examples of dual Lie bialgebras and dual Poisson-Lie groups

Example 2.4 (Trivial Poisson structure). Let G be a Lie group with trivial Poisson structure, and $\mathfrak{g}$ its Lie algebra with trivial cocommutator. Then $\mathfrak{g}^*$ is an abelian Lie algebra, but has a non-trivial cocommutator.

Example 2.5 (Two dimensional Lie bialgebras). Recall the notations of example 2.2. It is easy to check that $b_2(\beta)^* = b_2(\beta^{-1})$ and $\tilde{b}_2^* = \tilde{b}_2$.

Example 2.6 (Standard structure on $sl_2(\mathbb{C})$). In this case, the dual Lie bialgebra is generated by e^*, f^*, h^* with defining relations

$$[h^*, e^*] = -\frac{1}{2}e^*, \qquad [h^*, f^*] = -\frac{1}{2}f^*, \qquad [e^*, f^*] = 0,$$

$$\delta(e^*) = \frac{1}{2}h^* \wedge e^*, \qquad \delta(f^*) = \frac{1}{2}h^* \wedge f^*, \qquad \delta(h^*) = e^* \wedge f^*.$$

Notice that, as a Lie algebra, it is isomorphic to $T_2 \oplus T_2/x_1 - x_2$, where we denote by x_1, y_1 and x_2, y_2 the generators of the two copies of T_2 respectively.

Now let us construct the Poisson structure on the simply connected Lie groups corresponding to these Lie algebras. Recall that the Poisson bivector is obtained as the solution to the differential equation $\nabla_a(\tilde{\Pi}) = \delta(a) + ad(a)\tilde{\Pi}$ with initial condition $\tilde{\Pi}(e) = 0$. Applying formula (2.6) to the examples, we obtain:

Example 2.7 (Dual of the trivial structure). Since $\mathfrak{g}^*$ is abelian, $G^* = \mathfrak{g}^*$ as a commutative Lie group, but has a non-trivial Poisson structure. In this case, (2.6) yields $\tilde{\Pi}(a) = \Pi(a) = \delta(a)$. In other words, we obtain the same Poisson structure on $\mathfrak{g}^*$ as in Lecture 1. In particular, if $\Gamma \subset \mathfrak{g}^*$ is a lattice under which the Poisson structure is not invariant then the Lie group $H = \mathfrak{g}^*/\Gamma$ has a Lie bialgebra structure on its Lie algebra which doesn't lift to a Poisson structure on the group.

Example 2.8 (Two dimensional Lie bialgebras). Let us compute the Poisson structures on the simply connected Lie group $H = \left\{ \begin{pmatrix} p & q \\ 0 & 1 \end{pmatrix} | p > 0 \right\}$ induced by the Lie bialgebra structures $b_2(\beta)$ and $\tilde{b}_2$ on $T_2 = \mathrm{Lie}(H)$. Since $ad(ax+by)_{|\wedge^2 T_2} = a$, formula (2.6) yields

$$\tilde{\Pi}\left(exp\begin{pmatrix} a & b \\ 0 & 0 \end{pmatrix}\right) = \tilde{\Pi}\left(\begin{pmatrix} e^a & \frac{e^a-1}{a}b \\ 0 & 1 \end{pmatrix}\right) = \frac{1-e^{-a}}{a}\delta\left(\begin{pmatrix} a & b \\ 0 & 0 \end{pmatrix}\right).$$

Thus the Poisson structure has the following form:

1. $b_2(\beta)$: in this case, we have $\delta\left(\begin{pmatrix} a & b \\ 0 & 0 \end{pmatrix}\right) = \beta b x \wedge y$, whence

$$\tilde{\Pi}\left(\begin{pmatrix} p & q \\ 0 & 1 \end{pmatrix}\right) = \beta q p^{-1} x \wedge y, \qquad \Pi\left(\begin{pmatrix} p & q \\ 0 & 1 \end{pmatrix}\right) = \beta p q \partial_p \wedge \partial_q.$$

2. $\tilde{b}_2$: here, we have $\delta\Big(\begin{pmatrix} a & b \\ 0 & 0 \end{pmatrix}\Big) = ax \wedge y$, and we obtain

$$\tilde{\Pi}\Big(\begin{pmatrix} p & q \\ 0 & 1 \end{pmatrix}\Big) = (1 - p^{-1})x \wedge y, \qquad \Pi\Big(\begin{pmatrix} p & q \\ 0 & 1 \end{pmatrix}\Big) = p(p-1)\partial_p \wedge \partial_q.$$

Example 2.9 (Standard structure on $sl_2(\mathbb{C})$). The simply connected Lie group corresponding to $sl_2(\mathbb{C})$ is $SL_2(\mathbb{C})$. The Poisson bivector for $SL_2(\mathbb{C})$ is given by

$$\tilde{\Pi}\Big(\begin{pmatrix} 1 & t \\ 0 & 1 \end{pmatrix}\Big) = \frac{t}{2}\begin{pmatrix} 0 & 1 \\ 0 & 0 \end{pmatrix} \wedge \begin{pmatrix} 1 & 0 \\ 0 & -1 \end{pmatrix}, \quad \tilde{\Pi}\Big(\begin{pmatrix} 1 & 0 \\ t & 1 \end{pmatrix}\Big) = \frac{t}{2}\begin{pmatrix} 0 & 0 \\ 1 & 0 \end{pmatrix} \wedge \begin{pmatrix} 1 & 0 \\ 0 & -1 \end{pmatrix},$$

$$\Pi\Big(\begin{pmatrix} e^t & 0 \\ 0 & e^{-t} \end{pmatrix}\Big) = 0.$$

The Poisson-Lie group $SL_2(\mathbb{C})^*$ is the (simply connected covering of the) subgroup $\Big\{\Big(\begin{pmatrix} p & q \\ 0 & 1 \end{pmatrix}, \begin{pmatrix} p' & q' \\ 0 & 1 \end{pmatrix}\Big) \mid p = p'\Big\} \subset H_{\mathbb{C}} \times H_{\mathbb{C}}$, with the Poisson bivector field given on generators by

$$\Pi\Big(\begin{pmatrix} 1 & t \\ 0 & 1 \end{pmatrix}, \begin{pmatrix} 1 & 0 \\ 0 & 1 \end{pmatrix}\Big) = \frac{t}{2}\Big(\begin{pmatrix} 1 & 0 \\ 0 & 0 \end{pmatrix}, \begin{pmatrix} 1 & 0 \\ 0 & 0 \end{pmatrix}\Big) \wedge \Big(\begin{pmatrix} 0 & 1 \\ 0 & 0 \end{pmatrix}, \begin{pmatrix} 0 & 0 \\ 0 & 0 \end{pmatrix}\Big),$$

$$\Pi\Big(\begin{pmatrix} 1 & 0 \\ 0 & 1 \end{pmatrix}, \begin{pmatrix} 1 & t \\ 0 & 1 \end{pmatrix}\Big) = \frac{t}{2}\Big(\begin{pmatrix} 1 & 0 \\ 0 & 0 \end{pmatrix}, \begin{pmatrix} 1 & 0 \\ 0 & 0 \end{pmatrix}\Big) \wedge \Big(\begin{pmatrix} 0 & 0 \\ 0 & 0 \end{pmatrix}, \begin{pmatrix} 0 & 1 \\ 0 & 0 \end{pmatrix}\Big),$$

$$\Pi\Big(\begin{pmatrix} e^t & 0 \\ 0 & 1 \end{pmatrix}, \begin{pmatrix} e^t & 0 \\ 0 & 1 \end{pmatrix}\Big) = t\Big(\begin{pmatrix} 0 & 1 \\ 0 & 0 \end{pmatrix}, \begin{pmatrix} 0 & 0 \\ 0 & 0 \end{pmatrix}\Big) \wedge \Big(\begin{pmatrix} 0 & 0 \\ 0 & 0 \end{pmatrix}, \begin{pmatrix} 0 & 1 \\ 0 & 0 \end{pmatrix}\Big).$$

Notice that $SL_2(\mathbb{C})^*$ is solvable and hence contractible, while on the other hand $SL_2(\mathbb{C})$ is homotopically a 3-sphere.

We conclude this lecture with two problems:

Problem 2.1. Let G be a Poisson-Lie group, and let

$$G^{(0)} = \{x \in G \mid \Pi(x) = 0\}$$

be the subset of all points where the Poisson structure vanishes. Show that $G^{(0)}$ is a closed Lie subgroup of G.

For example, in the case of the Poisson structures on H (example 2.8), we have $H^{(0)}(b_2(\beta)) = \Big\{\begin{pmatrix} p & 0 \\ 0 & 1 \end{pmatrix}\Big\} \simeq \mathbb{R}^+$ degenerating to $H^{(0)}(\tilde{b}_2) = \Big\{\begin{pmatrix} 1 & q \\ 0 & 1 \end{pmatrix}\Big\} \simeq \mathbb{R}$.

Problem 2.2. Let G be a Poisson-Lie group with Lie bialgebra $(\mathfrak{g}, [\,,\,], \delta)$. Let $H \subset G$ be a closed Lie subgroup and let $\mathrm{Lie}(H) = \mathfrak{h} \subset \mathfrak{g}$. Show that the homogeneous space G/H inherits a (unique) Poisson structure such that $\pi : G \to G/H$ is a Poisson map if and only if $\delta(\mathfrak{h}) \subset \mathfrak{h} \otimes \mathfrak{g} + \mathfrak{g} \otimes \mathfrak{h}$.

Remark: The Poisson manifold G/H is an example of a *Poisson homogeneous space.* By definition, a Poisson homogeneous space for a Poisson-Lie group G is a homogeneous space G/H equipped with a Poisson structure such that the natural map $G \times G/H \to G/H$ is a Poisson map. However, not all Poisson homogeneous spaces are constructed as in Problem 2.2. Indeed, a transitive action of a Lie group G on a symplectic manifold M preserving the symplectic structure endows M with a Poisson homogeneous space structure, which is not of the above type. The Poisson homogeneous spaces have been classified by Drinfeld, [Dr7].

Lecture 3

Coboundary Lie bialgebras

In this lecture we focus on a very important class of Lie bialgebras $\mathfrak{g}$, for which the coalgebra structure has a simple expression in terms of a special element $r \in \mathfrak{g} \otimes \mathfrak{g}$ (*coboundary* Lie bialgebras). In the next lecture, we will see how to canonically embed any Lie bialgebra into one of this type.

3.1 Some Lie algebra cohomology

Let $\mathfrak{g}$ be a Lie algebra defined over a field k. We define here the cohomology groups $H^i(\mathfrak{g}, V)$ with values in a $\mathfrak{g}$-module V, as these will appear at various stages of these notes (see e.g [GG]).

Definition: Consider the vector spaces (of *cochains*): $C^n = \mathrm{Hom}_k(\Lambda^n \mathfrak{g}, V)$ for $n \in \mathbb{N}$ and the following *differential* maps $\partial_n : C^n \to C^{n+1}$ given by

$$\partial_n f(x_1 \wedge \ldots \wedge x_{n+1}) = \sum_{i=1}^{n+1} (-1)^{i+1} x_i . f(x_1 \wedge \ldots \wedge \widehat{x_i} \ldots \wedge x_{n+1})$$
$$+ \sum_{i<j} (-1)^{i+j} f([x_i, x_j] \wedge x_1 \ldots \wedge \widehat{x_i} \ldots \wedge \widehat{x_j} \ldots \wedge x_{n+1}).$$

The elements in $Z_n = \mathrm{Ker}\ \partial_n \subset C^n$ are called the *n-cocycles*, and the elements in $B_n = \mathrm{Im}\ \partial_{n-1} \subset C^n$ are called the *n-coboundaries*. It is easy to check that $B_n \subset Z_n$, i.e $\partial_{n+1}\partial_n = 0$. By definition, the *n-th cohomology* group is $H^n(\mathfrak{g}, V) = Z_n / B_n$ (a k-vector space).

The zero-th cohomology group is $H^0(\mathfrak{g}, V) = \{v \in V \mid \mathfrak{g}.v = 0\}$. Also, notice that the cocycle condition for the cocommutator $\delta : \mathfrak{g} \to \Lambda^2 \mathfrak{g}$ in the definition of a Lie bialgebra is indeed the condition that δ be a 1-cocycle of $\mathfrak{g}$ with values in $\Lambda^2 \mathfrak{g}$.

Remark: In the case of a finite dimensional Lie algebra $\mathfrak{g}$ defined over $\mathbb{R}$, there is another interpretation of these cohomology groups. Let us describe it in the case $V = \mathbb{R}$ (the trivial representation). Let G be the simply connected Lie group with Lie algebra $\mathfrak{g}$, and consider the cohomology $H^l_{DR,left}(G)$ of the subcomplex of the usual De Rham complex consisting of (say) left-invariant differential forms on G. A left invariant l-form λ is determined by its value at the identity of G, and one can check that it is closed if and only if $\lambda(e) : \Lambda^l\mathfrak{g} \to k$ is a cocycle, and exact if and only if $\lambda(e) : \Lambda^l\mathfrak{g} \to k$ is a coboundary. Thus $H^l_{DR,left}(G) = H^l(\mathfrak{g}, k)$.

3.2 Coboundary Lie bialgebras

Recall that a Lie bialgebra $\mathfrak{g}$ is a Lie algebra equipped with a 1-cocycle with values in $\Lambda^2\mathfrak{g}$, $\delta : \mathfrak{g} \to \Lambda^2\mathfrak{g}$, satisfying the coJacobi identity.

Definition: An element $r \in \Lambda^2\mathfrak{g}$ is said to be a *coboundary structure* of the Lie bialgebra $(\mathfrak{g}, [\,,\,], \delta)$ if $\delta = \partial r$, i.e if for every $a \in \mathfrak{g}$, $\delta(a) = \partial r(a) = [a\otimes 1 + 1\otimes a, r]$. A *coboundary Lie bialgebra* is a triple $(\mathfrak{g}, [\,,\,], r)$, $r \in \Lambda^2\mathfrak{g}$, such that $(\mathfrak{g}, [\,,\,], \partial r)$ is a Lie bialgebra.

A morphism of coboundary Lie bialgebras $\phi : (\mathfrak{g}, [\,,\,], r) \to (\mathfrak{g}', [\,,\,]', r')$ is a Lie bialgebra morphism ϕ such that $\phi(r) = r'$. Notice that a given Lie bialgebra can have several coboundary structures. Indeed, if r is a coboundary structure then coboundary structures r' are in one-to-one correspondence with elements $\alpha \in (\Lambda^2\mathfrak{g})^{\mathfrak{g}}$ via $r' = r + \alpha$.

3.3 The classical Yang-Baxter map

Not any cocycle defined by some $r \in \Lambda^2\mathfrak{g}$ will give rise to a Lie bialgebra structure on $\mathfrak{g}$, as the coJacobi identity may not be satisfied. However, there is the following characterization of such $r \in \Lambda^2\mathfrak{g}$, due to Drinfeld.

Let us first explain some notations, which we will constantly use throughout these notes. If $x = \sum_i y_i \otimes z_i \in \mathfrak{g}\otimes\mathfrak{g}$, we set $x_{12} = \sum_i y_i \otimes z_i \otimes 1 \in U(\mathfrak{g})^{\otimes 3}$, $x_{13} = \sum_i y_i \otimes 1 \otimes z_i \in U(\mathfrak{g})^{\otimes 3}$, etc. When it is more convenient for typographical reasons, we will sometimes write the indices as superscripts, i.e x^{12} for x_{12}, etc.

Definition: The *classical Yang-Baxter map* is the map

$$\begin{aligned} \mathrm{CYB} :\ & \mathfrak{g}^{\otimes 2} \to \mathfrak{g}^{\otimes 3}, \\ & r \mapsto [r_{12}, r_{13}] + [r_{12}, r_{23}] + [r_{13}, r_{23}]. \end{aligned}$$

It is easy to check that CYB restricts to a map $\Lambda^2\mathfrak{g} \to \Lambda^3\mathfrak{g}$.

Theorem 3.1 (Drinfeld). *Let $(\mathfrak{g}, [\,,\,])$ be a Lie algebra and let $r \in \Lambda^2\mathfrak{g}$. Then $(\mathfrak{g}, [\,,\,], \partial r)$ is a Lie bialgebra if and only if the element*

$$\mathrm{CYB}(r) := [r_{12}, r_{13}] + [r_{12}, r_{23}] + [r_{13}, r_{23}] \in \mathfrak{g}^{\otimes 3}$$

is $\mathfrak{g}$-invariant.

Proof: One can show by a rather long direct computation using the Jacobi identity of $\mathfrak{g}$ that

$$Alt(\delta \otimes Id)\delta(x) + [x, \mathrm{CYB}(r)] = 0 \tag{3.1}$$

which proves the claim (see [CP], section 2.1). We give a shorter, more geometric proof in the case $k = \mathbb{R}$ and $\mathfrak{g}$ is finite dimensional (which can be adapted to the infinite dimensional case and to any (characteristic zero) field k using the language of formal groups).

Let $r \in \Lambda^2\mathfrak{g}$, and let G be the simply connected Lie group with Lie algebra $\mathfrak{g}$. Then, by the main theorem of Poisson-Lie theory and formula (2.6), $\delta = \partial r$ defines a Lie bialgebra structure on $\mathfrak{g}$ if and only if the bivector field defined by

$$\begin{aligned}\Pi_r(e^a) &= (d_e\rho_{|e^a} \otimes d_e\rho_{|e^a})\Big(\frac{1 - e^{-ad(a)}}{ad(a)}\delta(a)\Big)\\ &= (d_e\rho_{|e^a} \otimes d_e\rho_{|e^a})\big((1 - e^{-ad(a)})r\big)\\ &= (d_e\rho_{|e^a} \otimes d_e\rho_{|e^a})r - (d_e\lambda_{|e^a} \otimes d_e\lambda_{|e^a})r\end{aligned}$$

defines a Poisson-Lie group structure on G. To verify that Π_r is indeed a Poisson bivector, we need to check the Jacobi identity. We first set a few notations. Let (e_i) be a basis of $\mathfrak{g}$, and let $r = \sum_{i,j} r_{ij} e_i \otimes e_j$. Also, let L_a (resp. R_a) be the left (resp. right)-invariant vector field on G such that $L_a(e) = R_a(e) = a \in \mathfrak{g}$. We let $R_i = R_{e_i}$ and $L_i = L_{e_i}$. We extend R to $U(\mathfrak{g})$ by the rule $R_{uv} = R_v R_u$ (this is again a right-invariant differential operator) and to $U(\mathfrak{g})^{\otimes n}$ by $R(a_1 \otimes \ldots \otimes a_n) = R(a) \otimes \ldots \otimes R(a_n)$. Similarly, we extend L to $U(\mathfrak{g})$ and to $U(\mathfrak{g})^{\otimes n}$ by the rule $L(a_1 \otimes \ldots \otimes a_n) = L(a_1) \otimes \ldots \otimes L(a_n)$.

Now, recall that by definition, for $f, g \in C^\infty(G)$, we have

$$\begin{aligned}\{f, g\} &= df \otimes dg(\Pi_r)\\ &= \sum_{i,j} r_{ij}(L_i f L_j g - R_i f R_j g).\end{aligned}$$

Using this, we see that

$$\begin{aligned}\mathrm{Jac}(f, g, h) := \{\{f, g\}, h\}_{cycl} &= \sum r_{ij} r_{kl}(R_i R_k f R_l g R_j h + R_k f R_i R_l g R_j h)_{cycl}\\ &\quad + A_{RL} + A_{LR} + A_{LL}\\ &= m_3 R(r_{12} r_{13} + r_{12} r_{23})(f \otimes g \otimes h)_{cycl}\\ &\quad + A_{RL} + A_{LR} + A_{LL}\end{aligned} \tag{3.2}$$

where the index *cycl* means sum over cyclic permutations, m_3 is the multiplication map, and A_{RL}, A_{LR} and A_{LL} are similar terms involving (R_i, R_j, L_k, L_l), (L_i, L_j, R_k, R_l), (L_i, L_j, L_k, L_l) respectively.

Now, it is clear that the l.h.s of (3.2) is skew-symmetric with respect to f, g, h, and hence $\mathrm{Jac}(f,g,h) = \frac{1}{2}(\mathrm{Jac}(f,g,h) - \mathrm{Jac}(f,h,g))$. Let us see what this yields on the r.h.s: the A_{RR} term gives

$$\begin{aligned}\frac{1}{2}m_3 R(r_{12}r_{13} + r_{23}r_{12} - r_{13}r_{12} - r_{13}r_{32})_{cycl}(f \otimes g \otimes h) \\ = \frac{1}{2}m_3 R([r_{12}, r_{13}] + [r_{12}, r_{23}] + [r_{13}, r_{23}])(f \otimes g \otimes h) \\ = \frac{1}{2}m_3 R(\mathrm{CYB}(r))(f \otimes g \otimes h).\end{aligned}$$

In a similar way, the A_{LL} term is given by $-\frac{1}{2}m_3 L(\mathrm{CYB}(r))(f \otimes g \otimes h)$, and the A_{RL} and A_{LR} terms vanish because $[L_i, R_j] = 0$ for all i, j. To sum up, we have shown that

$$\mathrm{Jac}(f,g,h) = \frac{1}{2}m_3(R(\mathrm{CYB}(r)) - L(\mathrm{CYB}(r)))(f \otimes g \otimes h).$$

Hence, the condition that Π_r be a Poisson bivector is equivalent to the condition $R(\mathrm{CYB}(r)) - L(\mathrm{CYB}(r)) = 0$, which exactly means that $\mathrm{CYB}(r)$ is $\mathfrak{g}$-invariant. This completes the proof of Theorem 3.1.

□

There is yet another, perhaps more conceptual way of writing the above proof of Theorem 3.1, using the Schouten bracket: let M be any manifold, and let $\Pi \in \Gamma(\Lambda^2 TM)$. As we saw in 1.4.1, the Jacobi identity for the bracket $\{\,,\,\} : \Lambda^2(\mathcal{C}^\infty(M)) \to \mathcal{C}^\infty(M)$ defined by $\{f, g\} = df \otimes dg(\Pi)$ is equivalent to the equation $[\Pi, \Pi]_s = 0$. If $M = G$ is a Poisson-Lie group and $\Pi = d\rho \cdot r - d\lambda \cdot r$, we have

$$\begin{aligned}[\Pi, \Pi]_s &= [d\rho \cdot r - d\lambda \cdot r, d\rho \cdot r - d\lambda \cdot r]_s \\ &= d\rho \cdot CYB(r) - d\lambda \cdot CYB(r),\end{aligned}$$

Thus Π is a Poisson bivector field if and only if $[r, r]_s$ is $\mathfrak{g}$-invariant.

3.4 Triangular Lie bialgebras and the classical Yang-Baxter equation

As we have seen, $(\mathfrak{g}, [\,,\,], \partial r)$ is a Lie bialgebra if and only if $\mathrm{CYB}(r)$ is $\mathfrak{g}$-invariant. This is the case in particular when $\mathrm{CYB}(r) = 0$.

Definition: The equation $CYB(r) = 0$ is called the *classical Yang-Baxter equation*, and will be henceforth denoted by CYBE. Any solution of the CYBE in $\mathfrak{g} \otimes \mathfrak{g}$ is called an *r-matrix*.

Definition: A coboundary Lie bialgebra $(\mathfrak{g}, [\,,\,], r)$ is *triangular* if $\mathrm{CYB}(r) = 0$. Likewise, a *triangular structure* on a Lie bialgebra $(\mathfrak{g}, [\,,\,], \delta)$ (respectively on a Lie algebra $\mathfrak{g}$) is an element $r \in \Lambda^2\mathfrak{g}$ such that $\partial r = \delta$ and $\mathrm{CYB}(r) = 0$ (respectively $\mathrm{CYB}(r) = 0$). The element r is called the *r-matrix* of $\mathfrak{g}$.

Thus there is a 1-1 correspondence between triangular Lie bialgebras $(\mathfrak{g}, [\,,\,], r)$ and solutions of the CYBE in $\Lambda^2\mathfrak{g}$.

In particular, if $\phi : \mathfrak{g} \to \mathfrak{a}$ is a Lie algebra homomorphism then any triangular structure r on $\mathfrak{g}$ induces a triangular structure $(\phi \otimes \phi)r$ on $\mathfrak{a}$. Note that this is not true for a general coboundary structure.

We will denote by $TLBA(k)$ the category of triangular Lie bialgebras defined over the field k.

The following proposition will enable us to classify all the skew solutions of the CYBE. We keep the notations of Theorem 3.1.

Proposition 3.1. *Let $r \in \Lambda^2\mathfrak{g}$ be an r-matrix. Then the bivectors*

$$\Pi_r^L(x) = (d_e\lambda_x \otimes d_e\lambda_x)r,$$
$$\Pi_r^R(x) = (d_e\rho_x \otimes d_e\rho_x)r,$$

are Poisson bivectors, i.e they define Poisson structures $\{,\}^L$ and $\{,\}^R$ respectively on G.

Proof: Let us set $\mathrm{Jac}^L(f,g,h) = \{\{f,g\}^L,h\}^L_{cycl}$. The computation in the proof of Theorem 3.1 implies that

$$\mathrm{Jac}^L(f,g,h) = -\frac{1}{2}m_3L(\mathrm{CYB}(r))(f \otimes g \otimes h) = 0$$

hence $\{,\}^L$ is a Poisson bracket. The same proof works for $\{,\}^R$.

□

Remark: These Poisson brackets are not Poisson-Lie structures on G (for example, $\Pi^L(e) = \Pi^R(e) = r \neq 0$ in general). The usual Poisson-Lie bivector is the *difference* $\Pi = \Pi^R - \Pi^L$.

3.5 Classification of triangular structures

Definition: A triangular Lie bialgebra $(\mathfrak{g}, [\,,\,], r)$ is *nondegenerate* if $r \in \Lambda^2\mathfrak{g}$ is nondegenerate (i.e induces a nondegenerate map $\mathfrak{g}^* \to \mathfrak{g}$).

The following construction will enable us to reduce any triangular structure to a nondegenerate one and will be often used:

Proposition 3.2. *Let $r \in \mathfrak{g} \otimes \mathfrak{g}$ be a solution to the CYBE. Let*

$$\mathfrak{g}_r^+ = \mathrm{Span}\{(Id \otimes f)r \mid f \in \mathfrak{g}^*\},$$
$$\mathfrak{g}_r^- = \mathrm{Span}\{(f \otimes Id)r \mid f \in \mathfrak{g}^*\}.$$

Then $\mathfrak{g}_r^+$ and $\mathfrak{g}_r^-$ are (finite dimensional) Lie subalgebras of $\mathfrak{g}$.

Proof: We will give the proof for $\mathfrak{g}_r^+$ (the case of $\mathfrak{g}_r^-$ is treated in a similar way). Let $r = \sum_1^l a_i \otimes b_i$ be a minimal expression for r (i.e r cannot be written as a sum of less than l terms). Then $\{a_i\}$ is a basis of $\mathfrak{g}_r^+$, and $\{b_i\}$ is a basis of $\mathfrak{g}_r^-$. Let us complete $\{b_i\}$ to a basis of $\mathfrak{g}$, and let $\{b_i^*\}$ be the dual basis of $\mathfrak{g}^*$. Applying the operator $1 \otimes b_i^* \otimes b_j^*$ to

$$\mathrm{CYB}(r) = \sum_{k,l} \left([a_k, a_l] \otimes b_k \otimes b_l + a_k \otimes [b_k, a_l] \otimes b_l + a_k \otimes a_l \otimes [b_k, b_l]\right) = 0$$

we obtain

$$[a_i, a_j] + \sum_k \lambda_k^{i,j} a_k = 0$$

for some constants $\lambda_k^{i,j}$. This concludes the proof of the proposition.

□

Remark: In the same fashion, one proves that $\mathfrak{g}_+$ and $\mathfrak{g}_-$ are naturally subcoalgebras of $\mathfrak{g}$ (with the cocommutator induced by r). This is based on the identities $(\delta \otimes 1)(r) = [r_{13}, r_{23}]$ and $(1 \otimes \delta)(r) = [r_{13}, r_{12}]$, which in turn follow from the CYBE.

Now let $(\mathfrak{g}, [\,,\,], r)$ be any triangular Lie bialgebra. Keeping the notations of the proposition, and using the fact that $r \in \Lambda^2\mathfrak{g}$, we see that $\mathfrak{h} := \mathfrak{g}_r^+ = \mathfrak{g}_r^-$ is a Lie subalgebra of $\mathfrak{g}$, and that $(\mathfrak{h}, [\,,\,], r)$ is a nondegenerate triangular Lie bialgebra. It thus remains to classify such structures. It turns out that these have a nice interpretation in terms of Lie algebra cohomology.

Proposition 3.3. *An element $r \in \Lambda^2\mathfrak{h}$ defines a nondegenerate triangular Lie bialgebra structure on a finite dimensional Lie algebra $\mathfrak{h}$ if and only if the induced form $r^{-1} \in \Lambda^2\mathfrak{h}^*$ is a nondegenerate 2-cocycle on $\mathfrak{h}$ with values in the trivial representation k.*

Proof: As for Theorem 3.1, we treat the case $k = \mathbb{R}$ and give a geometric proof, which easily generalizes to an arbitrary field k of characteristic zero using the theory of formal groups. Let $r \in \Lambda^2\mathfrak{h}$ be a nondegenerate element. Let H be the simply connected Lie group with Lie algebra $\mathfrak{h}$ and recall the notations from Proposition 3.1. The bivector Π^L induces a left-invariant Poisson structure on H if and only if r is an r-matrix. Moreover, the Poisson structure is then nondegenerate, and hence symplectic. But a left-invariant 2-form ω is closed if and only if $\omega(e)$ is a $2-$ *cocycle* of $\mathfrak{h}$ with trivial coefficients. Thus, the proposition follows from the fact that $\omega(e) = r^{-1}$.

□

Remark: Notice the difference between the CYBE and the condition that a map $\Lambda^2\mathfrak{h} \to k$ be a 2-cocycle. The CYBE is quadratic while the cocycle condition is linear. Therefore, the reformulation of the CYBE in terms of 2-cocycles is a useful simplification.

The results of this section can thus be summarized as follows

Theorem 3.2. *There is a 1-1 correspondence between triangular structures on a Lie algebra* $\mathfrak{g}$ *and pairs* $(\mathfrak{h}, B)$ *where* $\mathfrak{h} \subset \mathfrak{g}$ *is a finite dimensional Lie subalgebra and* $B \in \Lambda^2\mathfrak{h}^*$ *is a nondegenerate 2-cocycle with values in the trivial representation.*

3.6 Quasitriangular Lie bialgebras

The notion of a triangular Lie bialgebra is too restrictive, as we will see in the examples (and in the next lecture). We define here a slightly more general type of coboundary Lie bialgebras:

Definition: An element $\tilde{r} \in \mathfrak{g} \otimes \mathfrak{g}$ such that $\tilde{r} + \tilde{r}^{21}$ is $\mathfrak{g}$-invariant is a *quasitriangular structure* for a Lie bialgebra $(\mathfrak{g}, [\,,\,], \delta)$ if $\delta = \partial\tilde{r}$ and $\mathrm{CYB}(\tilde{r}) = 0$. A *quasitriangular Lie bialgebra* is a triple $(\mathfrak{g}, [\,,\,], \tilde{r})$ such that $\tilde{r}$ satisfies the above conditions and that $(\mathfrak{g}, [\,,\,], \partial\tilde{r})$ is a Lie bialgebra.

In other words, the cocommutator is given by an r-matrix whose symmetric part is $\mathfrak{g}$-invariant. In particular, a triangular Lie bialgebra is quasitriangular.

We will denote by $QTLBA(k)$ the category of quasitriangular Lie bialgebras defined over k.

Remarks: (i) It is easy to see that any quasitriangular Lie bialgebra has a coboundary structure. Indeed, $\delta = \partial\tilde{r} = \partial r$, where $r = \frac{\tilde{r}-\tilde{r}^{21}}{2}$ is the skew part of $\tilde{r}$. However, r is not an r-matrix (i.e $\mathrm{CYB}(r) \neq 0$) in general. Conversely, one can check that a coboundary Lie bialgebra $(\mathfrak{g}, [\,,\,], r)$ has a quasitriangular structure if and only if there exists $T \in (S^2\mathfrak{g})^{\mathfrak{g}}$ such that $\mathrm{CYB}(r) = \frac{1}{4}[T_{12}, T_{23}]$, in which case we can choose $\tilde{r} = r + \frac{T}{2}$.
(ii) As opposed to the case of triangular structures, the image of a quasitriangular structure is not necessarily quasitriangular and even does not necessarily define a coboundary structure.

3.7 Examples of coboundary, triangular and quasitriangular Lie bialgebras

Let us return to our favorite examples, and see which are coboundary.

Example 3.1 (Trivial Lie bialgebra). A trivial Lie bialgebra obviously has a zero triangular structure. Conversely, a Lie bialgebra structure on an abelian Lie algebra is coboundary if and only if it is trivial, so the dual $\mathfrak{g}^*$ of a Lie algebra $\mathfrak{g}$ has no coboundary structure unless $\mathfrak{g}$ is abelian.

Example 3.2 (Two dimensional Lie bialgebras). Recall the notations from Lecture 2. It is easy to see that $H^1(T_2, \Lambda^2 T_2) = k$, and that coboundaries are given by

$$\begin{aligned} \delta_\alpha : T_2 &\to \Lambda^2 T_2, \\ x &\mapsto \alpha x \wedge y, \\ y &\mapsto 0. \end{aligned}$$

Thus the Lie bialgebra $\tilde{b}_2$ has a coboundary structure (which is triangular since $\Lambda^3 T_2 = 0$) with $r = x \wedge y$, but the Lie bialgebras $b_2(\beta)$ do not (the cocycle δ_β has cohomology class $\beta \in H^1(T_2, \Lambda^2 T_2)$).

Example 3.3 (Standard structure on $\mathfrak{g} = sl_2(\mathbb{C})$). Recall that it is given by the cocommutator

$$\delta(e) = \frac{1}{2} e \wedge h, \qquad \delta(f) = \frac{1}{2} f \wedge h, \qquad \delta(h) = 0.$$

It has a quasitriangular structure, given by the r-matrix $\tilde{r} = e \otimes f + \frac{1}{4} h \otimes h$ and we have $\tilde{r} + \tilde{r}^{12} = e \otimes f + f \otimes e + \frac{1}{2} h \otimes h = \Omega \in (S^2\mathfrak{g})^{\mathfrak{g}}$ (the *Casimir* element). Note that this Lie bialgebra does not admit a triangular structure: it is well known that $(\mathfrak{g} \otimes \mathfrak{g})^{\mathfrak{g}} = \mathbb{C}\Omega$, hence $(\Lambda^2 \mathfrak{g})^{\mathfrak{g}} = 0$, and $r = \frac{\tilde{r} - \tilde{r}^{21}}{2} = \frac{1}{2}(e \otimes f - f \otimes e)$ is the only skew-symmetric coboundary structure (and is not an r-matrix).

We end this lecture with some problems:

Problem 3.1 (Kupersmidt-Stoyanov,[KS]). Show that any coboundary Lie bialgebra structure on the algebra $\mathfrak{a} = \mathbb{C}[[x]]\frac{d}{dx}$ of formal vector fields on $\mathbb{A}^1$ can be obtained as an image of the $\tilde{b}_2$ Lie bialgebra structure of T_2 under a homomorphism $T_2 \to \mathfrak{a}$, and classify all such coboundary structures up to isomorphism.

Problem 3.2 (Balinsky-Burman, [BB]). Let V be a finite-dimensional vector space. We say that a Poisson bracket on V is of *degree n* if the Poisson bivector is a polynomial on V of degree n. Poisson brackets of degree $0, 1, 2$ are called constant, linear and quadratic respectively.
(a) Let A be an associative, finite dimensional algebra with unit over $\mathbb{R}$, and let

$\{\,,\}$ be a Poisson bracket on A, such that the multiplication map $m : A \times A \to A$ is a Poisson map. Show that the Poisson bracket is quadratic.
(b) Let G_A be the connected component of the identity in the multiplicative group of A. Show that any coboundary Poisson-Lie structure on G_A extends to a quadratic one on A.

Lecture 4

Drinfeld's double construction

This lecture is dedicated to a fundamental construction due to Drinfeld (the *double* construction), which will be one of the key steps in the quantization of Poisson-Lie groups and Lie bialgebras.

4.1 Manin triples

As usual, we let k be any field of characteristic zero.

Definition: A *finite dimensional Manin triple* is a triple of finite dimensional Lie algebras $(\mathfrak{g}, \mathfrak{g}_+, \mathfrak{g}_-)$, where $\mathfrak{g}$ is equipped with a nondegenerate invariant bilinear form $<,>$ such that

1. $\mathfrak{g}_+$, $\mathfrak{g}_-$ are Lie subalgebras of $\mathfrak{g}$ such that $\mathfrak{g}_+ \oplus \mathfrak{g}_- = \mathfrak{g}$ as a vector space,

2. $\mathfrak{g}_+$ and $\mathfrak{g}_-$ are isotropic with respect to $<,>$.

In particular, since $<,>$ is nondegenerate, $\mathfrak{g}_+$ and $\mathfrak{g}_-$ are maximal isotropic subalgebras (*Lagrangian subalgebras*).

The importance of the notion of a Manin triple comes from the following construction:

Let $(\mathfrak{g}, \mathfrak{g}_+, \mathfrak{g}_-)$ be a (finite dimensional) Manin triple. Then the nondegenerate form $<,>$ induces a nondegenerate pairing $\mathfrak{g}_+ \otimes \mathfrak{g}_- \to k$, and hence a Lie algebra structure on $\mathfrak{g}_+^* \simeq \mathfrak{g}_-$. Let us denote the induced coalgebra structure on $\mathfrak{g}_+$ by δ. To show that $(\mathfrak{g}_+, [\,,\,], \delta)$ is a Lie bialgebra, we must check the cocycle condition

$$\delta([a,b]) = ad(a)\delta(b) - ad(b)\delta(a)$$

which we do by direct computation. Let $(e_i)_1^n$ be a basis of $\mathfrak{g}_+$ and let (e_i^*) be the dual basis in $\mathfrak{g}_- \simeq \mathfrak{g}_+^*$. Set $[e_i, e_j] = \sum_s \alpha_{ij}^s e_s$ and $[e_i^*, e_j^*] = \sum_s \beta_s^{ij} e_s^*$.

Then, by definition, $\delta(e_i) = \sum_{s,t} \beta_i^{st} e_s \otimes e_t$. We have

$$\begin{aligned}(e_r^* \otimes e_s^*, \delta[e_k, e_l]) &= ([e_r^*, e_s^*], [e_k, e_l]) \\ &= ([[e_r^*, e_s^*], e_k], e_l) \\ &= ([[e_r^*, e_k], e_s^*] + [e_r^*, [e_s^*, e_k]], e_l) \\ &= \sum_t \alpha_{kt}^r \beta_l^{ts} + \sum_t \alpha_{tl}^s \beta_k^{rt} + \sum_t \alpha_{kt}^s \beta_l^{rt} + \sum_t \alpha_{tl}^r \beta_k^{st}.\end{aligned} \tag{4.1}$$

In a similar way, we compute

$$\begin{aligned}(e_r^* \otimes e_s^*, ad(e_k)(\delta(e_l))) &= ([e_r^*, e_k] \otimes e_s^* + e_r^* \otimes [e_s^*, e_k], \delta(e_l)) \\ &= \sum_t \alpha_{kt}^r \beta_l^{ts} + \sum_t \alpha_{kt}^s \beta_l^{rt}\end{aligned} \tag{4.2}$$

and

$$\begin{aligned}-(e_r^* \otimes e_s^*, ad(e_l)(\delta(e_k))) &= -([e_r^*, e_l] \otimes e_s^* + e_r^* \otimes [e_s^*, e_l], \delta(e_k)) \\ &= \sum_t \alpha_{lt}^r \beta_k^{ts} + \sum_t \alpha_{lt}^s \beta_k^{rt}.\end{aligned} \tag{4.3}$$

Comparing (4.1), (4.2) and (4.3) gives the cocycle identity. Therefore $\mathfrak{g}_+$ is a Lie bialgebra, and $\mathfrak{g}_-$ is its dual Lie bialgebra.

Conversely, if $\mathfrak{a}$ is a Lie bialgebra, we construct a Manin triple in the following way: $\mathfrak{g} = \mathfrak{a} \oplus \mathfrak{a}^*$, with nondegenerate form given by

$$< x + y, x' + y' >= y(x') + y'(x) \qquad x, x' \in \mathfrak{a},\ y, y' \in \mathfrak{a}^*$$

To extend the Lie algebra bracket to the whole of $\mathfrak{g}$ in such a way that $<,>$ is invariant, i.e that

$$\begin{aligned}&< [y, x], x' > =< y, [x, x'] >, \qquad x, x' \in \mathfrak{a},\ y, y' \in \mathfrak{a}^*. \\ &< [x, y], y' > =< x, [y, y'] >,\end{aligned}$$

we must set

$$[y, x] = ad^*(x)(y) - ad^*(y)(x) \tag{4.4}$$

where ad^* denotes the coadjoint actions of $\mathfrak{a}$ on $\mathfrak{a}^*$ and $\mathfrak{a}^*$ on $\mathfrak{a}$. A computation similar to the one above shows that this is indeed a Lie algebra bracket (i.e satisfies the Jacobi identity). Hence $(\mathfrak{g}, \mathfrak{a}, \mathfrak{a}^*)$ is a Manin triple.

4.2 Drinfeld's double

The preceding construction shows that the notions of a (finite dimensional) Manin triple and a Lie bialgebra are equivalent.

Let $\mathfrak{a}$ be a Lie bialgebra and let $(\mathfrak{g}, \mathfrak{a}, \mathfrak{a}^*)$ be the associated Manin triple. We now define on the Lie algebra $\mathfrak{g}$ a Lie bialgebra structure. Set $\delta = \delta_{\mathfrak{a}} - \delta_{\mathfrak{a}^*}$. It is clear that it is a coalgebra structure on $\mathfrak{g}$. One can check directly that δ is a 1-cocycle of $\mathfrak{g}$ with coefficients in $\mathfrak{g}\otimes\mathfrak{g}$, but this also follows from the computation below and the results of the previous lecture (the minus sign is critical here). Let us keep the notations of the preceding section. Let $\tilde{r} = \sum e_i \otimes e_i^* \in \mathfrak{a}\otimes\mathfrak{a}^*$ be the canonical element (corresponding to $Id \in \mathrm{End}(\mathfrak{a}) \simeq \mathfrak{a}\otimes\mathfrak{a}^*$). For $x \in \mathfrak{a}$, we have

$$\begin{aligned}
\partial\tilde{r}(x) &= [x\otimes 1 + 1\otimes x, \tilde{r}] \\
&= \sum_i [x, e_i]\otimes e_i^* + e_i \otimes [x, e_i^*] \\
&= \sum_i [x, e_i]\otimes e_i^* - e_i \otimes ad^*(x)e_i^* + e_i \otimes ad^*(e_i^*)x \\
&= \sum_i e_i \otimes ad^*(e_i^*)x \\
&= \delta(x).
\end{aligned}$$

A similar computation shows that $\partial\tilde{r}(y) = \delta(y)$ for $y \in \mathfrak{a}^*$:

$$\begin{aligned}
\partial\tilde{r}(y) &= [y\otimes 1 + 1\otimes y, \tilde{r}] \\
&= \sum_i [y, e_i]\otimes e_i^* + e_i \otimes [y, e_i^*] \\
&= \sum_i ad^*(y)e_i \otimes e_i^* - ad^*(e_i)y \otimes e_i^* + e_i \otimes [y, e_i^*] \\
&= \sum_i -ad^*(e_i)y \otimes e_i^* \\
&= \delta(y).
\end{aligned}$$

Therefore $\delta = \partial\tilde{r}$. Furthermore, $\tilde{r} + \tilde{r}^{12} = \sum_i e_i \otimes e_i^* + e_i^* \otimes e_i = \Omega$, the Casimir element corresponding to the nondegenerate invariant form $<\,,\,>$ on $\mathfrak{g}$, and is thus $\mathfrak{g}$-invariant. Finally, we have

$$\begin{aligned}
\mathrm{CYB}(\tilde{r}) &= \sum_{i,j} \left([e_j, e_i]\otimes e_j^* \otimes e_i^* + e_j \otimes [e_j^*, e_i]\otimes e_i^* + e_j \otimes e_i \otimes [e_j^*, e_i^*]\right) \\
&= \Big(\sum_{i,j} [e_j, e_i]\otimes e_j^* \otimes e_i^* + e_j \otimes ad^*(e_i)e_j^* \otimes e_i^*\Big) \\
&\quad + \Big(\sum_{i,j} e_j \otimes e_i \otimes [e_j^*, e_i^*] - e_j \otimes ad^*(e_j^*)e_i \otimes e_i^*\Big) \\
&= 0.
\end{aligned}$$

Thus $\tilde{r}$ is an r-matrix and $(\mathfrak{g}, [\,,\,], \tilde{r})$ is quasitriangular.

The Lie bialgebra thus associated to $\mathfrak{a}$ is called the *Drinfeld double* of $\mathfrak{a}$ and is denoted $D\mathfrak{a}$. We have shown the following :

Theorem 4.1 (Drinfeld). *Let $(\mathfrak{a}, [\,,\,], \delta)$ be a finite dimensional Lie bialgebra. Then the Drinfeld double $D\mathfrak{a}$ of $\mathfrak{a}$ is a quasitriangular Lie bialgebra.*

Since $\delta = \delta_{\mathfrak{a}} - \delta_{\mathfrak{a}^*}$, the embeddings $\mathfrak{a} \hookrightarrow \mathfrak{g}$, $\mathfrak{a}^{*op} \hookrightarrow \mathfrak{g}$ are Lie bialgebra maps (where $\mathfrak{a}^{*op}$ is the Lie bialgebra $\mathfrak{a}^*$ with opposite cocommutator $-\delta_{\mathfrak{a}^*}$). In particular, we have a canonical embedding of any Lie bialgebra into a quasitriangular one, in which the r-matrix is explicitly described.

In fact, Drinfeld's double is in some sense a universal construction of quasitriangular Lie bialgebras. This statement is made more precise by the following proposition.

Proposition 4.1. *Let $(\mathfrak{g}, [\,,\,], r)$ be a finite dimensional quasitriangular Lie bialgebra. Set*

$$\mathfrak{g}_+ = \mathrm{Span}\{(Id \otimes f)r \mid f \in \mathfrak{g}^*\},$$
$$\mathfrak{g}_- = \mathrm{Span}\{(f \otimes Id)r \mid f \in \mathfrak{g}^*\}.$$

Suppose that r is such that $\mathfrak{g}_+ + \mathfrak{g}_- = \mathfrak{g}$. Let $D\mathfrak{g}_+$ be the Drinfeld double of $\mathfrak{g}_+$, and let $\tilde{r}$ be its r-matrix. Then $\mathfrak{g}$ is isomorphic, as a quasitriangular Lie bialgebra, to a quotient of $D\mathfrak{g}_+$.

Proof: By Proposition 3.2, $\mathfrak{g}_+$ and $\mathfrak{g}_-$ are Lie subalgebras of $\mathfrak{g}$. We have $r \in \mathfrak{g}_+ \otimes \mathfrak{g}_-$, and it is easy to see that the map $\mathfrak{g}_+^* \to \mathfrak{g}_-$, $f \to (f \otimes Id)r$ is an isomorphism. Notice that $\mathfrak{g}_+$ and $\mathfrak{g}_-$ may have nontrivial intersection. By construction, $D\mathfrak{g}_+ = \mathfrak{g}_+ \oplus \mathfrak{g}_+^* \simeq \mathfrak{g}_+ \oplus \mathfrak{g}_-$, where the commutator is given by $[x, y] = ad^*(x)y - ad^*(y)x$, and with quasitriangular structure given by the canonical element $\tilde{r} \in \mathfrak{g}_+ \otimes \mathfrak{g}_-$.

Consider the natural map $\pi : D\mathfrak{g}_+ \to \mathfrak{g}$ which is the sum of the two inclusions $i_\pm : \mathfrak{g}_\pm \hookrightarrow \mathfrak{g}$ (notice that π is surjective but not injective if $\mathfrak{g}_+ \cap \mathfrak{g}_- \neq 0$). Now, for $x = (Id \otimes g)r \in \mathfrak{g}_+, y = (f \otimes Id)r \in \mathfrak{g}_-$, we have

$$\begin{aligned}
\pi([x, y]) &= \pi(ad^*(x)y - ad^*(y)x) \\
&= -(f([x, .]) \otimes Id)r + (Id \otimes g([y, .]))r \\
&= f \otimes Id \otimes g([r_{12}, r_{13}]) + f \otimes Id \otimes g([r_{13}, r_{23}]) \\
&= -f \otimes Id \otimes g([r_{12}, r_{23}]) \\
&= [\pi(x), \pi(y)] \ (\in \mathfrak{g}).
\end{aligned}$$

Hence, π is a Lie algebra morphism. Moreover, $(\pi \otimes \pi)\tilde{r} = r$, which implies that $\pi : D\mathfrak{g}_+ \to \mathfrak{g}$ is a (surjective) quasitriangular Lie bialgebra map. The proposition is proved.

□

More generally, the above proof implies the following:

Proposition 4.2. *Let $\mathfrak{g}$ be a Lie algebra and $r \in \mathfrak{g}\otimes\mathfrak{g}$ an r-matrix. There exists a finite dimensional Lie bialgebra $(\mathfrak{g}_+, [\,,\,], \delta)$ and a Lie algebra map $\phi : D\mathfrak{g}_+ \to \mathfrak{g}$ such that $\phi_{|\mathfrak{g}_+}$ is injective and $\phi \otimes \phi(\tilde{r}) = r$ where $\tilde{r}$ is the usual quasitriangular structure on $D\mathfrak{g}_+$.*

4.3 Examples

Example 4.1 (Triangular Lie bialgebras). Let $(\mathfrak{g}, [\,,\,], \delta)$ be a triangular Lie bialgebra, with nondegenerate r-matrix $r \in \Lambda^2\mathfrak{g}$. Then, using the notations of Proposition 4.1, we have $\mathfrak{g} = \mathfrak{g}_+ = \mathfrak{g}_-$. In this case, the double of $\mathfrak{g}_+$ is $D\mathfrak{g}_+ = (\mathfrak{g}\oplus\mathfrak{g}, [\,,\,], \delta_1 - \delta_2)$, where the commutator is given by formula (4.4) and the projection is $\pi : D\mathfrak{g}_+ \twoheadrightarrow \mathfrak{g}\,, (x_1, x_2) \mapsto x_1 - x_2$.

Example 4.2 (The standard structure on $sl_2(\mathbb{C})$). In this case, we have $r = e \otimes f + \frac{1}{4}h \otimes h$, and $\mathfrak{g}_+ = \mathbb{C}e \oplus \mathbb{C}h$, $\mathfrak{g}_- = \mathbb{C}f \oplus \mathbb{C}h$. Hence, we have $D\mathfrak{g}_+ = \mathbb{C}e_1 \oplus \mathbb{C}e_2 \oplus \mathbb{C}h_1 \oplus \mathbb{C}h_2$, with quasitriangular structure given by $\tilde{r} = e_1 \otimes e_2 + \frac{1}{4}h_1 \otimes h_2$, and the projection is $\pi : D\mathfrak{g}_+ \twoheadrightarrow sl_2(\mathbb{C})$, $e_1 \mapsto e$, $h_1 \mapsto h$, $e_2 \mapsto f$, $h_2 \mapsto h$.

4.4 Standard Lie bialgebra structure on simple Lie algebras

Let $\mathfrak{g}$ be a simple Lie algebra over $\mathbb{C}$. We will now define on $\mathfrak{g}$ an analogue of the standard (quasitriangular) Lie bialgebra structure on $sl_2(\mathbb{C})$, by the method of Proposition 4.2, i.e as a quotient of a double Lie bialgebra.

4.4.1 Notations

Let us briefly recall the structure of simple Lie algebras, and fix notations to be used throughout these notes. Let $\mathfrak{h} \subset \mathfrak{g}$ be a Cartan subalgebra, $\Delta \subset \mathfrak{h}^*$ the root system, and $A = (a_{ij})$ the Cartan matrix of $\mathfrak{g}$. Let $\Delta = \Delta^+ \cup \Delta^-$ be a polarization of Δ ($\Delta^- = -\Delta^+$) and $\Gamma = \{\alpha_1, \dots \alpha_n\}$ be the set of positive simple roots. Let us write $\mathfrak{g}_\alpha = \{x \in \mathfrak{g} \mid [h, x] = \alpha(h)x \quad \forall h \in \mathfrak{h}\}$, $\mathfrak{n}_\pm = \bigoplus_{\alpha\in\Delta^\pm} \mathfrak{g}_\alpha$ and

$$\mathfrak{g} = \mathfrak{n}^+ \oplus \mathfrak{h} \oplus \mathfrak{n}^-$$

for the usual decomposition of $\mathfrak{g}$ into the Cartan subalgebra, the positive and the negative nilpotent subalgebras. The root subspaces $\mathfrak{g}_\alpha$ are one-dimensional, and $[\mathfrak{g}_\alpha, \mathfrak{g}_\beta] = \mathfrak{g}_{\alpha+\beta}$. Let $(\,,\,)$ be the nondegenerate invariant bilinear form on $\mathfrak{g}$ normalized so that $(\alpha, \alpha) = 2$ for long roots (we will also write $(\,,\,)$ for the induced form on $\mathfrak{g}^*$), and set $d_i = \frac{2}{(\alpha_i,\alpha_i)}$. We have $(\mathfrak{g}_\alpha, \mathfrak{g}_\beta) = 0$ if $\alpha + \beta \neq 0$, and $(\,,\,)_{|\mathfrak{h}}$ is nondegenerate. For $\alpha \in \Delta^+$, choose elements $e_\alpha \in \mathfrak{g}_\alpha$ and $f_\alpha \in \mathfrak{g}_{-\alpha}$ such that $(e_\alpha, f_\alpha) = 1$, and set $h_\alpha = [e_\alpha, f_\alpha] \in \mathfrak{h}$. With this normalization, we have

$$[h_{\alpha_i}, e_\alpha] = (\alpha_i, \alpha)e_\alpha, \qquad [h_{\alpha_i}, f_\alpha] = -(\alpha_i, \alpha)f_\alpha.$$

It is well known that $\mathfrak{g}$ admits the following presentation, due to Serre: the "Chevalley" generators are $(e_i, h_i, f_i)_{i=1,\dots n}$ and the relations are

$$[e_i, f_i] = h_i, \qquad [e_i, f_j] = 0, \qquad (i \neq j)$$

$$[h_i, e_j] = a_{ij}e_i, \qquad [h_i, f_j] = -a_{ij}f_j, \qquad [h_i, h_j] = 0,$$

$$ad^{1-a_{ij}}(e_i)e_j = 0, \qquad ad^{1-a_{ij}}(f_i)f_j = 0. \qquad (*)$$

The last two equations $(*)$ are called the *Serre relations*. Note that we now have $(e_i, f_i) = d_i^{-1}$.

4.4.2 Standard structure

Note that the invariant form establishes isomorphisms $\mathfrak{n}_- \tilde{\to} \mathfrak{n}_+^*$ and $\mathfrak{h} \tilde{\to} \mathfrak{h}^*$. Thus $\mathfrak{g}$ almost has the structure of a Manin triple. To really obtain one, we use the following trick to "double" the Cartan subalgebra:

Consider the Lie algebra $\tilde{\mathfrak{g}} = \mathfrak{n}_+ \oplus \mathfrak{h}^{(1)} \oplus \mathfrak{h}^{(2)} \oplus \mathfrak{n}_-$, where $\mathfrak{h}^{(1)} \simeq \mathfrak{h}^{(2)} \simeq \mathfrak{h}$, with slightly modified commutation relations:

$$[\mathfrak{h}^{(1)}, \mathfrak{h}^{(2)}] = 0, \qquad [h^{(i)}, e_\alpha] = \alpha(h)e_\alpha, \qquad [h^{(i)}, f_\alpha] = -\alpha(h)f_\alpha,$$

$$[e_\alpha, f_\alpha] = \frac{1}{2}(h_\alpha^{(1)} + h_\alpha^{(2)}).$$

Define the projection $\pi : \tilde{\mathfrak{g}} \twoheadrightarrow \mathfrak{g}$ by $\pi_{|\mathfrak{n}_+ \oplus \mathfrak{n}_-} = \mathrm{Id}$, $\pi(h_\alpha^{(1)}) = \pi(h_\alpha^{(2)}) = h_\alpha$. It is a Lie algebra map. Now define a nondegenerate invariant bilinear form $(,)_{\tilde{\mathfrak{g}}}$ on $\tilde{\mathfrak{g}}$ by the formula:

$$(x + h^{(1)} + h^{(2)}, x' + h^{'(1)} + h^{'(2)})_{\tilde{\mathfrak{g}}} = 2((h^{(1)}, h^{'(2)}) + (h^{(2)}, h^{'(1)})) + (x, x').$$

It is easy to check that $(\tilde{\mathfrak{g}}, \mathfrak{n}_+ \oplus \mathfrak{h}^{(1)}, \mathfrak{n}_- \oplus \mathfrak{h}^{(2)})$ is a Manin triple. Thus $\tilde{\mathfrak{g}}$ has a structure of quasitriangular Lie bialgebra with r-matrix given by

$$\tilde{r} = \sum_{\alpha \in \Delta^+} e_\alpha \otimes f_\alpha + \frac{1}{2} \sum_i k_i^{(1)} \otimes k_i^{(2)}$$

where (k_i) is an orthonormal basis of $\mathfrak{h}$ for $(,)$. The projection $\pi : \tilde{\mathfrak{g}} \twoheadrightarrow \mathfrak{g}$ therefore endows $\mathfrak{g}$ with a quasitriangular Lie bialgebra structure with r-matrix

$$r = \sum_{\alpha \in \Delta^+} e_\alpha \otimes f_\alpha + \frac{1}{2} \sum_i k_i \otimes k_i$$

Notice that $r + r^{21} = \Omega$, the Casimir of $\mathfrak{g}$. This structure is called the *standard structure*. The Lie subalgebras $\mathfrak{b}_\pm = \mathfrak{n}_\pm \oplus \mathfrak{h}$ are Lie subbialgebras, and the above construction shows that $\mathfrak{g}$ is "almost" the double of $\mathfrak{b}_+$ (or $\mathfrak{b}_-$).

Exercise 4.1. Show that $\tilde{\mathfrak{g}}$ is naturally isomorphic to the Lie algebra $\mathfrak{g} \oplus \mathfrak{h}$, and that the invariant bilinear form above is mapped under this isomorphism to the form $(,)_\mathfrak{g} - (,)_\mathfrak{h}$.

Let us compute the cocommutator for the standard structure. We have

$$\begin{aligned}\delta(k_i) &= ad(k_i)r = 0\\ \delta(e_\beta) &= ad(e_\beta)r = \sum_\alpha [e_\beta, e_\alpha] \otimes f_\alpha + \sum_\alpha e_\alpha \otimes [e_\beta, f_\alpha]\\ &\quad + \frac{1}{2}\Big(\sum_i [e_\beta, k_i] \otimes k_i + k_i \otimes [e_\beta, k_i]\Big)\\ &= e_\beta \otimes h_\beta - \frac{1}{2}(e_\beta \otimes h_\beta + h_\beta \otimes e_\beta)\\ &= \frac{1}{2} e_\beta \wedge h_\beta\end{aligned}$$

for *simple* roots β. A similar computation shows that

$$\delta(f_\beta) = \frac{1}{2} f_\beta \wedge h_\beta.$$

The cobracket of e_α, f_α for a general root α can be computed from these equalities using the cocycle identity.

The presentation in terms of Chevalley generators is slightly different because $(e_i, f_i) = d_i^{-1}$, so we have the following formulas:

$$\delta(h_i) = 0, \qquad \delta(e_i) = \frac{d_i}{2} e_i \wedge h_i, \qquad \delta(f_i) = \frac{d_i}{2} f_i \wedge h_i.$$

In conclusion, let us give the following straightforward problem:

Problem 4.1 (Extension of scalars). Let A be a finite dimensional commutative algebra which is *Frobenius*: there exists a *trace* map $tr : A \to k$ such that the bilinear form $(a, b)_A = tr(ab)$ is nondegenerate (for example, $A = \mathbb{C}[t]/(t^{n+1})$ and $(t^i, t^j) = \delta_{i+j,n}$). Let $(\mathfrak{g}, \mathfrak{g}_+, \mathfrak{g}_-)$ be a Manin triple with bilinear form $(\,,\,)_\mathfrak{g}$. Let $\mathfrak{g}_A = \mathfrak{g} \otimes_k A$, and $\mathfrak{g}_{\pm,A} = \mathfrak{g}_\pm \otimes_k A$. Then the bilinear form $(\,,\,)_\mathfrak{g} \otimes (\,,\,)_A$ on $\mathfrak{g}_A$ is invariant and nondegenerate, and $(\mathfrak{g}_A, \mathfrak{g}_{+,A}, \mathfrak{g}_{-,A})$ is a Manin triple.

Lecture 5

Belavin-Drinfeld classification (I)

The results of the last lecture (Proposition 4.1 and Proposition 4.2) and of lecture 3 (Proposition 3.3) are not very efficient for classifying r-matrices explicitly, and for a general Lie algebra $\mathfrak{g}$, there is no known classification or explicit construction. However, things improve when $\mathfrak{g}$ is a simple Lie algebra. In this case, Belavin and Drinfeld [BD1] gave in 1982 a complete description of *non-skew* r-matrices giving rise to Lie bialgebra structures on $\mathfrak{g}$. This lecture is devoted to explaining this classification.

5.1 Coboundary structure on simple Lie bialgebras

Recall the two following standard results about simple Lie algebras:

Theorem 5.1 (Whitehead). *Let $\mathfrak{g}$ be a simple Lie algebra over complex numbers, and V any finite dimensional $\mathfrak{g}$-module. Then $H^1(\mathfrak{g}, V) = H^2(\mathfrak{g}, V) = 0$.*

Theorem 5.2. *Let $\mathfrak{g}$ be a simple Lie algebra over $\mathbb{C}$ and let $\Omega \in \mathfrak{g} \otimes \mathfrak{g}$ be the Casimir element associated to a nondegenerate invariant form on $\mathfrak{g}$. Then*

$$(\Lambda^3 \mathfrak{g})^{\mathfrak{g}} = \mathbb{C}Z, \qquad Z = [\Omega_{12}, \Omega_{23}].$$

Using these, we can prove

Proposition 5.1. *Let $(\mathfrak{g}, [\, , \,], \delta)$ be a Lie bialgebra where $\mathfrak{g}$ is a simple Lie algebra. Then $(\mathfrak{g}, [\, , \,], \delta)$ is quasitriangular and admits two quasitriangular structures $\overline{r}_1$ and $\overline{r}_2$, related to each other by $\overline{r}_1 = -\overline{r}_2^{op}$ (which may or may not coincide). Moreover, any quasitriangular structure on $(\mathfrak{g}, [\, , \,], \delta)$ equals r_1 or r_2.*

Proof: By Theorem 5.1, $H^1(\mathfrak{g}, \Lambda^2 \mathfrak{g}) = 0$. Therefore $\delta : \mathfrak{g} \to \Lambda^2 \mathfrak{g}$ is a coboundary, i.e $\delta = \partial r$ for some $r \in \Lambda^2 \mathfrak{g}$. This r is unique since $\Lambda^2(\mathfrak{g}) \cap (\mathfrak{g} \otimes \mathfrak{g})^{\mathfrak{g}} = \{0\}$. Moreover, by Theorem 5.2 we have $CYB(r) \in (\Lambda^3 \mathfrak{g})^{\mathfrak{g}} = \mathbb{C}[\Omega_{12}, \Omega_{23}]$. Hence, $(\mathfrak{g}, [\, , \,], \delta)$ admits two quasitriangular structures given by $\overline{r}_1 = r + \lambda\Omega$ and

$\bar{r}_2 = r - \lambda\Omega$. These structures coincide for $\lambda = 0$ (in which case they are triangular structures) and are distinct otherwise.

□

In other terms, the classification of Lie bialgebra structures on $\mathfrak{g}$ reduces to the classification of r-matrices satisfying $r + r_{21} \in \mathbb{C}\Omega$.

5.2 Skew-symmetric r-matrices

As we have seen in Lecture 3 (Proposition 3.3), there is a one-to-one correspondence between triangular structures on $\mathfrak{g}$ and pairs $(\mathfrak{h}, r^{-1})$ where $\mathfrak{h}$ is a Lie subalgebra of $\mathfrak{g}$ and r^{-1} is a nondegenerate 2-cocycle on $\mathfrak{h}$ with values in the trivial representation $\mathbb{C}$. This classification is all but explicit as it involves computing $H^2(\mathfrak{h}, \mathbb{C})$ for *all* subalgebras $\mathfrak{h}$ of $\mathfrak{g}$ and, as we know, *any* (finite dimensional) Lie algebra can be embedded in $gl(N)$ for N large enough, and hence into $sl(N)$ for N large enough! The only thing we can say is

Proposition 5.2. *Let $\mathfrak{g}$ be a simple Lie algebra. There are no nondegenerate triangular structures on $\mathfrak{g}$.*

Proof: By Whitehead's theorem, we have $H^2(\mathfrak{g}, \mathbb{C}) = 0$. Hence every 2-cocycle $c : \Lambda^2\mathfrak{g} \to \mathbb{C}$ is a coboundary: $c(x \wedge y) = f([x, y])$ for some $f \in \mathfrak{g}^*$. Let us fix a nondegenerate invariant bilinear form $(\, , \,)$ on $\mathfrak{g}$. Then there exists $z \in \mathfrak{g}$ such that $c(x \wedge y) = ([x, y], z) = (x, [y, z])$. But then z is in the kernel of c (regarded as a bilinear form) and c is degenerate.

□

5.3 Non skew-symmetric r-matrices

We let $\mathfrak{g}$ denote a simple Lie algebra and $(\, , \,)$ an invariant nondegenerate form on $\mathfrak{g}$. Let $\mathfrak{h} \subset \mathfrak{g}$ be a Cartan subalgebra, $\Delta \subset \mathfrak{h}^*$ the associated root system, and $\Gamma \subset \Delta$ the system of simple roots. Also let $(e_\alpha, f_\alpha, h_\alpha)$, $\alpha \in \Gamma$ be the system of Chevalley generators defined in the previous lecture. Finally, let $\Omega \in \mathfrak{g} \otimes \mathfrak{g}$ be the Casimir corresponding to $(\, , \,)$, and $\Omega_0 \in \mathfrak{h} \otimes \mathfrak{h}$ its "$\mathfrak{h}$-part".

We are now looking for r-matrices $r \in \mathfrak{g} \otimes \mathfrak{g}$ such that

$$r + r^{21} = \epsilon\Omega \qquad (\epsilon \neq 0). \tag{5.1}$$

Rescaling if necessary, we will assume $\epsilon = 1$.
Let us first describe the set of solutions, as given by Belavin and Drinfeld. They are classified by the following combinatorial data:

Definition A *Belavin-Drinfeld triple* is a triple $(\Gamma_1, \Gamma_2, \tau)$ where $\Gamma_1, \Gamma_2 \subset \Gamma$ and $\tau : \Gamma_1 \to \Gamma_2$ such that

1. τ is a bijective map preserving the form $(\,,\,)$,
2. for any $\gamma_1 \in \Gamma_1$, there exists $n > 0$ such that $\tau^n(\gamma_1) \in \Gamma_2 \backslash \Gamma_1$.

A typical example of Belavin-Drinfeld triple is as follows (for $\mathfrak{g} = sl_n$):

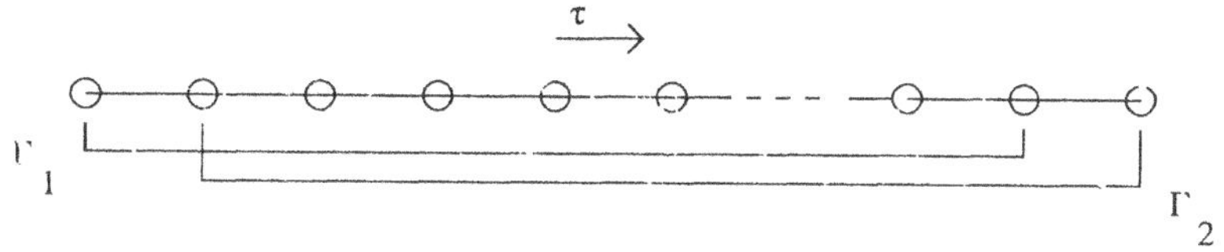

Figure 5.1: A Belavin-Drinfeld triple for sl_n (the Cremmer-Gervais triple)

Let $(\Gamma_1, \Gamma_2, \tau)$ be a Belavin-Drinfeld triple, and let $\mathfrak{g}_{\Gamma_i}$ be the subalgebra generated by $\mathfrak{g}_{\pm\alpha}$, $\alpha \in \Gamma_i$ (i.e the sum of root subspaces corresponding to roots generated by elements of Γ_i).

By 1., the isomorphism $\tau : \Gamma_1 \to \Gamma_2$ extends to a map $\mathbb{Z}\Gamma_1 \xrightarrow{\sim} \mathbb{Z}\Gamma_2$, and induces an isomorphism $\tau : \mathfrak{g}_{\Gamma_1} \xrightarrow{\sim} \mathfrak{g}_{\Gamma_2}$, defined on generators by $\tau(e_\alpha) = e_{\tau(\alpha)}$ and $\tau(f_\alpha) = f_{\tau(\alpha)}$ for $\alpha \in \Gamma_1$. We associate to it a basis of $\mathfrak{g}$ in the following way: we choose $x_\alpha \in \mathfrak{g}_\alpha$ in such a way that $(x_\alpha, x_{-\alpha}) = 1$ and $\tau(x_\alpha) = x_{\tau(\alpha)}$ for *all* $\alpha \in \mathbb{Z}\Gamma_1$. This is made possible by property 2. Note that the choice of (x_α) is not unique.

Define a partial order on Δ_+ by $\alpha < \beta$ if there exists $n > 0$ such that $\tau^n(\alpha) = \beta$ (in particular, $\alpha \in \mathbb{Z}\Gamma_1$ and $\beta \in \mathbb{Z}\Gamma_2$) (again, this uses property 2.).

Theorem 5.3 (Belavin-Drinfeld [BD1]). *Let $\mathfrak{g}$ be a simple Lie algebra, with invariant nondegenerate form $(\,,\,)$. Let $\mathfrak{g} = \mathfrak{n}_- \oplus \mathfrak{h} \oplus \mathfrak{n}_+$ be a triangular decomposition of $\mathfrak{g}$, and $(\Gamma_1, \Gamma_2, \tau)$ a Belavin-Drinfeld triple. Let $r_0 \in \mathfrak{h} \otimes \mathfrak{h}$ satisfy*

$$r_0 + r_0^{21} = \Omega_0, \tag{5.2}$$

$$(\tau\alpha \otimes 1)r_0 + (1 \otimes \alpha)r_0 = 0 \qquad (\alpha \in \Gamma_1). \tag{5.3}$$

Now define

$$r = r_0 + \sum_{\alpha\in\Delta_+} x_{-\alpha} \otimes x_\alpha + \sum_{\alpha,\beta\in\Delta_+,\alpha<\beta} x_{-\alpha} \wedge x_\beta. \tag{5.4}$$

Then

1. *r is an r-matrix satisfying $r + r^{21} = \Omega$,*
2. *Any r-matrix satisfying $r + r^{21} = \Omega$ is of the above form (for a suitable decomposition $\mathfrak{g} = \mathfrak{n}_- \oplus \mathfrak{h} \oplus \mathfrak{n}_+$, and a suitable choice of a basis (x_α)).*

5.4 Proof of the classification theorem

Let $r \in \mathfrak{g} \otimes \mathfrak{g}$, and consider the map $f : \mathfrak{g} \to \mathfrak{g}$ defined by $r = (f \otimes 1)\Omega$, i.e $r = \sum_\nu f(I_\nu) \otimes I_\nu$ for an orthonormal basis (I_ν) of $\mathfrak{g}$. Let us rewrite the CYBE and the condition (5.1) in terms of f:

Lemma 5.1. *The CYBE and condition (5.1) are equivalent to the system*

$$f + f^* = 1, \tag{5.5}$$

$$(f-1)[f(x), f(y)] = f([(f-1)x, (f-1)y]). \tag{5.6}$$

where f^* stands for the adjoint of f with respect to the form $(\, , \,)$.

Proof: By (5.1), we have

$$\Omega = r + r^{21} = (f \otimes 1 + 1 \otimes f)\Omega = ((f + f^*) \otimes 1)\Omega$$

from which (5.5) follows. To see (5.6), note that in terms of f, the CYBE for r is

$$0 = \sum_{\mu,\nu} \big([f(I_\mu), f(I_\nu)] \otimes I_\mu \otimes I_\nu + f(I_\mu) \otimes [I_\nu, f(I_\mu)] \otimes I_\nu$$
$$+ f(I_\mu) \otimes f(I_\nu) \otimes [I_\mu, I_\nu]\big).$$

Applying $Id \otimes Id \otimes I_\nu^*$, and using the invariance of $(\, , \,)$ and the orthonormality of (I_λ), we get

$$0 = \sum_\mu \big([f(I_\mu), f(I_\nu)] \otimes I_\mu + f(I_\mu) \otimes [I_\mu, f(I_\nu)] - f(I_\mu) \otimes f([I_\mu, I_\nu])\big).$$

Finally, applying $Id \otimes I_\mu^*$ and using $f + f^* = 1$, we obtain

$$[f(I_\mu), f(I_\nu)] - f([I_\mu, f(I_\nu)]) - f([f(I_\mu), I_\nu]) + f([I_\mu, I_\nu]) = 0$$

which we rewrite as (5.6).

□

5.4.1 The Cayley transform

The main idea of the proof of the classification theorem is to use the *Cayley transform* θ of f. Morally, $\theta = \frac{f}{f-1}$. Let us suppose for a moment that this expression makes sense (i.e $f - 1$ is invertible), and that f satisfies the system (5.5,5.6). Then $\theta\theta^* = \frac{ff^*}{(f-1)(f^*-1)} = \frac{f(1-f)}{-(f-1)f} = 1$ and from (5.6) we see that $[\theta(x), \theta(y)] = \theta([x, y])$. Hence, if it exists, θ is an orthogonal automorphism of $\mathfrak{g}$. Unfortunately, one can show that f (and thus $1 - f$) are never invertible (i.e r is degenerate, see [BD1]). We therefore need to make the following adjustments: we will consider θ as a map

$$\theta : \mathrm{Im}\,(f-1)/\mathrm{Ker}\, f \to \mathrm{Im}\, f/\mathrm{Ker}\,(f-1).$$

Notice that $\mathrm{Ker}\, f \subset \mathrm{Im}(f-1)$ for any linear operator f. Luckily, our above claim still holds in a modified form:

Lemma 5.2. *Let $f : \mathfrak{g} \to \mathfrak{g}$ satisfy (5.5). Then we have* $\mathrm{Ker}\, f = \mathrm{Im}\,(f-1)^{\perp}$, $\mathrm{Ker}\,(f-1) = \mathrm{Im}\, f^{\perp}$ *and the map θ is orthogonal. Furthermore, f satisfies (5.6) if and only if* $\mathrm{Im}\, f$ *and* $\mathrm{Im}\,(f-1)$ *are Lie subalgebras of $\mathfrak{g}$, and θ is a Lie algebra isomorphism.*

Proof: The first part of the lemma follows easily from $f + f^* = 1$. Now suppose that f satisfies (5.6). This implies

$$[f(x), f(y)] = f([x, f(y)] + [f(x), y] - [x, y]),$$
$$[(f-1)(x), (f-1)(y)] = (f-1)([x, f(y)] + [f(x), y] - [x, y])$$

from which we conclude that $\mathrm{Im}\, f$ and $\mathrm{Im}\,(f-1)$ are subalgebras. Also, it is easy to see from (5.6) that θ is an isomorphism. Conversely if $\mathrm{Im}\, f$ and $\mathrm{Im}\,(f-1)$ are Lie subalgebras and θ an isomorphism, then for any $x, y \in \mathfrak{g}$, we have $[(f-1)x, (f-1)y] = (f-1)z$ for some $z \in \mathfrak{g}$. Since θ is a Lie algebra map, $[f(x), f(y)] - f(z) \in \ker\,(f-1)$. Subtracting, we get

$$[f(x), y] + [x, f(y)] - [x, y] - z \in \ker\,(f-1).$$

Applying $(f-1)$ yields (5.6).

□

We will use the notation $C_1 = \mathrm{Im}\,(f-1), C_2 = \mathrm{Im}\, f$. The triple (C_1, C_2, θ) is called the Cayley transform of f.

5.4.2 Proof of part 1)

Starting from a Belavin-Drinfeld triple, we will define a triple (C_1, C_2, θ) satisfying the conditions of lemma 5.2, which is the Cayley transform of a map $f : \mathfrak{g} \to \mathfrak{g}$. Let us first explain the idea of the construction. Notice that, by lemma 5.2, the subalgebras C_1, C_2 will have to satisfy $C_i^{\perp} \subset C_i$. Examples of such subalgebras are constructed in the following way. Let $L \subset \Gamma$ be a subset of simple roots. Let $\hat{L} \subset \mathbb{Z}L$ be the set of all roots generated by L. We define the following subalgebras:

$$\mathfrak{h}_L = \bigoplus_{\alpha \in L} \mathfrak{h}_{\alpha}, \qquad \mathfrak{g}_L = \bigoplus_{\alpha \in \hat{L}} \mathfrak{g}_{\alpha} \oplus \mathfrak{h}_L, \qquad \mathfrak{n}_L^{\pm} = \bigoplus_{\alpha \in \Delta_+ \setminus \hat{L}} \mathfrak{g}_{\pm\alpha}.$$

The (almost parabolic) subalgebras associated to L are defined by $\mathfrak{p}_L^{\pm} = \mathfrak{g}_L \oplus \mathfrak{n}_L^{\pm}$. Notice that $\mathfrak{n}_L^{\pm} \subset \mathfrak{p}_L^{\pm}$ are ideals. It is clear that $\mathfrak{p}_L^{\pm\perp} = \mathfrak{n}_L^{\pm} \oplus \mathfrak{h}_L^{\perp} \subset \mathfrak{p}_L^{\pm}$.

We have $\mathfrak{h}_L^{\perp} \oplus \mathfrak{h}_L = \mathfrak{h}$. If $V \subset \mathfrak{h}_L^{\perp}$ is such that $V^{\perp} \cap \mathfrak{h}_L^{\perp} \subset V$, then $\mathfrak{p}_L^{\pm} \oplus V$ is a subalgebra, satisfying

$$\begin{aligned}(\mathfrak{p}_L^{\pm} \oplus V)^{\perp} &= \mathfrak{n}_L^{\pm} \oplus (V^{\perp} \cap \mathfrak{h}_L^{\perp}) \subset \mathfrak{p}_L^{\pm} \oplus V,\\ (\mathfrak{p}_L^{\pm} \oplus V)/(\mathfrak{p}_L^{\pm} \oplus V)^{\perp} &= \mathfrak{g}_L \oplus V/(V^{\perp} \cap \mathfrak{h}_L^{\perp}).\end{aligned}$$

In fact, one can show that, up to an automorphism of $\mathfrak{g}$, any subalgebra C satisfying $C^{\perp} \subset C$ is of this type.

It is thus natural to consider the case $C_1 = \mathfrak{p}_{\Gamma_1}^{+} + V_1$, $C_2 = \mathfrak{p}_{\Gamma_2}^{-} + V_2$ for some $\Gamma_1, \Gamma_2 \subset \Gamma$, and suitable $V_1, V_2 \subset \mathfrak{h}$. We now need an orthonormal Lie algebra isomorphism $C_1/C_1^{\perp} \xrightarrow{\sim} C_2/C_2^{\perp}$. The simplest such maps are those induced by a norm-preserving bijection $\tau : \Gamma_1 \xrightarrow{\sim} \Gamma_2$ and defined on generators by

$$\theta(e_\alpha) = e_{\tau(\alpha)} \qquad \theta(f_\alpha) = f_{\tau(\alpha)} \qquad \theta(h_\alpha) = h_{\tau(\alpha)} \qquad (\alpha \in \Gamma_1)$$

(and $\theta : V_1/V_1^{\perp} \xrightarrow{\sim} V_2/V_2^{\perp}$).

The last step is to construct a function $f : \mathfrak{g} \to \mathfrak{g}$ whose Cayley transform is (C_1, C_2, τ). We look for f of the form $f = f_+ + f_- + f_0$ with $f_\pm : \mathfrak{n}_\pm \to \mathfrak{n}_\pm$, $f_0 : \mathfrak{h} \to \mathfrak{h}$. It is easy to see that the conditions imposed on f_0 are $f_0 + f_0^* = 1$ and $f_0(h_\alpha) = (f_0 - 1)h_{\tau(\alpha)}$ for $\alpha \in \Gamma_1$ ($f_0 - 1$ is not necessarily invertible). In the case of f_+, we have $\mathrm{Ker}\,(f_+ - 1) = C_2^{\perp} \cap \mathfrak{n}^+ = \{0\}$, and $f_+ - 1$ is invertible. Similarly, $f_- - 1$ is invertible, and our conditions are now

$$\begin{aligned} f_+^* + f_- &= 1, & f_+ + f_-^* &= 1,\\ f_+/(f_+ - 1) &= \psi, & f_-/(f_- - 1) &= \phi,\end{aligned}$$

with

$$\psi(x) = \begin{cases} \theta(x) & x \in \mathfrak{g}_\alpha, \alpha \in \mathbb{Z}^+\Gamma_1 \\ 0 & x \in \mathfrak{g}_\alpha, \alpha \notin \mathbb{Z}^+\Gamma_1 \end{cases}, \qquad \phi(x) = \begin{cases} \theta(x) & x \in \mathfrak{g}_\alpha, \alpha \in \mathbb{Z}^-\Gamma_1 \\ 0 & x \in \mathfrak{g}_\alpha, \alpha \notin \mathbb{Z}^-\Gamma_1 \end{cases}.$$

It is easy to see that half of the above conditions imply the other half.

Lemma 5.3. *$\psi - 1$ is invertible if and only if for any $\alpha \in \Gamma_1$ there exists $k > 0$ such that $\tau^k(\alpha) \in \Gamma_2 \backslash \Gamma_1$.*

Proof: If τ satisfies the above condition, ψ is nilpotent and hence $\psi - 1$ is invertible. Conversely, if $\alpha \in \Gamma_1$ is such that $\tau^k(\alpha) \in \Gamma_1$ for all $k > 0$, then we must have $\alpha = \tau^n(\alpha)$ for some $n > 0$ and 1 is an eigenvalue of ψ.

□

The conditions defining a Belavin-Drinfeld triple are now clear, and we can proceed to show that formula (5.4) yields an r-matrix. Let $(\Gamma_1, \Gamma_2, \tau)$ be a Belavin-Drinfeld triple, and let $r_0 \in \mathfrak{h} \otimes \mathfrak{h}$ satisfying (5.2) and (5.3). From $r_0 \in \mathfrak{h} \otimes \mathfrak{h}$ we construct $f_0 : \mathfrak{h} \to \mathfrak{h}$ satisfying the above conditions. Define ψ by the formula above, and let $f_+ = \frac{\psi}{\psi - 1}$, $f_- = 1 - f_+^*$, and $f = f_- + f_0 + f_+$. Explicitly, f is given by

$$f = f_0 - \sum_{n \geq 1} \psi^n + \sum_{l \geq 0} \psi^{*l}$$

$$\text{where} \qquad \psi^*(x) = \begin{cases} \theta^{-1}(x) & x \in \mathfrak{g}_\alpha, \alpha \in \mathbb{Z}^-\Gamma_2 \\ 0 & x \in \mathfrak{g}_\alpha, \alpha \notin \mathbb{Z}^-\Gamma_2 \end{cases}.$$

Both sums are finite as ψ, ψ^* are nilpotent. The element $r \in \mathfrak{g} \otimes \mathfrak{g}$ associated to f is

$$r = r_0 + \sum_{\alpha \in \Delta_+} x_{-\alpha} \otimes x_\alpha + \sum_{\alpha, \beta \in \Delta_+, \alpha < \beta} x_{-\alpha} \wedge x_\beta.$$

We are thus left to check that the Cayley transform (C_1, C_2, θ) of f satisfies the conditions of lemma 5.2, which is clear from the above analysis.

5.4.3 Proof of part 2)

We now show that *every* r-matrix satisfying (5.1) arises from a Belavin-Drinfeld triple for a suitable splitting $\mathfrak{g} = \mathfrak{n}_- \oplus \mathfrak{h} \oplus \mathfrak{n}_+$. The first step is to identify such a splitting. Let r be an r-matrix satisfying (5.1) and let $f : \mathfrak{g} \to \mathfrak{g}$ denote the corresponding map. Recall that in our construction of part 1), we had $\mathfrak{n}^+ \subset (\mathfrak{p}_{\Gamma_1}^+ \oplus V_1)^\perp = \mathrm{Ker}\, f$ and $\mathfrak{n}^- \subset (\mathfrak{p}_{\Gamma_2}^- \oplus V_2)^\perp = \mathrm{Ker}\,(f-1)$. By analogy, we set

$$\mathfrak{g}^\lambda = \bigcup_{n > 0} \mathrm{Ker}\,(f - \lambda)^n$$

for $\lambda \in \mathbb{C}$ and we let $\mathfrak{g}' = \bigoplus_{\lambda \neq 0,1} \mathfrak{g}^\lambda$. We have $\mathfrak{g} = \mathfrak{g}^0 \oplus \mathfrak{g}' \oplus \mathfrak{g}^1$.

Lemma 5.4. *$\mathfrak{g}^0 \oplus \mathfrak{g}'$ and $\mathfrak{g}^1 \oplus \mathfrak{g}'$ are solvable Lie subalgebras in $\mathfrak{g}$.*

Proof: It is easy to see using (5.6) that $[\mathfrak{g}^\lambda, \mathfrak{g}^\mu] \subset \mathfrak{g}^{\lambda\mu/(\lambda+\mu-1)}$ when $\lambda + \mu \neq 1$, and $[\mathfrak{g}^\lambda, \mathfrak{g}^\mu] = 0$ if $\lambda + \mu = 1, \lambda\mu \neq 0$. Thus $\mathfrak{g}^0 \oplus \mathfrak{g}'$ and $\mathfrak{g}^1 \oplus \mathfrak{g}'$ are subalgebras in which $\mathfrak{g}^0$ and $\mathfrak{g}^1$ are ideals respectively. From $f + f^* = 1$, we deduce that $\mathfrak{g}^0 \subset (\mathfrak{g}^0)^\perp$ and $\mathfrak{g}^1 \subset (\mathfrak{g}^1)^\perp$. In particular, $(\mathfrak{g}^i, [\mathfrak{g}^i, \mathfrak{g}^i]) = 0$ for $i = 0, 1$, and by Cartan's criterion, $\mathfrak{g}^0$ and $\mathfrak{g}^1$ are solvable (see [Bou], chap. 1). Therefore, to prove the lemma, it remains to show that $\mathfrak{g}'$ is solvable. This is a consequence of the following fact: any finite dimensional Lie algebra with an automorphism ϕ such that $\phi - 1$ is invertible, is solvable (cf [BD1]). Here, $\phi = \frac{f}{f-1}$ is well defined on $\mathfrak{g}'$, and $\phi - 1 = (f-1)^{-1}$ is invertible.

□

Now, by lemma 5.4, we can find two Borel subalgebras $\mathfrak{b}_+ \supset \mathfrak{g}^0 \oplus \mathfrak{g}'$ and $\mathfrak{b}_- \supset \mathfrak{g}^1 \oplus \mathfrak{g}'$. Furthermore, $\mathfrak{b}_+ + \mathfrak{b}_- = \mathfrak{g}$ ($\mathfrak{b}_+$ and $\mathfrak{b}_-$ are complementary Borel subalgebras). From this, we can reconstruct a splitting $\mathfrak{g} = \mathfrak{n}_- \oplus \mathfrak{h} \oplus \mathfrak{n}_+$, namely $\mathfrak{h} = \mathfrak{b}_+ \cap \mathfrak{b}_-$, $\mathfrak{n}_\pm = [\mathfrak{b}_\pm, \mathfrak{b}_\pm]$. We have $\mathfrak{g}' \subset \mathfrak{h}$, and $\mathfrak{g}^0 = (\mathfrak{g}^0 \oplus \mathfrak{g}')^\perp \supset \mathfrak{b}_+^\perp = \mathfrak{n}_+$. Similarly, $\mathfrak{n}_- \subset \mathfrak{g}^1$.

Using (5.6) and the fact that $\mathfrak{b}_+ = \{x \in \mathfrak{g} \mid [x, \mathfrak{g}^0] \subset \mathfrak{g}^0\}$, it is easy to see that $f(\mathfrak{b}_+) \subset \mathfrak{b}_+$, and $f(\mathfrak{n}_+) \subset \mathfrak{n}_+$. Similarly $f(\mathfrak{b}_-) \subset \mathfrak{b}_-$ and $f(\mathfrak{n}_-) \subset \mathfrak{n}_-$. Hence f preserves the splitting $\mathfrak{g} = \mathfrak{n}_+ \oplus \mathfrak{h} \oplus \mathfrak{n}_-$. Now let (C_1, C_2, θ) be the Cayley transform of f. Recall that by definition $C_1 = \operatorname{Im}(f-1) \supset \mathfrak{g}^0 \oplus \mathfrak{g}'$ and $C_2 = \operatorname{Im} f \supset \mathfrak{g}^1 \oplus \mathfrak{g}'$.

Lemma 5.5. *There exists $\Gamma_1, \Gamma_2 \subset \Gamma$ and vector spaces $V_i \subset \mathfrak{h}_{\Gamma_i}^\perp$ such that*

$$C_1 = \mathfrak{p}_{\Gamma_1}^+ \oplus V_1, \qquad C_2 = \mathfrak{p}_{\Gamma_2}^- \oplus V_2.$$

Proof: Note that $\mathfrak{h} + C_i$ are Lie subalgebras of $\mathfrak{g}$ for $i = 1, 2$. Indeed, C_1 is a subalgebra so $[C_1, (\mathfrak{g}^0 \oplus \mathfrak{g}') \cap \mathfrak{h}] \subset C_1$. Furthermore, there are no coroots of $\mathfrak{h}$ orthogonal to $(\mathfrak{g}^0 \oplus \mathfrak{g}') \cap \mathfrak{h}$. This implies that $[\mathfrak{h}, C_1] \subset C_1$. A similar argument works for C_2. Since $\mathfrak{n}_+ \subset \mathfrak{g}^0 \subset C_1$, one has $\mathfrak{b}_+ \subset \mathfrak{h} + C_1$. It is well known that all subalgebras $\mathfrak{a} \supset \mathfrak{b}_\pm$ are of the form $\mathfrak{p}_L^\pm + \mathfrak{h}$ for some subset $L \subset \Gamma$. This proves the lemma.

□

Now let $\tau : \Gamma_1 \to \Gamma_2$ be the bijection defined by $\theta(\mathfrak{g}_\alpha) = \mathfrak{g}_{\tau(\alpha)}$ for $\alpha \in \Gamma_1$. Using the results in part 1), it is easy to check that $(\Gamma_1, \Gamma_2, \tau)$ is a Belavin-Drinfeld triple and that r is the r-matrix corresponding to it. This concludes the proof of the theorem.

Problem 5.1. Show that the standard structure on a simple Lie algebra $\mathfrak{g}$ is obtained when $\Gamma_1 = \Gamma_2 = \emptyset$ and that it is, up to a factor and adding r_0, the only Lie bialgebra structure compatible with the decomposition $\mathfrak{n}_- \oplus \mathfrak{h} \oplus \mathfrak{n}_+$ (i.e such that $\mathfrak{b}_\pm = \mathfrak{n}_\pm \oplus \mathfrak{h}$ are Lie subbialgebras).

Lecture 6

Infinite dimensional Lie bialgebras

In this lecture, we extend the concepts of a Manin triple and the Drinfeld's double to the infinite dimensional setting. This will lead us to the CYBE with spectral parameters, and to the fundamental examples of infinite dimensional Lie bialgebras.

6.1 Infinite Manin triples

We now generalize the definition of a Manin triple to the infinite dimensional setting:

Definition: A *Manin triple* is a triple of Lie algebras $(\mathfrak{g}, \mathfrak{g}_+, \mathfrak{g}_-)$ in which $\mathfrak{g} = \mathfrak{g}_+ \oplus \mathfrak{g}_-$ is equipped with an invariant nondegenerate bilinear form $(,)$ such that

1. $\mathfrak{g}_+, \mathfrak{g}_-$ are isotropic subalgebras,

2. The form $(,)$ induces an isomorphism $\mathfrak{g}_- \xrightarrow{\sim} \mathfrak{g}_+^*$.

With the above definition, the notions of a Manin triple and a Lie bialgebra are equivalent. Indeed, let $(\mathfrak{g}, \mathfrak{g}_+, \mathfrak{g}_-)$ be a Manin triple. We claim that the Lie bracket $[,]$ on $\mathfrak{g}_- \simeq \mathfrak{g}_+^*$ is continuous for the weak topology, and thus induces a genuine cocommutator $\delta : \mathfrak{g}_+ \to \Lambda^2\mathfrak{g}_+$. For simplicity, we will restrict ourselves to Lie algebras of countable dimension. If $x \in \mathfrak{g}$, we denote by x_+ and x_- the $\mathfrak{g}_+$ and $\mathfrak{g}_-$ components of x respectively. Let $x_n \in \mathfrak{g}_-$, $x_n \to 0$. This means that for every $a \in \mathfrak{g}_+$, $\lim (x_n, a) = 0$. Now if $y \in \mathfrak{g}_-$, we have

$$\lim ([x_n, y], a) = \lim (x_n, [y, a]) = \lim (x_n, [y, a]_+) = 0$$

for every $a \in \mathfrak{g}_+$. Thus $[x_n, y] \to 0$. One checks that the cocommutator is a 1-cocycle in the same way as in the finite-dimensional case.

Conversely, given a Lie bialgebra $\mathfrak{a}$, we define just as in 4.1 a Manin triple $(\mathfrak{a} \oplus \mathfrak{a}^*, \mathfrak{a}, \mathfrak{a}^*)$. The Lie algebra $\mathfrak{a} \oplus \mathfrak{a}^*$ is called the Drinfeld double of $\mathfrak{a}$.

Remark: Just as the notion of Lie bialgebra is not self dual in the infinite dimensional setting, the notion of Manin triple is not symmetric. That is, if $(\mathfrak{g}, \mathfrak{g}_+, \mathfrak{g}_-)$ is a Manin triple and $\dim \mathfrak{g}_+ = \infty$ then $(\mathfrak{g}, \mathfrak{g}_-, \mathfrak{g}_+)$ is *not* a Manin triple since $\mathfrak{g}_-^* = \mathfrak{g}_+^{**} \neq \mathfrak{g}_+$. In particular, in this case, $\mathfrak{g}_-$ is not a Lie bialgebra, but only a "topological Lie bialgebra" (an analogue of the notion of topological Lie algebra). Likewise, $\mathfrak{g} = \mathfrak{a} \oplus \mathfrak{a}^*$ is only a "topological Lie bialgebra". We will not need these notions in these notes and will not give a precise definition.

6.2 Examples

6.2.1 The standard structure on Kac-Moody algebras

Kac-Moody algebras are infinite dimensional analogues of simple Lie algebras, and share many of their properties. In particular, we will define a standard Lie bialgebra structure on any Kac-Moody algebra.

Let $A = (a_{ij})_{i,j=1\ldots n}$ be a *generalized Cartan matrix*, i.e an $n \times n$ matrix with $a_{ii} = 2$, $a_{ij} \in \mathbb{Z}^-(i \neq j)$ and $a_{ij} = 0 \Leftrightarrow a_{ji} = 0$. Suppose that A is symmetrizable, i.e there exists $d_i \in \mathbb{N}$ such that $d_i a_{ij} = d_j a_{ji} =: b_{ij}$. We associate to A a (Kac-Moody) Lie algebra $\mathfrak{g}_A$ which admits a Serre presentation by generators e_i, f_i, h_i for $i = 1, \ldots n$ and relations

$$[e_i, f_i] = h_i, \qquad [e_i, f_j] = 0, \qquad (i \neq j)$$

$$[h_i, e_j] = a_{ij} e_i, \qquad [h_i, f_j] = -a_{ij} f_j, \qquad [\mathfrak{h}, \mathfrak{h}] = 0,$$

$$ad^{1-a_{ij}}(e_i)e_j = 0, \qquad ad^{1-a_{ij}}(f_i)f_j = 0.$$

The subalgebra generated by (h_i) is called the Cartan subalgebra, and the subalgebra $\mathfrak{n}_+$ (resp $\mathfrak{n}_-$) generated by (e_i) (resp. (f_i)) is called the positive (resp. negative) nilpotent subalgebra. The algebra $\mathfrak{g}_A$ decomposes into a sum of root subspaces $\mathfrak{g}_A = \mathfrak{h} \oplus \bigoplus_{\alpha \in \Delta} \mathfrak{g}_\alpha$ with $\mathfrak{g}_\alpha = \{x \in \mathfrak{g} \mid [h, x] = \alpha(h)x \quad \forall h \in \mathfrak{h}\}$. As opposed to the finite dimensional case, the root subspaces are not in general one-dimensional (but are still finite dimensional).The set of roots Δ (in general infinite) has a polarization $\Delta = \Delta_+ \cup \Delta_-$ with $\Delta_- = -\Delta_+$ corresponding to the decomposition $\mathfrak{g}_A = \mathfrak{n}_- \oplus \mathfrak{h} \oplus \mathfrak{n}_+$, and is generated by the set of simple roots defined by $\alpha_i(h_j) = a_{ij}$.

Define a symmetric bilinear form on $\mathfrak{h}$ by $(h_i, h_j) = d_i^{-1} d_j^{-1} b_{ij}$. This form may be degenerate: let $\mathfrak{l}$ be its kernel, and $\mathfrak{l}'$ a complement to $\mathfrak{l}$ in $\mathfrak{h}$. Extend $(,)$ to $\tilde{\mathfrak{h}} = \mathfrak{h} \oplus \mathfrak{l}^*$ by $(h, x) = x(h)$ for $x \in \mathfrak{l}^*$ and $h \in \mathfrak{l}$, and $(\mathfrak{l}^*, \mathfrak{l}^*) = (\mathfrak{l}', \mathfrak{l}^*) = 0$. Then $(,)$ is nondegenerate on $\tilde{\mathfrak{h}}$. Furthermore, extend the simple roots α_i to $\tilde{\mathfrak{h}}^*$ by $\alpha_i(z) = z(h_i)$ for $z \in \mathfrak{l}$. Now define the extended Kac-Moody algebra, generated by $e_i, f_i, \tilde{\mathfrak{h}}$ satisfying

$$[e_i, f_i] = h_i, \qquad [e_i, f_j] = 0, \qquad (i \neq j)$$

$$[h, e_j] = \alpha_i(h) e_i, \qquad [h, f_j] = -\alpha_i(h) f_j, \qquad [\tilde{\mathfrak{h}}, \tilde{\mathfrak{h}}] = 0,$$

$$ad^{1-a_{ij}}(e_i)e_j = 0, \qquad ad^{1-a_{ij}}(f_i)f_j = 0.$$

As a vector space, $\tilde{\mathfrak{g}}_A = \mathfrak{n}_- \oplus \tilde{\mathfrak{h}} \oplus \mathfrak{n}_+$. Also, notice that the center of $\tilde{\mathfrak{g}}_A$ coincides with $\mathfrak{l}$. The reason for considering $\tilde{\mathfrak{g}}_A$ rather than $\mathfrak{g}_A$ is the following result.

Proposition 6.1 (Kac). *There is a unique nondegenerate invariant bilinear form on $\tilde{\mathfrak{g}}_A$ which extends the form $(,)$ on $\tilde{\mathfrak{h}}$.*

We refer the reader to [Kac] for a proof of this.

Since $(,)$ is invariant and nondegenerate, it induces a pairing $\mathfrak{g}_\alpha \times \mathfrak{g}_{-\alpha} \to \mathbb{C}$. Let us choose a basis (e^i_α) of $\mathfrak{g}_\alpha$ and let (f^i_α) be the dual basis of $\mathfrak{g}_{-\alpha}$.

Just as in the case of a simple Lie algebra, we can define the standard bialgebra structure $\tilde{\mathfrak{g}}_A$, by considering it as "almost" the double of $\mathfrak{n}_+ \oplus \tilde{\mathfrak{h}}$. The cocommutator is given by $\delta(x) = [x \otimes 1 + 1 \otimes x, r]$, where

$$r = \sum_{\alpha \in \Delta_+} \sum_i e^i_\alpha \otimes f^i_\alpha + \frac{1}{2} \sum_j k_j \otimes k_j$$

for some orthonormal basis (k_j) of $\tilde{\mathfrak{h}}$. By analogy with the finite dimensional case, we will say that r is a *pseudoquasitriangular structure* for $(\tilde{\mathfrak{g}}_A, [,], \delta)$. (The term *pseudo* means that r lives in a completion of $\tilde{\mathfrak{g}}_A \otimes \tilde{\mathfrak{g}}_A$). It is easy to check that the cocommutator is defined on the Chevalley generators by

$$\delta(h_i) = 0,$$
$$\delta(e_i) = \frac{d_i}{2} e_i \wedge h_i,$$
$$\delta(f_i) = \frac{d_i}{2} f_i \wedge h_i.$$

Remark: It makes sense to consider the Lie algebra defined as above *without* the Serre relations. In this case, the a_{ij}s can be any complex numbers rather than just integers. One can check that such a Lie algebra also has a Lie bialgebra structure, constructed in the same way. See [Var] for more details.

6.3 The CYBE with spectral parameter

Let us now turn to another type of infinite dimensional Lie algebras, namely algebras of (meromorphic) functions on a smooth curve with values in a Lie algebra $\mathfrak{g}$. Before making this more precise, let us consider an example.

6.3.1 An example: the Yangian and its dual

Let $\mathfrak{g}_0$ be a finite dimensional Lie algebra with a fixed nondegenerate invariant bilinear form $(,)$. Set $\mathfrak{g} = \mathfrak{g}_0((t^{-1}))$, $\mathfrak{g}_- = t^{-1}\mathfrak{g}_0[[t^{-1}]]$ and $\mathfrak{g}_+ = \mathfrak{g}_0[t]$. We have

$\mathfrak{g} = \mathfrak{g}_+ \oplus \mathfrak{g}_-$. Equip the Lie algebra $\mathfrak{g}$ with the nondegenerate invariant bilinear form

$$< a(t), b(t) >= \mathrm{Res}_{t=0}(a(t), b(t))$$

where $a(t), b(t) \in \mathfrak{g}((t^{-1}))$. The Lie subalgebras $\mathfrak{g}_+$, $\mathfrak{g}_-$ are clearly isotropic. Furthermore

$$\mathfrak{g}_+^* = (\bigoplus_{n\geq 0} \mathfrak{g}_0 t^n)^* = \prod_{n\geq 0} (\mathfrak{g}_0 t^n)^* = \prod_{n\geq 1} \mathfrak{g}_0 t^{-n} = \mathfrak{g}_-$$

hence $(\mathfrak{g}, \mathfrak{g}_+, \mathfrak{g}_-)$ is an infinite Manin triple. We thus obtain a Lie bialgebra structure on $\mathfrak{g}_0[t]$, which is called the *Yangian Lie bialgebra*, in honor of Yang, who was the first to find the corresponding solution to the Yang-Baxter equation.

Explicitly, the cocommutator is given by

$$\delta(at^n) = \sum_{0\leq r\leq n-1} \sum_j [x_j, a]t^r \otimes x_j t^{n-1-r}$$

where (x_i) is an orthonormal basis of $\mathfrak{g}_0$.

The *(graded) dual Yangian* is obtained in a similar fashion, starting from the Manin triple given by $\mathfrak{g} = \mathfrak{g}_0((t))$, $\mathfrak{g}_+ = t^{-1}\mathfrak{g}_0[t^{-1}]$ and $\mathfrak{g}_- = \mathfrak{g}_0[[t]]$. In this case, we obtain cocommutation relations

$$\delta(at^n) = \sum_{1\leq r\leq n} \sum_j [x_j, a]t^{-r} \otimes x_j t^{r-n+1}.$$

As we have seen in the case of Kac-Moody algebras, it is tempting to associate to an infinite Manin triple $(\mathfrak{g}, \mathfrak{g}_+, \mathfrak{g}_-)$ some kind of (pseudo)quasitriangular structure on $\mathfrak{g}$, which is morally given by $r_\mathfrak{g} = \sum_i g_+^i \otimes g_-^i$ for some basis (g_+^i) of $\mathfrak{g}_+$ and the dual (topological) basis (g_-^i) of $\mathfrak{g}_-$. In the case of the Yangian Lie bialgebra, it is convenient to choose $g_+^{i,n} = x_i t^n \in \mathfrak{g}_0[t]$ and $g_-^{i,n} = x_i u^{-n-1} \in u^{-1}\mathfrak{g}_0[[u^{-1}]]$ for $n \geq 0$. We use two different variables t and u to distinguish between the two algebras $\mathfrak{g}_+$ and $\mathfrak{g}_-$. Hence

$$r_\mathfrak{g} = \sum_{i,n\geq 0} x_i t^n \otimes x_i u^{-n-1} \in \mathfrak{g}_+ \hat{\otimes} \mathfrak{g}_-.$$

Identifying $\mathfrak{g}_+ \hat{\otimes} \mathfrak{g}_-$ with $u^{-1}\mathfrak{g}_0 \otimes \mathfrak{g}_0[t][[u^{-1}]]$, we can write

$$r_\mathfrak{g} = \sum_{i,n\geq 0} x_i \otimes x_i t^n u^{-n-1} = \frac{\sum_i x_i \otimes x_i}{u-t} = \frac{\Omega}{u-t}$$

where $\Omega \in \mathfrak{g}_0 \otimes \mathfrak{g}_0$ is the Casimir corresponding to $(\,,\,)$ and where we expand $\frac{1}{u-t}$ as $\sum_{n\geq 0} t^n u^{-n-1}$, i.e "in the region $|t| < |u|$".

In the case of the dual Yangian, we obtain similarly

$$r_{\mathfrak{g}} = \sum_{i,n\geq 0} x_i t^{-n-1} \otimes x_i u^n = \frac{\Omega}{t-u}$$

where this time we expand $\frac{1}{t-u}$ as $\sum_{n\geq 0} t^{-n-1}u^n$ ("in the region $|u| < |t|$").

Another way of saying this is to consider the evaluation maps

$$\pi_x : \mathfrak{g}_0[t,t^{-1}] \to \mathfrak{g}_0, \quad at^n \mapsto x^n a$$

where $x \in \mathbb{C}^*$. Then for the Yangian Lie bialgebra, the Laurent series in z, w $(\pi_z \otimes \pi_w)(r_{\mathfrak{g}})$ converges in the region $|z| < |w|$ to $\frac{\Omega}{w-z}$. In the case of the dual Yangian Lie bialgebra, $(\pi_z \otimes \pi_w)(r_{\mathfrak{g}})$ converges in the region $|w| < |z|$ to $\frac{\Omega}{z-w}$.

Now, $r_{\mathfrak{g}}$ satisfies the CYBE in $\mathfrak{g}\otimes\mathfrak{g}$. Thus, $r_{\mathfrak{g}}(z-w) = (\pi_z \otimes \pi_w)(r_{\mathfrak{g}})$ satisfies the following equation:

$$[r_{\mathfrak{g}}^{12}(z_1-z_2), r_{\mathfrak{g}}^{13}(z_1-z_3)] + [r_{\mathfrak{g}}^{12}(z_1-z_2), r_{\mathfrak{g}}^{23}(z_2-z_3)] + [r_{\mathfrak{g}}^{13}(z_1-z_3), r_{\mathfrak{g}}^{23}(z_2-z_3)] = 0.$$

Notice that this equation makes sense in the region $|z_1| < |z_2| < |z_3|$ in the case of the Yangian, and in the region $|z_1| > |z_2| > |z_3|$ in the case of the dual Yangian.

6.3.2 The CYBE with spectral parameter

Motivated by the preceding example, we make the following definition. Let $\mathfrak{g}$ be any finite dimensional Lie algebra over $k = \mathbb{C}$.

Definition: Let r be a germ at 0 of a meromorphic function $\mathbb{C} \to \mathfrak{g}\otimes\mathfrak{g}$. The *CYB map with spectral parameter* is the map defined by

$$CYB(r)(z_1,z_2,z_3) = [r_{12}(z_1-z_2), r_{13}(z_1-z_3)] + [r_{12}(z_1-z_2), r_{23}(z_2-z_3)] + [r_{13}(z_1-z_3), r_{23}(z_2-z_3)].$$

The variable z in $r(z)$ is called the spectral parameter. The *CYBE with spectral parameter* is the equation

$$CYB(r)(z_1,z_2,z_3) = 0$$

for a function $r(z)$ with values in $\mathfrak{g}\otimes\mathfrak{g}$, which is meromorphic around the origin. Solutions of this equation are called *r-matrices with spectral parameter*. Notice that the usual CYBE is the CYBE with spectral parameter for constant functions.

6.3.3 Construction of a Lie bialgebra from an r-matrix

We have seen how the Yangian Lie bialgebra gives rise to the solution $r(z) = \frac{\Omega}{z}$ of the CYBE with spectral parameter (Yang's solution). Let us now explain how to conversely define a Lie bialgebra from an r-matrix with spectral parameter, when $\mathfrak{g}$ has an invariant nondegenerate form (in particular, if $\mathfrak{g}$ is simple). This construction is due to Semenov-Tian Shansky [Sem] and Belavin-Drinfeld [BD2].

Let $r(z)$ be an r-matrix, and assume that it satisfies the following conditions:

$$r(z) = \frac{\Omega}{z} + O(1) \qquad \text{(around 0)}, \tag{6.1}$$

$$r(z) + r^{21}(-z) = 0 \qquad \text{(unitarity condition)}, \tag{6.2}$$

(we will see in the next lecture that this includes all r-matrices that are nondegenerate in a suitable sense). Let us expand $r^{21}(t-u)$ in the region $|t| > |u|$, i.e replacing $\frac{1}{t-u}$ by $\sum_{m\geq 0} t^{-m-1}u^m$. We obtain a series

$$r^{21}(t-u) = \sum_{n\geq 0} c_n u^n \in \mathfrak{g}\otimes\mathfrak{g}((t))[[u]],$$

where $c_n \in \mathfrak{g}\otimes\mathfrak{g}((t))$. Let

$$\mathfrak{g}(r) = \text{Span}\{(f\otimes Id)c_n \mid f\in\mathfrak{g}^*,\ n\geq 0\} \subset \mathfrak{g}((t)).$$

Notice the analogy with Proposition 3.2. We have the following result:

Proposition 6.2. *We have $\mathfrak{g}((t)) = \mathfrak{g}[[t]] \oplus \mathfrak{g}(r)$. Furthermore, $\mathfrak{g}(r)$ is a Lie subalgebra of $\mathfrak{g}((t))$, isotropic under the usual invariant nondegenerate form on $\mathfrak{g}((t))$*

$$< a(t), b(t) >= \text{Res}_0(a(t), b(t))$$

and $(\mathfrak{g}((t)), \mathfrak{g}(r), \mathfrak{g}[[t]])$ is an infinite Manin triple.

Proof: It is easy to see from (6.1) that $c_n - t^{-n-1}\Omega \in \mathfrak{g}[[t]]$, and the first assertion follows. Now let us write $c_n(t) = \sum_i d^i \otimes c_n^i(t)$ where $c_n^i(t) \in \mathfrak{g}((t))$ and (d_i) is a basis of $\mathfrak{g}$. The CYBE with spectral parameters is

$$\sum_{i,j,n,m} [c_n^i(z_1), c_m^j(z_1)] \otimes d^i z_2^n \otimes d^j z_3^m + \sum_{i,j,n,m} c_n^i(z_1) \otimes [d^i, c_m^j(z_2)] z_2^n \otimes d^j z_3^m$$
$$+ \sum_{i,j,n,m} c_n^i(z_1) \otimes c_m^j(z_2) \otimes [d^i, d^j] z_3^{m+n} = 0.$$

Application of linear forms $Id\otimes f\otimes g$ (as in Proposition 3.3) shows that $\mathfrak{g}(r)$ is a Lie subalgebra. Since $c_n(t) = t^{-n-1}\Omega + O(1)$, $\mathfrak{g}(r)$ is isotropic if and only if we have

$$(\Omega, \text{coef. of } t^m \text{ in } c_n) + (\Omega, \text{coef. of } t^n \text{ in } c_m) = 0 \tag{6.3}$$

for $n, m \geq 0$. Let us write

$$r^{21}(t-u) = \frac{\Omega}{t-u} + \sum_i a_i (t-u)^i.$$

The coefficient of t^n in c_n is $(-1)^m a_{n+m} \binom{m+n}{m}$. By the unitarity condition, $a_i + (-1)^i a_i^{21} = 0$. This implies equation (6.3).
Finally, it is easy to see that $\mathfrak{g}[[t]] = \mathfrak{g}(r)^*$. This shows that $(\mathfrak{g}((t)), \mathfrak{g}(r), \mathfrak{g}[[t]])$ is a Manin triple and that $\mathfrak{g}(r)$ is a Lie bialgebra.

□

This construction is particularly convenient as it gives us an explicit description of the commutation and cocommutation relations in $\mathfrak{g}(r)$; Namely, consider the generating function $\tau(u) = r^{21}(t-u) \in \mathfrak{g} \otimes \mathfrak{g}(r)[[u]]$. The CYBE with spectral parameters implies the following relations:

$$[\tau^{13}(u), \tau^{23}(v)] = [r^{21}(u-v), \tau^{13}(u) + \tau^{23}(v)], \tag{6.4}$$

$$\delta(\tau(u)) = [\tau^{12}(u), \tau^{13}(u)]. \tag{6.5}$$

Notice that (6.4) makes sense: the pole part $\frac{\Omega}{z}$ in $r^{21}(u-v)$ disappears on the RHS of (6.4) because Ω is $\mathfrak{g}$-invariant.

6.4 Solutions of the CYBE with spectral parameters

As is the case of the CYBE, solutions of the CYBE with spectral parameter for a general $\mathfrak{g}$ are not classified. However, in their remarkable work in 1982 [BD1], Belavin and Drinfeld obtained an almost complete classification of r-matrices when $\mathfrak{g}$ is a simple Lie algebra (which was later completed by Stolin [Sto]) under some nondegeneracy condition. We state their result, describe the simplest solutions, and devote the next lecture to the proof of this deep theorem. We let $\mathfrak{g}$ be a simple algebra over $k = \mathbb{C}$, and $\Omega \in (S^2\mathfrak{g})^{\mathfrak{g}}$ a nonzero Casimir element.

Definition: An r-matrix $r \in \mathfrak{g} \otimes \mathfrak{g}((z))$ is called *nondegenerate* if the map $r^* : \mathfrak{g}^* \to \mathfrak{g}((z))$ extends to a linear isomorphism $\mathfrak{g}^*((z)) \tilde{\to} \mathfrak{g}((z))$.

Definition: Two r-matrices $r(z)$ and $r'(z)$ defined in a neighborhood U of the origin are *equivalent* if there exists a meromorphic function $\phi : U \to \operatorname{Aut} \mathfrak{g}$ such that $r(z-w) = (\phi(z) \otimes \phi(w)) r'(z-w)$.

Theorem 6.1 (Belavin-Drinfeld). *Let $r(z) \in \mathfrak{g} \otimes \mathfrak{g}((z))$ be a nondegenerate r-matrix. Then*

1. *$r(z)$ extends to a meromorphic function on the whole complex plane $\mathbb{C}$,*

2. *The poles of $r(z)$ are simple and form a lattice $\Gamma \subset \mathbb{C}$,*

3. *$r(z) = \frac{\epsilon\Omega}{z} + O(1)$ around 0,*

4. *$r(z)$ satisfies the unitarity condition $r(z) + r^{21}(-z) = 0$.*

Furthermore, solutions are of three types:

1. *$\Gamma = 0$: rational r-matrices,*

2. *$\Gamma \simeq \mathbb{Z}$: trigonometric r-matrices,*

3. *$\Gamma \simeq \mathbb{Z}^2$: elliptic r-matrices (only when $\mathfrak{g} = sl_n$).*

More precisely, a rational r-matrix is equivalent to one of the form $r(z) = P(z)$, for a rational function P, a trigonometric r-matrix is equivalent to one of the form $r(z) = P(e^z)$ for a rational function P and an elliptic r-matrix is of the form $r(z) = \theta(z)$ where θ is an elliptic function.

In other terms, the r-matrix $r(z)$ extends (after an equivalence if necessary) to a rational function on a one-dimensional algebraic group G ($G = \mathbb{C}$ for rational solutions, $G = \mathbb{C}^*$ for trigonometric solutions and G is an elliptic curve for elliptic solutions).
Remark: For elliptic solutions, there are no nonconstant equivalences.

We now proceed to describe the simplest r-matrices in each class, and give a geometric realization of the associated Manin triple.

Rational solutions: The simplest is Yang's r-matrix $r(z) = \frac{\Omega}{z}$ with which we have started, and the associated Manin triple is the usual splitting of $\mathfrak{g}((t))$ into regular and singular parts. Other solutions include $r(z) = \frac{\Omega}{z} + \rho$ where ρ is a skew-symmetric constant r-matrix, but there are many more (see [Sto]).

Trigonometric solutions: Let r_0 be any quasitriangular structure on $\mathfrak{g}$ such that $r_0 + r_0^{21} = \Omega$. Then

$$r(z) = \frac{r_0^{21} e^z + r_0}{e^z - 1} \tag{6.6}$$

is a (trigonometric) r-matrix. When we take for r_0 the standard structure, the associated Manin triple has the following description:

Choose the local coordinate $x = e^z$. Then $r = \frac{r_0^{21} x + r_0}{x - 1}$ and $\mathfrak{g}(r) \subset \mathfrak{g}((z))$ is the space of (Laurent expansions with respect to z of) rational functions f of x on $\mathbb{P}^1$, regular on $\mathbb{P}^1 \backslash \{1\}$, such that $f(0) \in \mathfrak{b}_+$, $f(\infty) \in \mathfrak{b}_-$ and $f(0) + f(\infty) \in \mathfrak{n}_- \oplus \mathfrak{n}_+$.

Elliptic solutions: They exist only when $\mathfrak{g} = sl_n$. Consider the two $n \times n$ matrices:

$$A = \begin{pmatrix} 0 & 0 & \dots & 0 & 1 \\ 1 & 0 & \dots & 0 & 0 \\ \vdots & \ddots & \vdots & \vdots & \vdots \\ 0 & 0 & \ddots & 0 & 0 \\ 0 & 0 & \dots & 1 & 0 \end{pmatrix} \qquad B = \begin{pmatrix} 1 & 0 & \dots & 0 & 0 \\ 0 & \xi & \dots & 0 & 0 \\ \vdots & \vdots & \ddots & \vdots & \vdots \\ 0 & 0 & \dots & \xi^{n-2} & 0 \\ 0 & 0 & \dots & 0 & \xi^{n-1} \end{pmatrix}$$

where ξ is a primitive n^{th}-root of unity. These matrices satisfy $A^n = B^n = I$, $AB = \xi^{-1}BA$, and generate the *Heisenberg group*. Notice that $\mathrm{Ad}A$ and $\mathrm{Ad}B$ are two commuting autormorphisms of $\mathfrak{g}$. Let $\tau \in \mathbb{C}$, $\mathrm{Im}(\tau) > 0$. Then $r(z)$ is the unique meromorphic function $\mathbb{C} \to \mathfrak{g} \otimes \mathfrak{g}$ with simple poles on $\mathbb{Z} \oplus \tau\mathbb{Z}$, residue Ω at the origin, and such that

$$r(z + m + l\tau) = (\mathrm{Ad}(A)^m \mathrm{Ad}(B)^l \otimes I) r(z)$$

There is an explicit formula for $r(z)$: introduce the Weierstraß ζ function:

$$\zeta(u) = \lim_{N\to\infty}\Big[\frac{1}{u} + \sum_{\substack{m,l=-N \\ (m,l)\neq(0,0)}}^{N} \Big(\frac{1}{u+m+l\tau} - \frac{1}{m+l\tau} + \frac{u}{(m+l\tau)^2}\Big).$$

Then we have

$$r(u) = \sum_{m,l=0}^{n-1} \frac{1}{n}(\mathrm{Ad}A^m \mathrm{Ad}B^l \otimes I)(\Omega)\zeta\Big(\frac{u}{n} - m - l\tau\Big). \tag{6.7}$$

The geometric interpretation of the corresponding Manin triple is the following: let us denote by $\mathcal{E} = \mathbb{C}/\mathbb{Z} \oplus \tau\mathbb{Z}$ the elliptic curve with modulus τ, and let P be a holomorphic $sl_n(\mathbb{C})$-bundle with a flat connection, with monodromies given by

$$h_1 = \mathrm{Ad}A\ (z \mapsto z+1), \qquad h_2 = \mathrm{Ad}B\ (z \mapsto z+\tau).$$

Then $\mathfrak{g}(r) \subset \mathfrak{g}((z))$ is the space of Laurent expansions of meromorphic sections of P with a unique pole at 0.

6.5 Affine Lie algebras

The picture would not be complete without mentioning *affine Lie algebras.* These infinite dimensional Lie algebras hold a special place in the theory as they can both be realized in a geometric way and as Kac-Moody algebras.

6.5.1 Definition

We use the geometric approach: let $\mathfrak{g}$ be a simple Lie algebra with a non-degenerate invariant form $(\ ,\ )$. Denote by $L(\mathfrak{g}) = \mathfrak{g}[t, t^{-1}]$ the *Loop algebra* of

$\mathfrak{g}$ consisting of $\mathfrak{g}$-valued Laurent polynomials (with pointwise bracket). It can be shown (cf [Kac]) that $L(\mathfrak{g})$ has a unique up to scaling 1-dimensional central extension:

$$0 \to \mathbb{C}c \to \hat{\mathfrak{g}} \to L(\mathfrak{g}) \to 0$$

given by the cocycle

$$c(f.x, g.y) = (x, y)\mathrm{Res}_0 f dg, \qquad f, g \in \mathbb{C}[t, t^{-1}],\ x, y \in \mathfrak{g}.$$

In other terms, $\hat{\mathfrak{g}} = \mathfrak{g}[t, t^{-1}] \oplus \mathbb{C}c$ as a vector space, with commutation relations

$$[c, f.x] = 0,$$
$$[f.x, g.y] = fg[x, y] + (x, y)\mathrm{Res}_0(fdg)c.$$

It is sometimes convenient to introduce a grading operator in $\hat{\mathfrak{g}}$: let $\tilde{\mathfrak{g}} = \hat{\mathfrak{g}} \oplus \mathbb{C}d$ with commutation relations as above and

$$[d, c] = 0, \qquad [d, f.x] = t\frac{df}{dt}.x.$$

Definition: The Lie algebra $\hat{\mathfrak{g}}$ is called the *affine Lie algebra* associated to $\mathfrak{g}$ and $\tilde{\mathfrak{g}}$ is called the *extended affine Lie algebra* associated to $\mathfrak{g}$.

As we mentioned earlier, there is a realization of $\tilde{\mathfrak{g}}$ as a Kac-Moody algebra. Let $\mathfrak{h} \subset \mathfrak{g}$ be a Cartan subalgebra, $\Delta \subset \mathfrak{h}^*$ the root system and $\Delta = \Delta_+ \cup \Delta_-$ a polarization. Let $\Gamma = \{\alpha_1, \dots \alpha_n\}$ be the set of simple roots, θ the maximal root and $\check{\theta}$ the associated coroot. Recall that the Cartan matrix of $\mathfrak{g}$ is defined as $A = (a_{ij})_{i,j=1}^n$ where $a_{ij} = \frac{2(\alpha_i, \alpha_j)}{(\alpha_i, \alpha_i)}$. Define the *extended Cartan matrix* as $\hat{A} = (a_{ij})_{i,j=0}^n$ where we set

$$a_{0i} = -(\theta, \alpha_i), \qquad a_{i0} = \frac{-2(\theta, \alpha_i)}{(\alpha_i, \alpha_i)}, \qquad a_{00} = 2.$$

and let $\mathfrak{g}_{\hat{A}}$ be the associated Kac-Moody algebra.

Theorem 6.2 (Gabber-Kac). *We have* $\mathfrak{g}_{\hat{A}} \simeq \tilde{\mathfrak{g}}$.

We refer the reader to [GK] for a proof. We will just describe the isomorphism here. Notice that $\hat{A}$ is of rank n, and let Z be its one-dimensional kernel. Let $z^*, f_i, e_i, h_i,\ i = 0, \dots n$ be the Chevalley generators of $\mathfrak{g}_{\hat{A}}$, and $\overline{e_i}, \overline{f_i}, \overline{h_i}$, $i = 1, \dots n$ be the Chevalley generators of $\mathfrak{g}$. Choose $e_\theta \in \mathfrak{g}_\theta$ and $f_\theta \in \mathfrak{g}_{-\theta}$ such that $(e_\theta, f_\theta) = 1$. The Lie algebra isomorphism is then given by:

$$e_i \mapsto \overline{e_i}, \qquad f_i \mapsto \overline{f_i}, \qquad h_i \mapsto \overline{h_i}, \qquad i = 1, \dots n,$$
$$e_0 \mapsto f_\theta t, \qquad f_0 \mapsto e_\theta t^{-1}, \qquad h_0 \mapsto c - \check{\theta}, \qquad z^* \mapsto d.$$

6.5.2 Lie bialgebra structure

As $\tilde{\mathfrak{g}}$ is a Kac-Moody algebra, it is equipped with the standard Lie bialgebra structure. However, it is interesting to note that this Lie bialgebra structure also has a geometric origin.

Let $X = \mathbb{P}^1$ with coordinate t and let $\omega = t^{-1}dt$ be a meromorphic one-form (with poles at 0 and ∞). Let $k_0 = \mathbb{C}((t))$, $k_\infty = \mathbb{C}((t^{-1}))$ be the local fields at 0 and ∞ respectively. Finally, let $\gamma : \mathcal{O}(X\backslash\{0,\infty\}) \to k_0 \oplus k_\infty$ be the canonical embedding. Fix a polarization $\mathfrak{g} = \mathfrak{n}_- \oplus \mathfrak{h} \oplus \mathfrak{n}_+$.

Consider the following Manin triple: set

$$\mathfrak{a} = L(\mathfrak{g}) \otimes_{\mathbb{C}[t,t^{-1}]} (k_0 \oplus k_\infty)$$

and equip $\mathfrak{a}$ with the following nondegenerate invariant scalar product

$$< (f_0, f_\infty), (g_0, g_\infty) >= \mathrm{Res}_0(f_0, g_0)\omega + \mathrm{Res}_\infty(f_\infty, g_\infty)\omega.$$

Now, define

$$\mathfrak{a}_+ = L(\mathfrak{g}) \subset L(\mathfrak{g}) \otimes_{\mathbb{C}[t,t^{-1}]} (k_0 \oplus k_\infty),$$
$$\mathfrak{a}_- = \{(x,y) \in L(\mathfrak{g}) \oplus L(\mathfrak{g}) \,|x \in \mathfrak{b}_- + t^{-1}\mathfrak{g}[t^{-1}], y \in \mathfrak{b}_+ + t\mathfrak{g}[t], x(\infty) + y(0) =\in \mathfrak{n}_+ \oplus \mathfrak{n}_-\},$$

where $\mathfrak{a}_+ \subset \mathfrak{a}$ via the embedding γ, and $\mathfrak{a}_- \subset \mathfrak{a}$ by the trivial embedding. It is easy to see that $(\mathfrak{a}, \mathfrak{a}_+, \overline{\mathfrak{a}_-})$ is a Manin triple where $\overline{\mathfrak{a}_-}$ is the closure with respect to the "adic" topology. Thus $\mathfrak{a}_+$ has a Lie bialgebra structure.

Problem 6.1. Check that this induces a Lie bialgebra structure on $L(\mathfrak{g}) = \mathfrak{a}_+$ which coincides with the structure induced by the standard one on $\tilde{\mathfrak{g}}$ on the quotient $L(\mathfrak{g}) = \tilde{\mathfrak{g}}/\mathbb{C}c \ominus \mathbb{C}d$.

Remark: Notice that this Manin triple involves two points rather than one as in the previous examples of the Yangian and elliptic Lie bialgebras.

Lecture 7

Belavin-Drinfeld classification (II)

The goal of this lecture is to explain Theorem 6.1, i.e why nondegenerate solutions of the CYBE with spectral parameter come in three kinds: rational, trigonometric and elliptic. We follow the paper [BD1]. However, we will not carry out the complete classification in each class here. The interested reader is referred to the original article [BD1] for the elliptic and trigonometric cases, and to the papers by Stolin [Sto] where the rational case is discussed.

Let $\mathfrak{g}$ be a simple Lie algebra over $\mathbb{C}$ with a fixed invariant nondegenerate form $(\ ,\ )$. Let (I_ν) be an orthonormal basis of $\mathfrak{g}$, and $\Omega = \sum_\nu I_\nu \otimes I_\nu \in \mathfrak{g} \otimes \mathfrak{g}$ denote the Casimir element. We are looking for solutions of the equation

$$[r_{12}(u), r_{13}(u+v)] + [r_{12}(u), r_{23}(v)] + [r_{13}(u+v), r_{23}(v)] = 0 \qquad (7.1)$$

for a function $r : U \to \mathfrak{g} \otimes \mathfrak{g}$ meromorphic in a (connected) neighborhood U of $0 \in \mathbb{C}$, and satisfying the following nondegeneracy condition: there exists $v \in U$ for which $r(v)$ is nondegenerate as an operator $\mathfrak{g}^* \to \mathfrak{g}$.

7.1 Properties of nondegenerate solutions

In this section, we prove that a nondegenerate solution necessarily has a simple pole at 0 with residue $t\Omega$ ($t \neq 0$), and satisfies the unitarity condition $r(u) + r^{21}(-u) = 0$.

Lemma 7.1. *The function $r(u)$ has a pole at 0, whose order is maximal among poles of $r(u)$. The residue at that pole is nondegenerate.*

Proof: We will first prove that $r(u)$ has a pole in U. Reasoning by contradiction, let us suppose that r is holomorphic in U. Let us show that $r(0)$ is nondegenerate. Define $f_u : \mathfrak{g} \to \mathfrak{g}$ by $r(u) = (f_u \otimes 1)\Omega$. Substituting $v = 0$, we rewrite (7.1) as

$$\sum_{\mu,\nu} \left([f_u(I_\mu), f_u(I_\nu)] \otimes I_\mu \otimes I_\nu + f_u(I_\nu) \otimes [I_\nu \otimes 1 + 1 \otimes I_\nu, r(0)] \right) = 0. \qquad (7.2)$$

Letting $[I_\lambda \otimes 1 + 1 \otimes I_\lambda, r(0)] = \sum c_{\mu,\nu}^\lambda I_\mu \otimes I_\nu$, we have

$$[f_u(I_\mu), f_u(I_\nu)] = \sum_\lambda c_{\mu,\nu}^\lambda f_u(I_\lambda). \tag{7.3}$$

Since $r(v)$ is nondegenerate for generic v, f_v is invertible. Let us choose such a v and consider $\phi_u = f_u f_v^{-1}$. It follows from (7.3) that ϕ_u is a Lie algebra endomorphism of $\mathfrak{g}$. In particular, $|\det \phi| \in \{0,1\}$. But since the assignment $u \mapsto \phi_u$ is holomorphic, we have $\det \phi_u = \det \phi_v = 1$ for all $u \in U$. In particular, $r(0)$ is nondegenerate.

From this and from the equation $[f_0(I_\mu), f_0(I_\nu)] = \sum c_{\mu,\nu}^\lambda f_0(I_\lambda)$, we deduce $c_{\mu,\nu}^\lambda + c_{\nu,\mu}^\lambda = 0$. Thus $[r(0) + r^{21}(0), I_\lambda \otimes 1 + 1 \otimes I_\lambda] = 0$ for all λ. This implies

$$r(0) + r^{21}(0) = t\Omega \tag{7.4}$$

for some $t \in \mathbb{C}$. Moreover, it follows from (7.1) that $r(0)$ satisfies the (constant) CYBE. But by the results of Lecture 5 (Proposition 5.2 and Theorem 5.3), there are no nondegenerate solutions satisfying (7.4). Hence $r(u)$ is not holomorphic. Let γ be a pole of order k: $r(\gamma + h) = \frac{\tau}{h^k} + O(h^{1-k})$. For $u \in U$ such that r is holomorphic at $u, u + \gamma$, expanding (7.1) around $v = \gamma$ yields

$$[r^{12}(u), \tau^{23}] + [r^{13}(u+\gamma), \tau^{23}] = 0 \qquad [r^{23}(u), \tau^{12}] + [r^{13}(u+\gamma), \tau^{12}] = 0. \tag{7.5}$$

We claim that τ is nondegenerate. To see this, write $\tau = \sum f(I_\mu) \otimes I_\mu$ and let $V = \operatorname{Im} f \subset \mathfrak{g}$. Let $\mathfrak{g}'$ be the normalizer of V in $\mathfrak{g}$. It is easy to see, using (7.5), that $[r^{12}(u), \tau^{23}] = -[r^{13}(u+\gamma), \tau^{23}] \subset \mathfrak{g} \otimes V \otimes \mathfrak{g}$, hence $r(u) \in \mathfrak{g} \otimes \mathfrak{g}'$. Similarly, we have $r(u) \in \mathfrak{g}' \otimes \mathfrak{g}$ and thus $r(u) \in \mathfrak{g}' \otimes \mathfrak{g}'$. As this is true for generic u, our nondegeneracy condition implies $\mathfrak{g}' = \mathfrak{g}$. Thus V is a nontrivial ideal, and $V = \mathfrak{g}$ by simplicity.

To conclude the proof of the lemma, notice that r has a pole of order at least k at 0. Indeed, if not, expanding (7.5) around $u = 0$ yields $[\tau^{13}, \tau^{23}] = 0$, which contradicts the nondegeneracy of τ.

□

Lemma 7.2. *All the poles of r are simple and $r(u) = \frac{t\Omega}{u} + \mathcal{O}(1)$ for some $t \in \mathbb{C}$.*

Proof: By lemma 7.1, it is enough to show that $r(u) = \frac{t\Omega}{u} + \mathcal{O}(1)$. Let us write $r(u) = \frac{\theta}{u^l} + \frac{\theta'}{u^{l-1}} + O(u^{2-l})$. Considering the coefficient of u^{1-l} in the expansion of (7.1) around $u = 0$, we obtain

$$[\theta'_{12}, r^{13}(v) + r^{23}(v)] + [\theta_{12}, \frac{dr^{13}(v)}{dv}] + [r^{13}(v), r^{23}(v)] = 0, \ (l = 1)$$

$$[\theta'_{12}, r^{13}(v) + r^{23}(v)] + [\theta_{12}, \frac{r^{13}(v)}{v}] = 0, \ (l > 1)$$

for $v \in U$ not a pole of r. Suppose that $l > 1$. Then letting $v \to 0$, and considering the coefficient of v^{-l-1}, we obtain $[\theta_{12}, \theta_{13}] = 0$, which is a contradiction (see lemma 7.1). Hence r has a simple pole at 0.

Let us now determine the residue θ. From (7.5), we have

$$[r^{12}(u) + r^{13}(u), \theta^{23}] = [r^{13}(u) + r^{23}(u), \theta^{12}] = 0$$

(for $u \in U$ not a pole of r). Let $\mathfrak{g}'$ be the centralizer of θ in $\mathfrak{g}$. As in lemma 7.1, we see that $r(u) \in \mathfrak{g}' \otimes \mathfrak{g}'$, which implies that $\mathfrak{g}' = \mathfrak{g}$, i.e θ is an invariant element. Thus $\theta = t\Omega$, $t \neq 0$.

□

Rescaling if necessary, we will assume for simplicity that $t = 1$.

Lemma 7.3. *The r-matrix r satisfies the unitary condition.*

Proof: By definition, r satisfies the equation

$$[r_{12}(z_{12}), r_{13}(z_{13})] + [r_{12}(z_{12}), r_{23}(z_{23})] + [r_{13}(z_{13}), r_{23}(z_{23})] = 0 \qquad (7.6)$$

where we use the notation $z_{ij} = z_i - z_j$. Interchanging z_1 and z_2 and switching the first two components of the tensor product yields

$$[r_{21}(z_{21}), r_{23}(z_{23})] + [r_{21}(z_{21}), r_{13}(z_{13})] + [r_{23}(z_{23}), r_{13}(z_{13})] = 0 \qquad (7.7)$$

Summing (7.6) and (7.7), we arrive at

$$[r_{12}(z_{12}) + r_{21}(z_{21}), r_{13}(z_{13}) + r_{23}(z_{23})] = 0.$$

Letting $z_3 \to z_2$, we obtain $[r_{12}(z_{12}) + r_{21}(z_{21}), \Omega_{23}] = 0$. Since Ω is nondegenerate, this implies that $r_{12}(z_{12}) + r_{21}(z_{21}) = 0$.

□

7.2 Meromorphic continuation of $r(z)$ to $\mathbb{C}$

We will show in this section that $r(z)$ can be extended to a meromorphic function on $\mathbb{C}$, whose poles are simple and form a (discrete) subgroup of $\mathbb{C}$. The basic result is the following:

Lemma 7.4. *For $u, v \in U$ in a neighborhood of 0, $r(u+v)$ is a rational function of $r(u)$ and $r(v)$*

In other terms, the components of $r(u + v)$ are rational functions of the components of $r(u)$ and $r(v)$.

Proof: We consider the equation (7.1) as an (inhomogeneous) system of linear equations in the unknown $r(u+v)$ and whose coefficients are given by $r(u)$ and $r(v)$. A solution to such a system is expressed as a rational function of the coefficients if the associated homogeneous system is nondegenerate (i.e it has only the trivial solution). In our case, this homogeneous system is

$$[r_{12}(u) - r_{23}(v), X_{13}] = 0. \tag{7.8}$$

We now show that this system has only the trivial solution for generic u, v in a neighborhood of 0. Consider (7.8) for $u = v \neq 0$. Thus, our system is equivalent to $[ur_{12}(u) - ur_{23}(u), X_{13}] = 0$. Letting $u \to 0$, we arrive at the system $[\Omega_{12} - \Omega_{23}, X_{13}] = 0$. This system only has the trivial solution: indeed, it is equivalent to the system

$$[g \otimes 1 - 1 \otimes g, X] = 0 \qquad \text{for all } g \in \mathfrak{g} \tag{7.9}$$

But then, for $g_1, g_2 \in \mathfrak{g}$,

$$\begin{aligned} &[[g_1, g_2] \otimes 1 + 1 \otimes [g_1, g_2], X] \\ &\qquad = [[g_1 \otimes 1 - 1 \otimes g_1, g_2 \otimes 1 - 1 \otimes g_2], X] = 0. \end{aligned} \tag{7.10}$$

Since $\mathfrak{g}$ is simple, this implies $[g \otimes 1 + 1 \otimes g, X] = 0$ for all $g \in \mathfrak{g}$, which, together with (7.9) implies $X = 0$.
Nondegeneracy of this system can be expressed as the nonvanishing of certain minors, which are meromorphic functions in u and v. Since this is an open condition, the system (7.8) is nondegenerate in a neighborhood of 0.

□

This result, together with lemma 7.1 has the following immediate corollary:

Corollary 7.1. *The function $r(u)$ extends to a meromorphic function on $\mathbb{C}$ with only simple poles.*

Our next task is to show that the set Γ of poles of r forms a subgroup of $\mathbb{C}$. This will be a consequence of the following:

Lemma 7.5. *Let $\gamma \in \Gamma$. There is an automorphism $A_\gamma \in \operatorname{Aut}\mathfrak{g}$ such that*

$$r(u+\gamma) = (A_\gamma \otimes 1) r(u). \tag{7.11}$$

Proof: Let τ be the residue of r at γ, and set $\tau = (A_\gamma \otimes 1)\Omega$. As we have seen in section 1, τ is nondegenerate (and thus A_γ invertible). Taking the residue of (7.1) at $u = \gamma$, we have $[\tau_{12}, r_{13}(v+\gamma)] + [\tau_{12}, r_{23}(v)] = 0$. Thus

$$\begin{aligned} [(A_\gamma \otimes 1)\Omega_{12}, r_{13}(v+\gamma)] &= -(A_\gamma \otimes 1 \otimes 1)[\Omega_{12}, r_{23}(v)] \\ &= (A_\gamma \otimes 1 \otimes 1)[\Omega_{12}, r_{13}(v)]. \end{aligned} \tag{7.12}$$

Taking now the residue at $v = 0$, we see that

$$[(A_\gamma \otimes 1)\Omega_{12}, (A_\gamma \otimes 1)\Omega_{13}] = (A_\gamma \otimes 1 \otimes 1)[\Omega_{12}, \Omega_{13}]$$

which implies that $A_\gamma \in \operatorname{Aut} \mathfrak{g}$. We can rewrite (7.12) as

$$[(A_\gamma \otimes 1)\Omega_{12}, r_{13}(v + \gamma) - (A_\gamma \otimes 1)r_{13}(v)] = 0$$

from which we conclude that $r(\gamma + v) = (A_\gamma \otimes 1)r(v)$.

□

Corollary 7.2. *The set Γ is a discrete subgroup of $\mathbb{C}$ and the map $\Gamma \to \operatorname{Aut} \mathfrak{g}$, $\gamma \mapsto A_\gamma$ is a group homomorphism.*

7.3 Proof of the classification theorem

By lemma 7.4 and the unitarity condition, there exists a rational function P such that

$$r(u + v) = P(r(u), r(v)). \tag{7.13}$$

A classical and well-known theorem of Weierstraß asserts that any meromorphic function $f : \mathbb{C} \to \mathbb{C}$ satisfying (7.13) is rational, trigonometric or elliptic. In the present case of vector-valued functions, we will use Myberg's generalization of Weierstraß's result.

7.3.1 Myberg's theorem

Quasiabelian function: A meromorphic function $\mathbb{C}^n \to \mathbb{C}^m$ is called *abelian* if it admits $2n$ periods which are linearly independent over $\mathbb{R}$. *Quasiabelian* functions are degenerate forms of abelian functions in the following sense:

Definition: A meromorphic function $f : \mathbb{C}^n \to \mathbb{C}^m$ is *quasiabelian* if there exists a linear coordinate system $(z_1, \dots z_n)$ on $\mathbb{C}^n$, integers $p, q, r \geq 0$, $p+q+r = n$ and vectors $\gamma_1, \dots \gamma_{2r} \in \mathbb{C}^n$ such that

1. For fixed $z_{p+q+1}, \dots z_n$, $f(z_1, \dots z_n)$ is a rational function of the variables $z_1, \dots z_p, e^{z_{p+1}}, \dots e^{z_{p+q}}$,

2. The vectors γ_i are periods of f,

3. The $2r$ vectors $\bar{\gamma}_i \in \mathbb{C}^r$ formed by the last r coordinates of the vectors γ_i are linearly independent over $\mathbb{R}$.

For example, when $n = 1$, abelian functions are elliptic and quasiabelian functions include rational and trigonometric functions. A proof of the following theorem can be found in [BD1].

Theorem 7.1 (Myberg). *Let $f : \mathbb{C} \to \mathbb{C}^m$ be a meromorphic function satisfying*

$$f(u+v) = P(f(u), f(v)) \tag{7.14}$$

for some (fixed) rational function P. Then there exists $n > 0$, a quasiabelian function $\tilde{f} : \mathbb{C}^n \to \mathbb{C}^m$ and a vector $\mathbf{a} \in \mathbb{C}^n$ such that $f(u) = \tilde{f}(u\mathbf{a})$. Furthermore, $\tilde{f}$ can be chosen in such a way that for every $\mathbf{x}, \mathbf{y} \in \mathbb{C}^n$, $\tilde{f}(\mathbf{x}+\mathbf{y}) = P(\tilde{f}(\mathbf{x}), \tilde{f}(\mathbf{y}))$.

7.3.2 Elliptic solutions

Proposition 7.1. *Suppose* $\operatorname{rank}\Gamma = 2$. *Then $r(u)$ is an elliptic function.*

Proof: Let γ_1, γ_2 be generators of Γ. We need to prove that there exists $m, n > 0$ such that $A_{\gamma_1}^n = A_{\gamma_2}^m = 1$, or, equivalently, that the subgroup H of $\operatorname{Aut}\mathfrak{g}$ generated by $A_{\gamma_1}, A_{\gamma_2}$ is finite. We argue by contradiction, using the following result:

Fact: Let H be an infinite commutative subgroup of $\operatorname{Aut}\mathfrak{g}$. Then there exists $a \in \mathfrak{g}$ such that $ha = a$ for every $h \in H$.

Proof: Let $\bar{H}$ be the smallest algebraic subgroup containing H. Since $\bar{H}$ is infinite, $\bar{\mathfrak{h}} = \operatorname{Lie}\bar{H} \neq 0$. Moreover $\bar{H}$ is commutative. But since $\mathfrak{g}$ is simple, $\bar{\mathfrak{h}} \subset \operatorname{Lie}(\operatorname{Aut}\mathfrak{g}) = \mathfrak{g}$ and we can pick $a \in \bar{\mathfrak{h}}$.

Suppose that H is infinite, and let $a \in \mathfrak{g}$ be as above. Let $\theta(u) = (a^* \otimes 1)r(u)$, where a^* is the linear form $(a, \cdot)$. By (7.11), the function θ is Γ-periodic. But $\theta(u)$ has a single, simple pole (mod Γ), which is a contradiction. Thus H is finite, and $r(u)$ an elliptic function.

□

Remark: Elliptic r-matrices exist only for $\mathfrak{g} = sl_n$. This is a consequence of the following result (see [BD1]):

Proposition 7.2. *Let A_1, A_2 be two commuting automorphisms of $\mathfrak{g}$ of finite order, with no nonzero common fixed point. Then $\mathfrak{g} \simeq sl_n$.*

One can further show that all elliptic r-matrices are equivalent to one defined in Lecture 6.

7.3.3 Rational and trigonometric r-matrices

It follows from lemma 7.4 and from Myberg's theorem that there exists $n > 0$, a quasiabelian function $\bar{r} : \mathbb{C}^n \to \mathfrak{g} \otimes \mathfrak{g}$ and a vector $\mathbf{a} \in \mathbb{C}^n$ such that $r(u) =$

$\bar{r}(u\mathbf{a})$. Furthermore, $\bar{r}$ can be chosen so as to satisfy the CYBE with spectral parameters

$$[r_{12}(\mathbf{u}), r_{13}(\mathbf{u}+\mathbf{v})] + [r_{12}(\mathbf{u}), r_{23}(\mathbf{v})] + [r_{13}(\mathbf{u}+\mathbf{v}), r_{23}(\mathbf{v})] = 0$$

where $\mathbf{u}, \mathbf{v} \in \mathbb{C}^n$. Indeed, as we have seen, the CYBE with spectral parameter can be regarded as an "addition formula": $r(u+v) = P(r(u), r(v))$ for some rational function P. To conclude the proof of theorem 6.1, it remains to prove that r is equivalent to a quasiabelian function of *one* variable. This is achieved by the following technical result:

Lemma 7.6. *There exists a hyperplane $V \subset \mathbb{C}^n$ with the following property: for every $\mathbf{b} \in \mathbf{a} + V$, the r-matrices $\tilde{r}(u) = \bar{r}(u\mathbf{b})$ and $r(u) = \bar{r}(u\mathbf{a})$ are equivalent.*

Proof: We first construct V. Let σ be a singular point of $\bar{r}$ and write $\bar{r}(z) = \frac{Y(z)}{f(z)}$ for holomorphic functions $Y : U \to \mathfrak{g} \otimes \mathfrak{g}$ and $f : U \to \mathbb{C}$ defined in a neighborhood $U \ni \sigma$, such that $Y(\sigma) \neq 0$. As usual, let us consider $k(z) : \mathfrak{g} \to \mathfrak{g}$ defined by $\bar{r}(z) = (k(z) \otimes 1)\Omega$. By the arguments of lemma 7.1, we have (for generic z) $[Y_{12}(\sigma), (k(z+\sigma) \otimes 1 + 1 \otimes k(z))\Omega^{13}] = 0$ and thus

$$[Y(\sigma), k(z+\sigma)x \otimes 1 + 1 \otimes k(z)x] = 0 \qquad \text{for any } x \in \mathfrak{g}. \tag{7.15}$$

Let $\mathfrak{k}(z) \subset \mathfrak{g} \oplus \mathfrak{g}$ be the subalgebra generated by $(\{k(z+\sigma)x, k(z)x\})$. For generic values of z, $\bar{r}(z+\sigma)$ and $\bar{r}(z)$ are both nondegenerate, and the projections $\pi_{1,2} : \mathfrak{k} \to \mathfrak{g}$ are surjective. Notice that since π_2 is surjective, $\operatorname{Ker} \pi_1 \subset (0, \mathfrak{g})$ is an ideal. By simplicity of $\mathfrak{g}$, we have $\operatorname{Ker} \pi_1 = 0$ or $(0, \mathfrak{g})$. But $Y(\sigma) \neq 0$, and $(0, \mathfrak{g}) \subset \mathfrak{k}$ is impossible. Thus, there exists $L(z, \sigma) \in \operatorname{Aut} \mathfrak{g}$ such that $\mathfrak{k} = \{L(z,\sigma)x \otimes 1 + 1 \otimes x \mid x \in \mathfrak{g}\}$. In particular,

$$k(z+\sigma) = L(z,\sigma)k(z) \tag{7.16}$$

This and (7.15) in turn imply

$$Y(\sigma) = c(z,\sigma)(L(z,\sigma) \otimes 1)\Omega \tag{7.17}$$

for some scalar $c(z,\sigma)$. This last equation shows that both $c(z,\sigma)$ and $L(z,\sigma)$ are in fact independent of z (notice that $c(z,\sigma)$ and $L(z,\sigma)$ are both determined by (7.17) since $L(z,\sigma) \in \operatorname{Aut} \mathfrak{g}$).

Now let S be the set of singular points of $\bar{r}$ and let Σ be the subgroup of $\mathbb{C}^n$ generated by S. By (7.16), we have just constructed a group homomorphism $\phi : \Sigma \to \operatorname{Aut} \mathfrak{g}$, $\sigma \mapsto L(\sigma)$ such that

$$\bar{r}(z+\sigma) = (\phi(\sigma) \otimes 1)\bar{r}(z). \tag{7.18}$$

Notice that this equality now holds for $z \in \mathbb{C}^n$ by analytic continuation. The unitarity condition for $\bar{r}$ (lemma 7.3) then implies

$$\bar{r}(z) = (\phi(\sigma) \otimes \phi(\sigma))\bar{r}(z). \tag{7.19}$$

We have $\Sigma \neq \mathbb{C}^n$ since the set of singular points of $\bar{r}$ is invariant under translation by Σ. Since S is an analytic subset of codimension 1, Σ is a union of (at most) countably many parallel affine hyperplanes. In particular, it contains a hyperplane V going through 0. Notice that by (7.17) the map $\phi : V \to \mathrm{Aut}\,\mathfrak{g}$ is holomorphic.

Now let $\mathbf{b} \in \mathbf{a}+V$ and $\tilde{r}(u) = \bar{r}(u\mathbf{b})$. The equivalence of $r(u)$ and $\tilde{r}(u)$ follows from (7.18) and (7.19):

$$\tilde{r}(u_1 - u_2) = \bar{r}((u_1 - u_2)\mathbf{b}) = \phi(u_1(\mathbf{b} - \mathbf{a}) \otimes \phi(u_2(\mathbf{b} - \mathbf{a}))\bar{r}((u_1 - u_2)\mathbf{a}).$$

□

Remark: In particular, this means that every solution of the CYBE with spectral parameters in $\mathbb{C}^n$ arises from a solution of the CYBE with spectral parameters in $\mathbb{C}$.

We now conclude the proof of the Belavin-Drinfeld classification theorem.

Proposition 7.3. *Suppose* $\operatorname{rank}\Gamma = 1$. *Then $r(u)$ is equivalent to an r-matrix of the form $f(e^{ku})$ for some $k \in \mathbb{C}$ and some rational function f. If* $\operatorname{rank}\Gamma = 0$ *then $r(u)$ is equivalent to a rational r-matrix.*

Proof: Let $p, q, r, (z_1, \dots z_n), \gamma_1, \dots \gamma_{2r}$ denote the usual data for the quasi-abelian function $\bar{r}$. Let $e_1, \dots e_n$ be the basis corresponding to $(z_1, \dots z_n)$ $(z = \sum z_i e_i)$ and let $W = \mathbb{C}e_1 \oplus \dots \mathbb{C}e_{p+q}$. We claim that $W \not\subset V$. To see this, notice that by (7.12) the complex numbers δ_i, $i = 1, \dots 2r$ defined by $\gamma_i \in \delta_i\mathbf{a} + V$ are in Γ. Moreover, $W \oplus \bigoplus_i \mathbb{R}\gamma_i = \mathbb{C}^n$, so that

$$\mathbb{C} \simeq \mathbb{C}^n/V \simeq (W \bigoplus_i \mathbb{R}\gamma_i)/V.$$

But by assumption $\operatorname{rank}\Gamma \leq 1$ so that the map $\bigoplus_i \mathbb{R}\gamma_i \to \mathbb{C}^n/V \simeq \mathbb{C}$ is not surjective. Hence $W \not\subset V$, i.e there exists $1 \leq i \leq p+q$ such that $e_i \notin W$. Then $\mathbf{a} \in ke_i + V$ for some $k \in \mathbb{C}$, and by lemma 7.6 $r(u) \sim \bar{r}(uke_i)$. Thus if $i \leq p$, $r(u) \sim P(u)$ for some rational function P, and if $i > p$, $r(u) \sim P(e^{ku})$ for some rational function P. It is easy to see that the first case happens when $\operatorname{rank}\Gamma = 0$ while the second occurs when $\operatorname{rank}\Gamma = 1$.

□

Lecture 8

Hopf algebras

We now turn to the quantum picture.

8.1 Definition of Hopf algebras

Let us first make a digression and explain where the term "quantum group" comes from:

8.1.1 Finite groups revisited

Recall that we have defined quantization of a Poisson manifold M as a quantization of the Poisson algebra $\mathcal{C}^\infty(M)$. Thus the process of quantization replaces a commutative algebra ($\mathcal{C}^\infty(M)$-the classical observables) by a noncommutative one (the quantum observables). Let us see what this means when, in addition, M has the structure of a group; in this case we write G for M: we have multiplication and inversion maps $m : G \times G \to G$, $inv : G \to G$ and a unit element $e \in G$ satisfying the usual axioms. For simplicity, let us consider the case of a finite group. Let $\mathcal{F}_k(G)$ be the algebra of k-valued functions on G.

The multiplication map m endows $\mathcal{F}_k(G)$ with an additional structure: define

$$\Delta : \mathcal{F}_k(G) \to \mathcal{F}_k(G \times G) \simeq \mathcal{F}_k(G) \otimes \mathcal{F}_k(G),$$
$$\Delta(f)(x, y) = f(m(x, y)).$$

Likewise, to the inversion map and the unit element there correspond maps

$$S : \mathcal{F}_k(G) \to \mathcal{F}_k(G), \qquad S(f)(x) = f(x^{-1}),$$
$$\epsilon : \mathcal{F}_k(G) \to k, \qquad \epsilon(f) = f(e).$$

Thus $\mathcal{F}_k(G)$ has altogether five structures: the unit (the function $f \equiv 1$) and the multiplication m, the *comultiplication* Δ and the *counit* ϵ, and the *antipode* S. Of course, there are many compatibility conditions which we will axiomatize in the next section when we define *Hopf algebras*. However, here is the fundamental observation: to (finite) groups we associate Hopf algebras in which the multiplication map m is commutative (i.e classical observables commute). Thus the Hopf algebras for which m and Δ are not necessarily commutative correspond to "quantum groups" rather than usual groups.

8.1.2 Coalgebras

Definition: A (coassociative) *coalgebra* defined over a field k is a k-vector space A equipped with two linear maps $\Delta : A \to A \otimes A$ (the comultiplication) and $\epsilon : A \to k$ (the counit) satisfying the following relations:

1. The map Δ is coassociative:

$$(\Delta \otimes 1)\Delta(x) = (1 \otimes \Delta)\Delta(x).$$

2. The counit axiom:

$$(\epsilon \otimes 1)\Delta(x) = (1 \otimes \epsilon)\Delta(x) = x.$$

We say that A is cocommutative if $\Delta = \Delta^{op}$, i.e if $\Delta(x) = (\Delta(x))^{21}$ for all $x \in A$. A homomorphism of coalgebras between two coalgebras $(A, \Delta_A, \epsilon_A)$ and $(B, \Delta_B, \epsilon_B)$ is a linear map $f : A \to B$ such that $\Delta_B(f(x)) = (f \otimes f)(\Delta_A(x))$ and $\epsilon_B(f(x)) = \epsilon_A(x)$ for all $x \in A$. Note that k itself is a (cocommutative) coalgebra with $\Delta : k \to k \otimes k$ and $\epsilon : k \to k$ being the identity maps.

8.1.3 Hopf algebras

Definition: A *Hopf algebra* defined over a field k is a k-vector space A with the following additional structures:

1.multiplication :	$m : A \otimes A \to A,$
2.unit :	$i : k \to A,$
3.comultiplication :	$\Delta : A \to A \otimes A,$
4.counit :	$\epsilon : A \to k,$
5.antipode :	$S : A \to A,$

satisfying the following axioms:

1. the map m defines a structure of an associative algebra on A with unit $i(1)$,
2. the maps Δ and ϵ define a structure of a coassociative coalgebra on A,
3. the maps $\Delta : A \to A \otimes A$ and $\epsilon : A \to k$ are homomorphisms of algebras,
4. the map S is a linear isomorphism and satisfies the following two relations:

$$m(1 \otimes S)\Delta(x) = m(S \otimes 1)\Delta(x) = i(\epsilon(x)).$$

Similarly, one defines the notion of a Hopf algebra over a commutative ring R.

A Hopf algebra is called commutative if $m(x,y) = m(y,x)$ for every $x, y \in A$. It is called *cocommutative* if $\Delta(x) = \Delta^{op}(x)$ for every $x \in A$. A Hopf subalgebra of A is a subalgebra B such that $\Delta(B) \subset B \otimes B$ and $S(B) \subset B$. A Hopf ideal of A is an ideal I such that $\Delta(I) \subset I \otimes A + A \otimes I$ and $S(I) \subset I$. As usual, by map of Hopf algebras, we will mean a map preserving all the structures. We will denote by $\mathrm{HA}(k)$ (resp. $\mathrm{HA}_f(k)$) the category of Hopf algebras (resp. finite dimensional Hopf algebras).

The above definition of Hopf algebra implies the following relations, which are constantly used:

5. the maps $m : A \otimes A \to A$ and $i : k \to A$ are coalgebra maps,

6. the map S is an algebra and coalgebra antihomomorphism:

$$S(xy) = S(y)S(x), \qquad \Delta(S(x)) = (S \otimes S)(\Delta^{op}(x)),$$

7. the antipode is compatible with the units:

$$\epsilon(S(x)) = \epsilon(x), \qquad S(i(1)) = i(1).$$

Proof of these results are not too difficult, especially if one uses the pictorial technique (see Exercise 8.1).

It also makes sense to consider the above axioms without the antipode:

Definition: A *bialgebra* is a k-vector space, equipped with the structures 1-4 satisfying axioms 1-3 (property 5 is then a consequence of these axioms).

In other terms, it is both an algebra and a coalgebra and these two structures are compatible.

8.2 Pictorial representation

Let us translate axioms 1.-4. in the language of pictures, which will make them easier to use, and (hopefully) more intuitive. Let us represent the basic operations on Hopf algebras as follows:

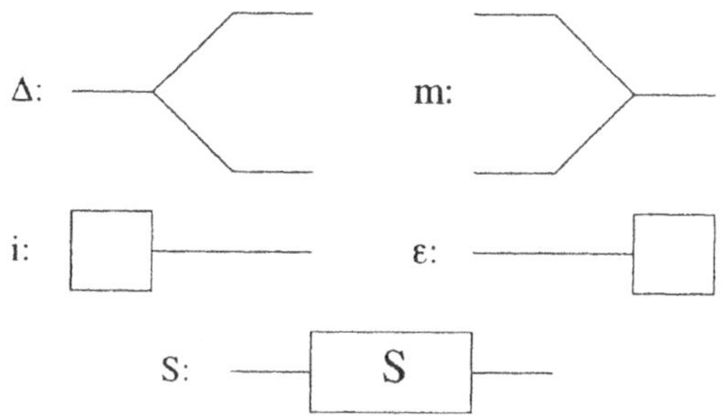

For example, if $x \in A$, $\Delta(x) \in A \otimes A$ has two components (more precisely, it is a sum of terms having two components), which stand at the ends of the two branches in the diagram representing Δ. The symbol $\square$ stands for the ground field k. More generally, a map $T : A \to A$ is represented as below:

A bare line represents the identity map. Finally, composition of morphisms is obtained by adjoining pictures (from left to right).

The relation defining Hopf algebras thus have the following pictorial representation: the first four lines correspond to axioms $1-4$, and the others correspond to the essential properties derived from them.

1.

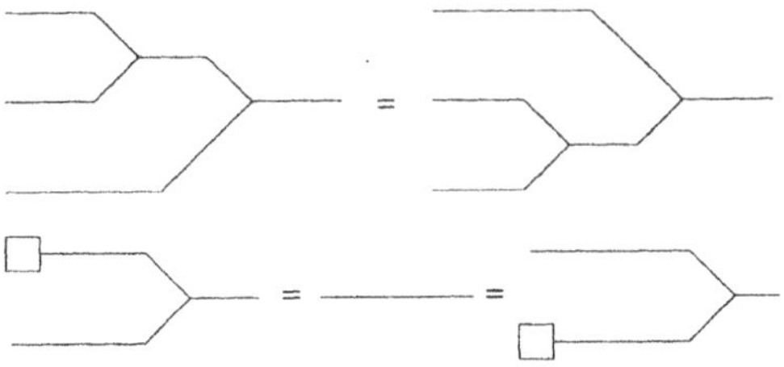

2.

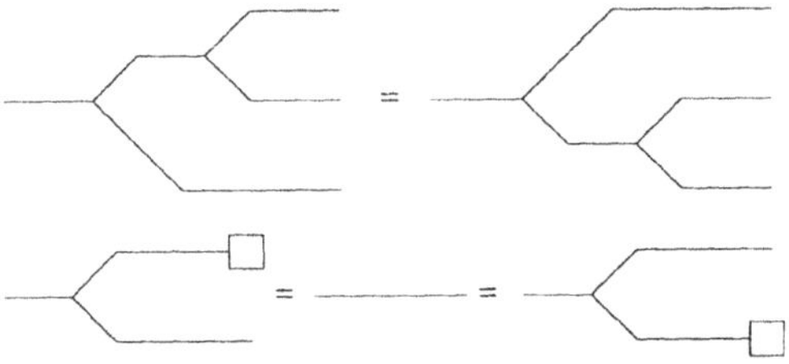

3.

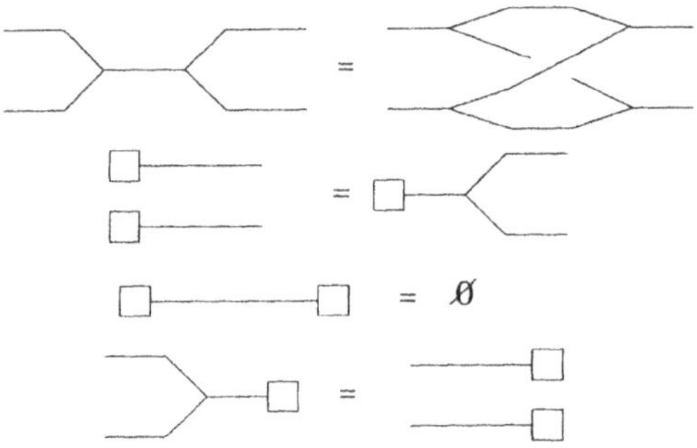

4.

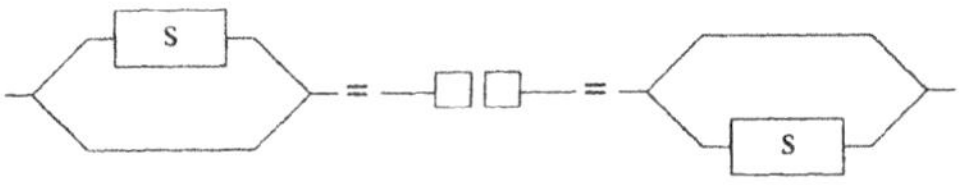

5. = 3.

6.

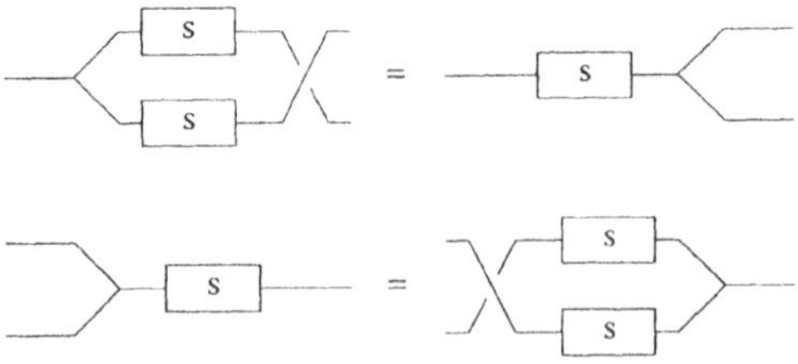

7.

In the above, the symbol $\emptyset$ stands for the identity map $Id : k \to k$.

Let us use this technique to prove a result which will be useful later:

Proposition 8.1 (uniqueness of antipode). *Let $(A, m, \Delta, i, \epsilon, S)$ be a Hopf algebra. Suppose that $S' \in \operatorname{End} A$ satisfies axiom 4., i.e $(A, m, \Delta, i, \epsilon, S')$ is a Hopf algebra. Then $S = S'$.*

Proof: The proof is based on the following observation. Let us define a map $\eta : \operatorname{End} A \to \operatorname{End} A$ by the picture:

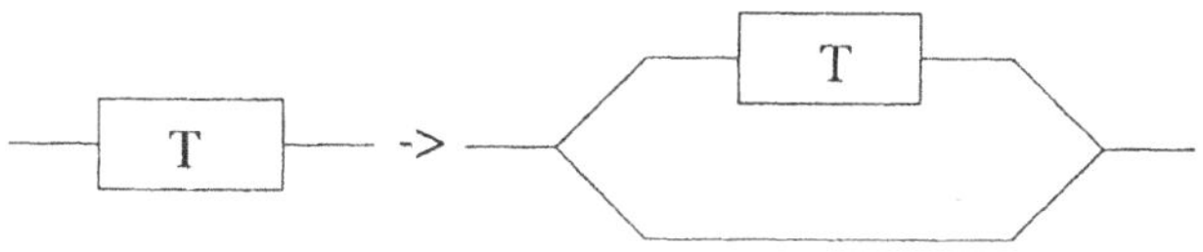

Lemma 8.1. *The map $\eta : \operatorname{End} A \to \operatorname{End} A$ is invertible, and its inverse is given by*

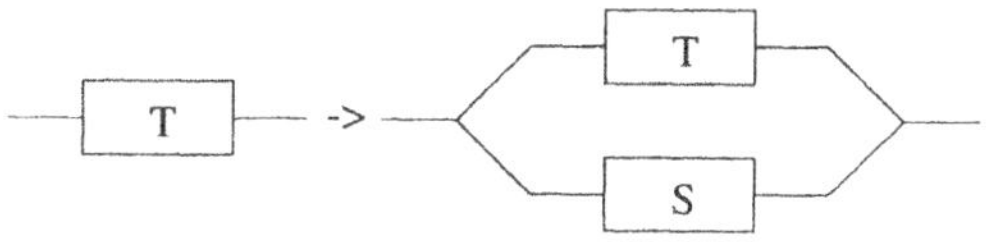

This is the proof:

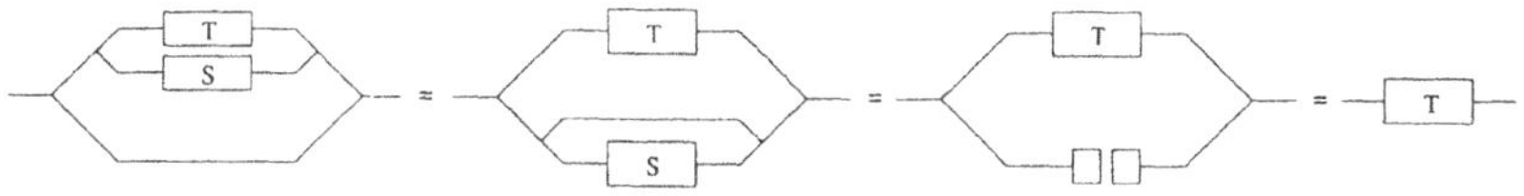

To conclude the proof of Proposition 8.1, it suffices to notice that

$$\eta(S) = \eta(S') = i \circ \epsilon.$$

$\square$

Exercise 8.1. Use axioms 1.–4. to derive relations 5.–7. This should convince you of the usefulness of the pictorial technique!

8.3 Examples of Hopf algebras

It is a worthwhile exercise to check that the following are Hopf algebras:

Example 8.1 (Functions on a finite group). This is what we started with. The same construction works for affine algebraic groups over k (and algebraic functions)-the important point here is that $\mathcal{F}_k(G \times G) \simeq \mathcal{F}_k(G) \otimes \mathcal{F}_k(G)$.

Example 8.2 (Group algebra). Let G be any group, and $k[G]$ its group algebra. Then $k[G]$ is a Hopf algebra with $\Delta(g) = g \otimes g$, $S(g) = g^{-1}$ and $\epsilon(g) = 1$ for $g \in G$.

Example 8.3 (Universal enveloping algebra). Let $\mathfrak{g}$ be a Lie algebra over k and $U(\mathfrak{g})$ its enveloping algebra. Define a coproduct on $U(\mathfrak{g})$ by $\Delta(x) = 1 \otimes x + x \otimes 1$ for $x \in \mathfrak{g}$ and extend it by $\Delta(xy) = \Delta(x)\Delta(y)$. Similarly, define the antipode by $S(x) = -x$ for $x \in \mathfrak{g}$ and $S(xy) = S(y)S(x)$, and the counit by $\epsilon(x) = 0$ for $x \in \mathfrak{g}$ and $\epsilon(xy) = \epsilon(x)\epsilon(y)$. This makes $U(\mathfrak{g})$ into a Hopf algebra.

The last two examples are essentially dual to the first. Notice that in 8.2, $\Delta(x) = x \otimes x \Leftrightarrow x \in G$ while in 8.3, $\Delta(x) = 1 \otimes x + x \otimes 1 \Leftrightarrow x \in \mathfrak{g}$. We thus make the following definition:

Definition: Let A be a Hopf algebra. An element $x \in A$ is called *grouplike* if $\Delta(x) = x \otimes x$. It is called *primitive* if $\Delta(x) = 1 \otimes x + x \otimes 1$.

Notice that example 8.1 is commutative, while the examples 8.2 and 8.3 are cocommutative. In fact, a converse statement is true:

Proposition 8.2 (see e.g. [Mum]). *Let A be a finitely generated commutative Hopf algebra defined over a field k of characteristic 0. Then $A \simeq \mathcal{F}_k(G)$ for some affine algebraic group G defined over k.*

Proof(sketch): It can be shown that the algebra A has no nilpotent elements. Thus A is identified with the algebra of (polynomial) functions on the affine variety Spec A. Furthermore, $\Delta : A \to A \otimes A$ defines a map m : Spec $A \times$ Spec $A \to$ Spec A. Likewise, $S : A \to A$ is now an algebra homomorphism, and defines a map inv : Spec $A \to$ Spec A. It remains to check that these endow Spec A with an (algebraic) group structure: the associativity of m is a consequence of the coassociativity of Δ, and the identity $m(inv \otimes \mathrm{Id})(x) = 1$ follows from the antipode axiom.

□

Proposition 8.3. *Let A be a cocommutative Hopf algebra over a field of characteristic zero generated by its primitive elements. Then $A = U(\mathfrak{g})$ (with its usual Hopf algebra structure) for some Lie algebra $\mathfrak{g}$.*

Proof: Let $\mathfrak{g} \subset A$ be the set of primitive elements. It follows from

$$\begin{aligned}\Delta([x,y]) &= [\Delta(x), \Delta(y)] \\ &= [1 \otimes x + x \otimes 1, y \otimes 1 + 1 \otimes y] \\ &= [x,y] \otimes 1 + 1 \otimes [x,y]\end{aligned}$$

that $\mathfrak{g}$ is a Lie algebra. Since $\mathfrak{g}$ generates A, we have a Hopf algebra map $\phi : U(\mathfrak{g}) \twoheadrightarrow A$. Let $I = \operatorname{Ker}\phi$. It is a Hopf ideal of $U(\mathfrak{g})$. Let $I = \bigcup_{n \geq 0} I_n$ be the filtration on I induced by the standard filtration of $U(\mathfrak{g})$. We have

$I_0 = I_1 = 0$. Suppose $I \neq \{0\}$, and let $x \in I_m$ be a nonzero element of minimal degree. It is easy to check that

$$\Delta(x) - 1 \otimes x - x \otimes 1 \in U(\mathfrak{g})_{m-1} \otimes U(\mathfrak{g})_{m-1}$$

(this is true for any element of $U(\mathfrak{g})_m$). By minimality of m, we must have $\Delta(x) = 1 \otimes x + x \otimes 1$, i.e x is primitive. This contradicts the fact that $m > 1$, since in characteristic zero the set of primitive elements of $U(\mathfrak{g})$ equals $\mathfrak{g}$. Hence $I = \{0\}$ and $A = U(\mathfrak{g})$.

□

Remark: Proposition 8.3 has the following well known generalization, referred to as Kostant's theorem: any cocommutative Hopf algebra over a field k of characteristic zero is of the form $k[G]\#U(\mathfrak{g})$, where G is a finite group and $\mathfrak{g}$ a Lie algebra. We will not give a proof of this result.

Example 8.4 (The quantum group $U_q(sl_2(\mathbb{C}))$. As promised, here is our first example of a neither commutative nor cocommutative Hopf algebra. Let $q \in \mathbb{C}\backslash\{0, \pm 1\}$, and let $U_q(sl_2(\mathbb{C}))$ be the Hopf algebra defined over $\mathbb{C}$ and generated by elements e, f, K, K^{-1} satisfying the following relations: K and K^{-1} are inverse to each other and

$$KeK^{-1} = q^2 e, \qquad KfK^{-1} = q^{-2} f, \qquad [e, f] = \frac{K - K^{-1}}{q - q^{-1}},$$

$$\Delta(e) = e \otimes K + 1 \otimes e, \qquad \Delta(f) = f \otimes 1 + K^{-1} \otimes f, \qquad \Delta(K) = K \otimes K,$$

$$S(e) = -eK^{-1}, \qquad S(f) = -Kf, \qquad S(K) = K^{-1},$$

$$\epsilon(e) = \epsilon(f) = 0, \qquad \epsilon(K) = 1.$$

In some sense, $U_q(sl_2(\mathbb{C}))$ is a deformation of $U(sl_2(\mathbb{C}))$ and tends to $U(sl_2(\mathbb{C}))$ as $q \to 1$. This will be formulated more precisely in the next lecture. According to our discussion, we should think of $U_q(sl_2(\mathbb{C}))$ as associated to a "quantum group". By abuse of language, it is often called a quantum group.

Remarks i) In examples 8.1-3, we have $S^2 = Id$ (because inversion is an involution of a group), but this is not true for $U_q(sl_2(\mathbb{C}))$.
ii) In $U_q(sl_2(\mathbb{C}))$, we have $S^2(u) = KuK^{-1}$, i.e the square of the antipode is an inner automorphism. We will encounter this situation again (see Problem 9.1).

Exercise 8.2. Let $(A, m, \Delta, i, \epsilon, S)$ be a commutative or cocommutative Hopf algebra. Show that $S^2 = Id$.

8.4 Duality in Hopf algebras

The notion of a Hopf algebra is self dual, at least if we restrict ourselves to finite dimensional Hopf algebras. This situation is reminiscent of that of Lie bialgebras.

Proposition 8.4. *Let $(A, m, \Delta, i, \epsilon, S)$ be a finite dimensional Hopf algebra. Then $(A^*, \Delta^*, m^*, \epsilon^*, i^*, S^*)$ is a Hopf algebra. The map*

$$\mathrm{HA}_f(k) \to \mathrm{HA}_f(k),$$
$$A \mapsto A^*$$

is an involution and an antiequivalence of categories.

Proof: By definition the dual of the multiplication in A is a comultiplication for A^*, the dual of the comultiplication in A a multiplication in A^*, the dual of the counit a unit and the dual of the unit a counit, and that the dual of the antipode is the antipode for A^*. In our pictorial technique, this amounts to exchanging left and right. We thus have to check that the axioms 1.-4. are symmetric, which is obvious from the pictures. The second part of the proposition is clear.

□

Exercise 8.3. Let G be a finite group. Show that the dual of $k[G]$ is $\mathcal{F}_k(G)$ (as a Hopf algebra).

8.5 Deformation Hopf algebras

Let $K = k[[h]]$. By a topologically free Hopf algebra over K we will understand a topologically free K-module equipped with a product, coproduct, unit, counit, and antipode, which satisfy the axioms of a Hopf algebra with tensor products understood in a completed sense. Note that a topologically free Hopf algebra is, strictly speaking, not a Hopf algebra, because of a somewhat different nature of tensor products. However, this difference is insignificant, at least for our purposes. Therefore, we will adopt a convenient abuse of terminology and will refer to topologically free Hopf algebras over K simply as "Hopf algebra over K". In particular, the category of topologically free Hopf algebras over K will be denoted by $HA(K)$. In accordance with Lecture 1, we make the following definition:

Definition: A *deformation Hopf algebra* is a topologically free Hopf algebra A. We say that A is a *deformation* of a Hopf algebra A_0 defined over k if $A/hA \simeq A_0$ as a Hopf algebra.
In the same fashion, we define deformation bialgebras.

We prove the following proposition, which we will use later:

Proposition 8.5. *Let A be a bialgebra which is a deformation of A_0. Suppose that A_0 has the structure of a Hopf algebra (i.e A_0 is equipped with an antipode $S_0 : A_0 \to A_0$ satisfying axiom 4.). Then A has a (unique) Hopf algebra structure compatible with its bialgebra structure and is a deformation of A_0 as a Hopf algebra.*

Proof: Recall the notation of Proposition 8.1. Let $\eta : \mathrm{End}(A) \to \mathrm{End}(A)$ and $\eta_0 : \mathrm{End}(A_0) \to \mathrm{End}(A_0)$ be as defined there. Since A_0 is a Hopf algebra, η_0 is invertible. Moreover, $\eta \equiv \eta_0 \pmod{h}$, so η is also invertible. Now set $S = \eta^{-1}(i \circ \epsilon)$. It is easy to check that S satisfies all the required axioms, and that $A/hA \simeq A_0$ as a Hopf algebra.

□

We can even say more: as we have seen before, any associative multiplication $*$ on $A_0[[h]]$ compatible with the multiplication in A_0 automatically has a unit. The same is true of the counit. In other words, to have a Hopf algebra deformation of A_0, we only need to construct an algebra deformation A of A_0 and a map $\Delta : A \to A \otimes A$ which is coassociative and which is an algebra homomorphism.

Lecture 9

Quantized universal enveloping algebras

We formulate in this lecture the notion of quantization of a Lie bialgebra in terms of Hopf algebras. We also state one of the main results of these notes: the existence of a functorial quantization of Lie bialgebras.

9.1 Quantized enveloping algebras

We let k be any characteristic zero field and set $K = k[[h]]$. The following definition is due to Drinfeld:

Definition: A deformation Hopf algebra H is called a *quantized universal enveloping algebra* if $H/hH \simeq U(\mathfrak{g})$ for some Lie algebra $\mathfrak{g}$, where $U(\mathfrak{g})$ has the usual Hopf algebra structure.
We will often drop the word "universal" for brevity.

In particular, a quantized enveloping algebra is cocommutative up to the first order. Of course, it may be noncocommutative. We denote by QUE(K) the full subcategory of HA(K) whose objects are the quantized universal enveloping algebras (where $HA(K)$ denotes the category of deformation Hopf algebras over K).

Recall the situation with deformation algebras: if A is a deformation of a commutative algebra A_0, then A_0 acquires from A a Poisson structure. In our case, we have:

Proposition 9.1. *Let H be a quantized enveloping algebra : $H/hH \simeq U(\mathfrak{g})$. Then the Lie algebra $\mathfrak{g}$ is naturally equipped with a Lie bialgebra structure defined by*

$$\delta(x) = \frac{\Delta(\tilde{x}) - \Delta^{op}(\tilde{x})}{h} \pmod{h} \tag{9.1}$$

where $\tilde{x}$ is any lifting of x to H.

Proof: Notice that (9.1) makes sense because H is cocommutative up to first order, i.e we take the first nontrivial term in $\Delta(\tilde{x}) - \Delta^{op}(\tilde{x})$. Let us check that δ is well defined: let $x \in \mathfrak{g}$, $\tilde{x}, \tilde{x} + hu$ be two liftings of x to H. Then

$$\begin{aligned} \frac{\Delta(\tilde{x}+hu) - \Delta^{op}(\tilde{x}+hu)}{h} &\equiv \frac{\Delta(\tilde{x}) - \Delta^{op}(\tilde{x})}{h} + (\Delta(u) - \Delta^{op}(u)) \pmod h \\ &\equiv \frac{\Delta(\tilde{x}) - \Delta^{op}(\tilde{x})}{h} \pmod h. \end{aligned}$$

Let us now show that $\delta(x) \in \Lambda^2\mathfrak{g} \subset U(\mathfrak{g}) \otimes U(\mathfrak{g})$. Since $\delta(x)$ is clearly skew, it is enough to show that $\delta(x) \in \mathfrak{g} \otimes \mathfrak{g}$, or that its two components are primitive elements of $U(\mathfrak{g})$. This is a consequence of the (co)Leibniz Identity: let Δ_0 denote the usual coproduct in $U(\mathfrak{g})$. Then

$$\begin{aligned} (\Delta_0 \otimes 1)\delta(x) &= \frac{1}{h}(\Delta \otimes 1)(\Delta(\tilde{x}) - \Delta^{op}(\tilde{x})) \pmod h \\ &\equiv \frac{1}{h}\{((1 \otimes \Delta) - (1 \otimes \Delta^{op}))\Delta(\tilde{x}) + \sigma_{23}(\Delta \otimes 1 - \Delta^{op} \otimes 1)\Delta(\tilde{x})\} \\ &= (1 \otimes \delta)\Delta_0(x) + \sigma_{23}(\delta \otimes 1)\Delta_0(x). \end{aligned}$$

It remains to check that δ satisfies the coJacobi identity and the cocycle condition. This follows from the coassociativity and homomorphism property of Δ respectively.

□

We are thus led to the following definition:

Definition: The quasiclassical limit of a quantized enveloping algebra H is the Lie bialgebra $(\mathfrak{g} = \mathrm{Prim}\, H/hH, [\,,\,], \delta)$ where $\mathrm{Prim}\, H/hH$ stands for the set of primitive elements of H/hH and where δ is defined as above. Conversely, we say that H is a quantization of the Lie bialgebra $(\mathfrak{g}, [\,,\,], \delta)$.

Remark: As it is the case for Poisson algebras, taking the quasiclassical limit of a quantized enveloping algebra is a well defined-operation, while there are in general an infinite number of distinct quantizations of a Lie bialgebra. For example, any Hopf algebra deformation of $U(\mathfrak{g})$ in which the coproduct is cocommutative up to second order is a quantization of the trivial Lie bialgebra.

Exercise 9.1 (CoPoisson Hopf algebras). Let A be a cocommutative Hopf algebra. A *CoPoisson* structure on A is a map $\delta : A \to \Lambda^2 A$ satisfying the following axioms: the coJacobi identity

$$(1 \otimes \delta)\delta + \text{cycl.} = 0,$$

the coLeibniz identity

$$(\Delta \otimes 1)\delta = (1 \otimes \delta)\Delta + \sigma_{23}(\delta \otimes 1)\Delta,$$

and the cocycle condition

$$\delta(ab) = \delta(a)\Delta(b) + \Delta(a)\delta(b).$$

Show that a coPoisson structure on $U(\mathfrak{g})$ naturally induces a Lie bialgebra structure on $\mathfrak{g}$, and that conversely, any Lie bialgebra structure on $\mathfrak{g}$ extends to a coPoisson structure on $U(\mathfrak{g})$.

9.2 The quantization theorem

The question of existence of a quantization for any Lie bialgebra is settled by the following theorem

Theorem 9.1 ([EK1, EK2]). *There exists a functor*

$$\tilde{Q} : LBA(k) \to QUE(K)$$

such that $\tilde{Q}((\mathfrak{g}, [\,,\,], \delta)$ *is a quantization of* $(\mathfrak{g}, [\,,\,], \delta)$ *for any* $(\mathfrak{g}, [\,,\,], \delta) \in \mathrm{Ob}(LBA(k))$. *In particular, any Lie bialgebra can be quantized.*

This answers a question of Drinfeld (see [Dr3]). In this paper, Drinfeld showed how to construct a quantization up to order h^4.

9.3 Examples

We can now explain the link between $U(sl_2(\mathbb{C}))$ and the quantum group $U_q(sl_2(\mathbb{C}))$ defined in the last lecture.

Let $U_h(sl_2(\mathbb{C}))$ be the Hopf algebra defined over $\mathbb{C}[[h]]$ as follows:
Set $q = e^h = \sum \frac{h^n}{n!} \in \mathbb{C}[[h]]$. Then $U_h(sl_2(\mathbb{C}))$ is generated by E, F, H and relations:

$$[H, E] = 2E, \qquad [H, F] = -2F, \qquad [E, F] = \frac{q^H - q^{-H}}{q - q^{-1}},$$

$$\begin{aligned} &\Delta(E) = E \otimes q^H + 1 \otimes E, && \Delta(F) = F \otimes 1 + q^{-H} \otimes F, && \Delta(H) = H \otimes 1 + 1 \otimes H, \\ &S(E) = -Eq^{-H}, && S(F) = q^H F, && S(H) = -H, \\ &\epsilon(E) = 0, && \epsilon(F) = 0, && \epsilon(H) = 0. \end{aligned}$$

It is easy to check that $U_h(sl_2(\mathbb{C}))/hU_h(sl_2(\mathbb{C})) \simeq U(sl_2(\mathbb{C}))$ as a Hopf algebra. Furthermore, we compute

$$\begin{aligned} \delta(H) &= 0, \\ \delta(E) &= \frac{1}{h}(\Delta(E) - \Delta^{op}(E)) \equiv E \wedge H \ (\mathrm{mod}\ h), \\ \delta(F) &= \frac{1}{h}(\Delta(F) - \Delta^{op}(F)) \equiv F \wedge H \ (\mathrm{mod}\ h). \end{aligned}$$

We recognize the standard Lie bialgebra structure on $sl_2(\mathbb{C})$ (up to a constant).

Remark: This Hopf algebra looks like $U_q(sl_2(\mathbb{C}))$ with $q = e^h$, but it is now a deformation Hopf algebra-the variable h is formal. This is what we mean when we say that $U_q(sl_2(\mathbb{C}))$ tends to $U(sl_2(\mathbb{C}))$ as $q \to 1$ (or $h \to 0$).

Remark: For the relation between the QUE algebra $U_h(sl_2(\mathbb{C}))$ and the algebra $U_q(sl_2(\mathbb{C}))$ over $\mathbb{C}[q, q^{-1}, (q - q^{-1})^{-1}]$, see [Kass].

Drinfeld-Jimbo quantum groups: Let $\mathfrak{g}$ be a Kac-Moody algebra. Let $A = (a_{ij})_{i,j=1,\ldots n}$ be its Cartan matrix. Suppose A is symmetrizable and choose $d_i \in \mathbb{N}$ such that $d_i a_{ij} = d_j a_{ji}$ and the d_i are coprime. Drinfeld and Jimbo independently introduced the following *quantum groups*, quantizations of $(\mathfrak{g}, [\,,\,], 2\delta)$ where δ is the standard structure:

We first introduce the *q-integers*:

$$[n]_q = \frac{q^n - q^{-n}}{q - q^{-1}} = q^{n-1} + q^{n-3} + \ldots + q^{-n+1}.$$

The *q-factorial* and *q-binomial coefficients* are then defined as:

$$[n]_q! = \prod_{k=1}^{n} [k]_q, \qquad \begin{bmatrix} n \\ k \end{bmatrix}_q = \frac{[n]_q!}{[n-k]_q![k]_q!}.$$

Notice that $[n]_q \in \mathbb{Z}[q, q^{-1}]$. In fact the same is true for $\begin{bmatrix} n \\ k \end{bmatrix}_q$ (see exercise 9.2). It is clear that $[n]_q \to n$ as $q \to 1$.

The quantum group $U_h(\mathfrak{g})$ is the Hopf algebra defined over $\mathbb{C}[[h]]$ generated by E_i, F_i, H_i for $i = 1, \ldots n$, and relations

$$[H_i, H_j] = 0, \qquad [H_j, E_i] = a_{ij} E_i, \qquad [H_j, F_i] = -a_{ij} F_i, \tag{9.2}$$

$$[E_i, F_j] = \delta_{ij} \frac{q^{d_i h H_i} - q^{-d_i h H_i}}{q^{d_i h} - q^{-d_i h}}, \tag{9.3}$$

$$\sum_{k=0}^{1-a_{ij}} (-1)^k \begin{bmatrix} 1 - a_{ij} \\ k \end{bmatrix}_{q^{d_i}} E_i^{1-a_{ij}-k} E_j E_i^k = 0, \tag{9.4}$$

$$\sum_{k=0}^{1-a_{ij}} (-1)^k \begin{bmatrix} 1 - a_{ij} \\ k \end{bmatrix}_{q^{d_i}} F_i^{1-a_{ij}-k} F_j F_i^k = 0, \tag{9.5}$$

$$\Delta(E_i) = E_i \otimes q^{d_i H_i} + 1 \otimes E_i, \qquad S(E_i) = -E_i q^{-d_i H_i}, \qquad \epsilon(E_i) = 0,$$
$$\Delta(F_i) = F_i \otimes 1 + q^{-d_i H_i} \otimes F_i, \qquad S(F_i) = -q^{d_i H_i} F_i, \qquad \epsilon(F_i) = 0,$$
$$\Delta(H_i) = H_i \otimes 1 + 1 \otimes H_i, \qquad S(H_i) = -H_i, \qquad \epsilon(H_i) = 0.$$

Notice the resemblance with the Serre presentation of $\mathfrak{g}$ in terms of Chevalley generators. In particular, the relations (9.4) and (9.5) are called the *quantum Serre relations.* For a proof of the fact that this is indeed a Hopf algebra, see for example [CP].

Remark: In the same way as we did for sl_2, we can easily define $U_q(\mathfrak{g})$ (by setting $q_i = e^{hd_i}$ and $K_i = q_i^{H_i}$), and consider the Hopf algebra defined over $\mathbb{C}$ by specializing $q = e^h$. The properties of this Hopf algebra strongly depend on the values of q, in particular on whether it is a root of unity or not. However, this is not the subject at hand. We refer the interested reader to [J] and [Lu] where this is discussed.

Exercise 9.2. Show that $\begin{bmatrix} n \\ k \end{bmatrix}_q \in \mathbb{Z}[q, q^{-1}]$ for $0 \le k \le n$.

Exercise 9.3. Show that $U_h(\mathfrak{g})$ is a quantization of $(\mathfrak{g}, [\,,\,], 2\delta)$ where δ is the standard Lie bialgebra structure on $\mathfrak{g}$.

9.4 Coboundary, quasitriangular, triangular Hopf algebras

As we have seen, the quasiclassical limit of a quantized enveloping algebra is a Lie bialgebra. We will now say what the counterparts of coboundary and (quasi)triangular structures are.

9.4.1 Coboundary Hopf algebras

We make the following definition:

Definition: Let H be a Hopf algebra. An invertible element $R \in H \otimes H$ is a coboundary structure on H if

1. $\Delta^{op} = R\Delta R^{-1}$,
2. $RR_{21} = 1$,
3. $R_{12}(\Delta \otimes 1)R = R_{23}(1 \otimes \Delta)R$.

We will call the pair (H, R) a coboundary Hopf algebra.

The meaning of axiom 1. is clear: the defect of cocommutativity is measured by an element $R \in H \otimes H$. Axiom 2. says that the map $\sigma_R \in \mathrm{End}(H \otimes H)$, $a \otimes b \mapsto \sigma R(a \otimes b)$ satisfies $\sigma_R^2 = 1$. Axiom 3. may seem obscure: it will become clear in the context of tensor categories.

The link with coboundary Lie bialgebras is given by

Proposition 9.2. *Let H be a quantized enveloping algebra which is a quantization of a Lie bialgebra $(\mathfrak{g}, [\,,\,], \delta)$. If R is a coboundary structure on H, then the element $r \in U(\mathfrak{g}) \otimes U(\mathfrak{g})$ defined by*

$$R \equiv 1 + hr \pmod{h^2}$$

is in fact in $\Lambda^2\mathfrak{g}$ and is a coboundary structure on $\mathfrak{g}$.

Proof: Let $a \in \mathfrak{g}$ and $\tilde{a}$ be a lifting of a to H. We have

$$\begin{aligned} \delta(a) &\equiv \frac{\Delta(\tilde{a}) - \Delta^{op}(\tilde{a})}{h} && \pmod{h} \\ &\equiv \frac{\Delta(\tilde{a}) - R\Delta(\tilde{a})R^{-1}}{h} && \pmod{h} \\ &= [a \otimes 1 + 1 \otimes a, r] \end{aligned}$$

since $\Delta(\tilde{a}) \equiv a \otimes 1 + 1 \otimes a \pmod{h}$. Furthermore, from $RR_{21} = 1$ it follows $r + r_{21} = 0$. The only thing that remains to be checked is that we have in fact $r \in \Lambda^2\mathfrak{g} \subset \Lambda^2(U(\mathfrak{g}))$. From axiom 3. we deduce

$$r_{12} + (\Delta_0 \otimes 1)r = r_{23} + (1 \otimes \Delta_0)r. \tag{9.6}$$

where Δ_0 is the usual (cocommutative) coproduct in universal enveloping algebras. We claim that $(\Delta_0 \otimes 1)r = r_{23} + r_{13}$. Indeed, let $Q = (\Delta_0 \otimes 1)r - r_{23} - r_{13} \in U(\mathfrak{g})^{\otimes 3}$; a quick computation shows that $Q_{123} = -Q_{312}$ which implies $Q = 0$ (the cyclic permutation is of odd order). Thus the first component of r is primitive and $r \in \Lambda^2\mathfrak{g}$.

□

9.4.2 (Quasi)triangular Hopf algebras

Here is the quantum analogue of the notion of a (quasi)triangular Lie bialgebra:

Definition: Let H be a Hopf algebra. An invertible element $R \in H \otimes H$ is called a quasitriangular structure on H if

1. $\Delta^{op} = R\Delta R^{-1}$,
2. the *Hexagon relations* are satisfied

$$(1 \otimes \Delta)R = R_{13}R_{12}, \qquad (\Delta \otimes 1)R = R_{13}R_{23}.$$

The element R is called a triangular structure of H if in addition $RR_{21} = 1$. As usual, we will call the pair (H, R) a (quasi)triangular Hopf algebra. The element R is called the *universal R-matrix* of H. The property $RR_{21} = 1$ is called unitarity of the R-matrix.

As usual, by a map of (quasi)triangular Hopf algebras we mean a map of Hopf algebras sending the R-matrix to the R-matrix. We will denote the category of (quasi)triangular Hopf algebras defined over k by (Q)THA(k) and the category of (quasi)triangular Hopf algebras defined over K as (Q)THA(K).

As in the quasiclassical picture of Lie bialgebras, the R-matrix satisfies an equation with values in $H^{\otimes 3}$:

Definition: Let H be a Hopf algebra. The Quantum Yang-Baxter equation (in short, QYBE) is the following equation for an element $R \in H \otimes H$:

$$R_{12}R_{13}R_{23} = R_{23}R_{13}R_{12}.$$

Proposition 9.3. *Let H be a Hopf algebra and $R \in H \otimes H$ a quasitriangular structure on H. Then R is a solution of the QYBE.*

Proof: We have

$$\begin{aligned} R_{12}R_{13}R_{23} &= R_{12}(\Delta \otimes 1)R \\ &= (\Delta^{op} \otimes 1)(R)R_{12} \\ &= R_{23}R_{13}R_{12}. \end{aligned}$$

□

Let us demonstrate how to prove Proposition 9.3 using the pictorial language. Let (H, R) be a quasitriangular Hopf algebra. Represent the R-matrix R as a map $k \to H \otimes H$ in the following way:

R :
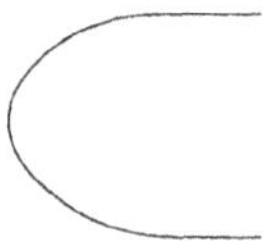

The relation $R\Delta = \Delta^{op}R$ can then be drawn as:

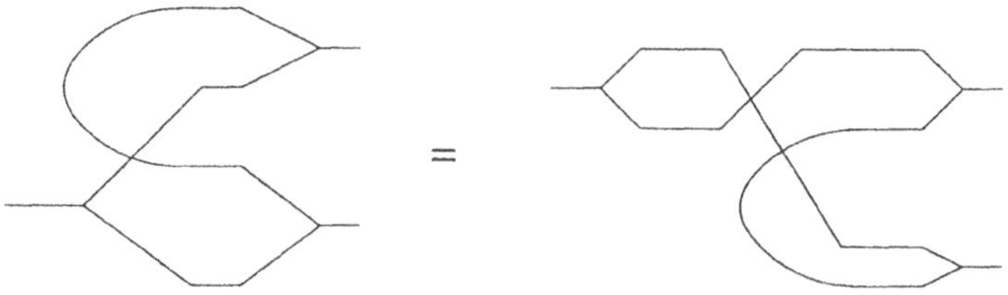

(9.7)

and the Hexagon relations are represented as:

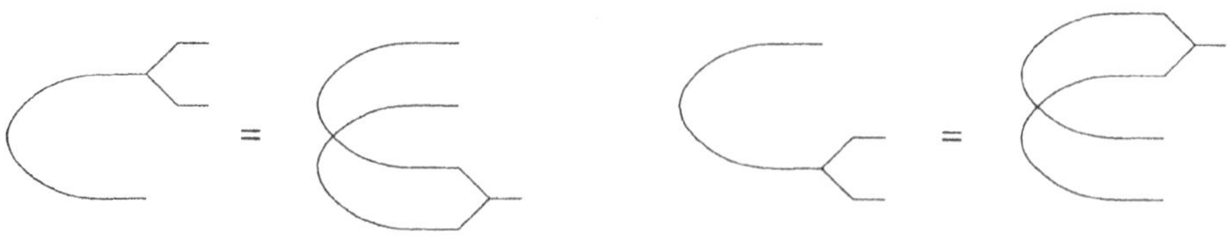

These pictures should be considered as expressions and the moves defined above encode the relation between these expressions. It gives us a way to visualize a computation: two pictures which can be deformed one into another by a sequence of moves represent equal expressions. In particular, the proof of the Quantum Yang Baxter equation is now given by the following picture:

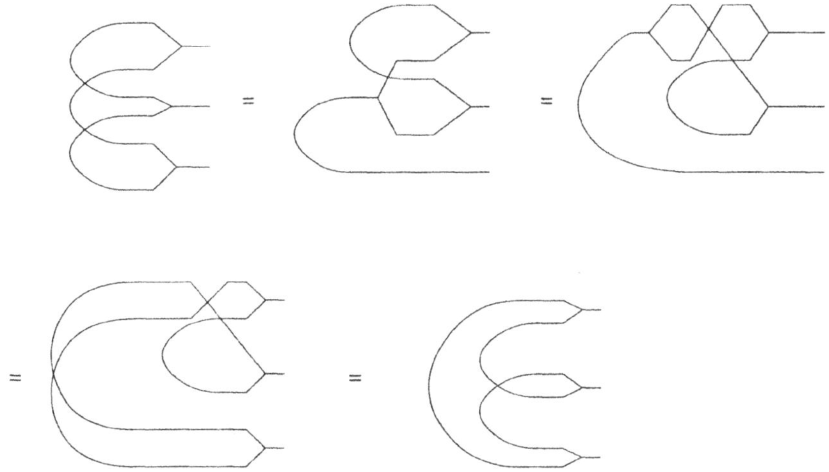

where we used relation (9.7) and the Hexagon relations.

This leads us to make the following definition:

Definition: Let V be a finite dimensional vector space. An element $R \in \mathrm{End}(V) \otimes \mathrm{End}(V)$ is called an *R-matrix* if it satisfies the QYBE (in $\mathrm{End}(V)^{\otimes 3}$). It is called unitary if in addition $RR_{21} = 1$.
In particular, if (H, R) is a (quasi)triangular Hopf algebra and V an H-module then the image of R in $\mathrm{End}(V) \otimes \mathrm{End}(V)$ is an R-matrix.

As the names suggest, the CYBE is the quasiclassical limit of the QYBE in the following sense:

Proposition 9.4. *Let A be an associative algebra and $R \in A \otimes A[[h]]$ a solution of the QYBE such that $R \equiv 1 \pmod h$. Define $r \in A \otimes A$ by the congruence $R \equiv 1 + hr \pmod{h^2}$. Then r is a solution of the CYBE.*

Proof: Expand $R_{12}R_{13}R_{23} = R_{23}R_{13}R_{12}$ in powers of h: the first nontrivial term is the coefficient of h^2 and it reads

$$r_{12}r_{13} + r_{12}r_{23} + r_{13}r_{23} = r_{23}r_{13} + r_{23}r_{12} + r_{13}r_{12}$$

which is exactly the CYBE.

□

The following problem is best solved using the pictorial technique:

Problem 9.1 (Drinfeld, [Dr6]). Let (H, m, Δ, S, R) be a quasitriangular finite-dimensional Hopf algebra and set

$$u = m((S \otimes 1)R^{21})$$

Show that u is invertible and that we have

$$S^2(x) = uxu^{-1} \qquad \text{for any } x \in H \tag{9.8}$$

Show that $S(u)u$ is a central element of H.

The link between (quasi)triangular Hopf algebras and (quasi)triangular Lie bialgebras is made complete by the following proposition:

Proposition 9.5. *Let (H, R) be a (quasi)triangular quantized enveloping algebra and $H/hH \simeq U(\mathfrak{g})$. Define $r \in U(\mathfrak{g}) \otimes U(\mathfrak{g})$ by $R \equiv 1 + hr \pmod{h^2}$. Then $r \in \mathfrak{g} \otimes \mathfrak{g}$ and is a (quasi)triangular structure for the Lie bialgebra $\mathfrak{g}$.*

Proof: We have already proved that $\delta(a) = [a \otimes 1 + 1 \otimes a, r]$ for $a \in \mathfrak{g}$. Furthermore, it follows from the Hexagon axioms that $(\Delta_0 \otimes 1)r = r_{13} + r_{23}$ and $(1 \otimes \Delta_0)r = r_{12} + r_{13}$, hence $r \in \mathfrak{g} \otimes \mathfrak{g}$. Since δ is skew-symmetric, we have $r + r_{21} \in (\mathfrak{g} \otimes \mathfrak{g})^{\mathfrak{g}}$. Finally, r satisfies the CYBE since R satisfies the QYBE. Thus the quasiclassical limit of a quasitriangular structure is again a quasitriangular structure. To conclude, notice that if, in addition, $RR_{21} = 1$ then $r + r_{21} = 0$, so the quasiclassical limit of a triangular structure is a triangular structure.

□

Example: Let $H = U_h(sl_2(\mathbb{C}))$. It is quasitriangular, and the universal R-matrix is given by

$$R = q^{\frac{1}{2}H \otimes H} \sum_{n \geq 0} q^{\frac{n(n-1)}{2}} \frac{(q - q^{-1})^n}{[n]_q!} E^n \otimes F^n \tag{9.9}$$

where $q = e^h$.

In fact, we will see in Lecture 12 that the quantum groups $U_h(\mathfrak{g})$ for $\mathfrak{g}$ a simple Lie algebra are all quasitriangular. Formula (9.9) was generalized to this case by Y. Soibelman ([So]) and by Levendorskii-Soibelman ([LS1], [LS2]).

9.4.3 Modifications of the quantization theorem

The preceding discussion leads us to make the following definition:

Definition: Let A be an associative algebra and let $R \in A \otimes A[[h]]$ be a solution of the QYBE such that $R \equiv 1(\text{mod } h)$. The *quasiclassical limit* of R is $r \in A \otimes A$ defined by $R \equiv 1 + hr \ (\text{mod } h^2)$ (and is a solution of the CYBE). Conversely, R is called a *quantization* of r.

The question of existence of a functorial quantization of (quasi)triangular Lie bialgebras is settled in [EK1] and [EK2]. We will say that a (quasi)triangular Hopf algebra (H, R) is a quantization of a (quasi)triangular Lie bialgebra $(\mathfrak{g}, r)$ if H is a quantization of $\mathfrak{g}$ and R is a quantization of r.

Theorem 9.2. *There exists a functor*

$$\tilde{Q}^{(q)t} : (Q)TLBA(k) \to (Q)THA(K)$$

such that $\tilde{Q}^{(q)t}(\mathfrak{g}, r)$ is a quantization of $(\mathfrak{g}, r)$.

Similarly, there is a functorial quantization of r-matrices. Let us denote by QYBA(K) the category whose objects are pairs (A, R) where A is an associative k-algebra and $R \in A \otimes A[[h]]$ is a solution of the QYBE. Morphisms are given by

$$\mathrm{Hom}_{\mathrm{QYBA}(K)}((A_1, R_1), (A_2, R_2)) = \{\phi : A_1[[h]] \to A_2[[h]] \mid (\phi \otimes \phi)R_1 = R_2\}$$

(we only consider homomorphisms of algebras).

In the same way, we define the category CYBA(k) whose objects are pairs (A, r), A an associative k-algebra and $r \in A \otimes A$ a solution of the CYBE. Morphisms are given by

$$\mathrm{Hom}_{\mathrm{CYBA}(k)}((A_1, r_1), (A_2, r_2)) = \{\phi : A_1 \to A_2 \mid (\phi \otimes \phi)r_1 = r_2\}.$$

Theorem 9.3. *There exists a functor*

$$\tilde{Q}^{YB} : \mathrm{CYBA}(k) \to \mathrm{QYBA}(K)$$

such that $\tilde{Q}^{YB}(A, r)$ is a quantization of (A, r).

9.5 Quantization by twists

This is a method for constructing new quantized enveloping algebras from old ones. It will appear in the construction of the functorial quantization.

Let $(H, m, \Delta, i, \epsilon, S)$ be a quantized enveloping algebra, $H/hH \simeq U(\mathfrak{g})$. Introduce the following notation: if $X \in H \otimes H$, we let $X_{12,3} = (\Delta \otimes 1)X$, $X_{1,23} = (1 \otimes \Delta)X$, etc.

Definition: An element $J \in H \otimes H$ is called a *twist* if

1. $J \equiv 1 \pmod h$,
2. $(\epsilon \otimes 1)J = (1 \otimes \epsilon)J = 1$,
3. $J_{12,3}(J_{12} \otimes 1)(1 \otimes J_{23})^{-1} J_{1,23}^{-1} \in (H^{\otimes 3})^H$,

where $(H^{\otimes 3})^H$ is the space of H-invariant elements of $H^{\otimes 3}$, i.e elements commuting with $(\Delta \otimes 1)\Delta(x)$ for all $x \in H$.

We define a new quantized enveloping algebra in the following way: as an algebra, $\tilde{H} = H$ and we set

$$\tilde{\Delta} = J^{-1}\Delta J,$$
$$\tilde{S} = Q^{-1}SQ,$$
$$\tilde{\epsilon} = \epsilon,$$

where $Q = m((S \otimes 1)(J))$. Notice that by 1., J and Q are invertible.

Proposition 9.6. $(\tilde{H}, m, \tilde{\Delta}, i, \tilde{\epsilon}, \tilde{S})$ *is a quantized enveloping algebra.*

Proof: Let us check the coassociativity of $\tilde{\Delta}$:

$$\begin{aligned}(1 \otimes \tilde{\Delta})\tilde{\Delta}(x) &= (1 \otimes J^{-1}\Delta J)(J^{-1}\Delta(x)J) \\ &= J_{23}^{-1} J_{1,23}^{-1}(1 \otimes \Delta)\Delta(x) J_{1,23} J_{23}.\end{aligned}$$

Similarly,

$$\begin{aligned}(\tilde{\Delta} \otimes 1)\tilde{\Delta}(x) &= (J^{-1}\Delta J \otimes 1)(J^{-1}\Delta(x)J) \\ &= J_{12}^{-1} J_{12,3}^{-1}(\Delta \otimes 1)\Delta(x) J_{12,3} J_{12}.\end{aligned}$$

These two expressions are equal by coassociativity of Δ and Property 3. In a similar way, one checks the compatibility conditions for $\tilde{S}$ and ϵ. Since $J \equiv 1 \pmod h$, it is clear that $\tilde{H}/h\tilde{H} \simeq U(\mathfrak{g})$ (but it may induce a different Lie coalgebra structure on $\mathfrak{g}$ than H, of course).

□

Example: Let $\mathfrak{g}$ be any Lie algebra and $\mathfrak{h} \subset \mathfrak{g}$ an abelian subalgebra. Then any $r \in \Lambda^2\mathfrak{h}$ defines a triangular structure on $\mathfrak{g}$. This structure can be easily quantized. Namely, let $H = U(\mathfrak{g})[[h]]$ with its usual Hopf algebra structure. Then $J = e^{hr/2}$ is a twist, and $(\tilde{H}, e^{hr})$ is a triangular Hopf algebra and a quantization of $(\mathfrak{g}, r)$.

Problem 9.2. Check the unproved assertions in Proposition 9.6.

Problem 9.3. Let (H, R) be a coboundary (resp. quasitriangular, resp. triangular) quantized enveloping algebra and $J \in H \otimes H$ a twist such that

$$J_{12,3}(J_{12} \otimes 1)(1 \otimes J_{23})^{-1} J_{1,23}^{-1} = 1$$

Show that $J_{21}^{-1} R J$ is a coboundary (resp. quasitriangular, resp. triangular) structure for H.

Problem 9.4. Let H be a cocommutative quantized enveloping algebra with $H/hH = U\mathfrak{g}$ and $J \in H \otimes H$ a twist as in Problem 9.3. Define $r \in U(\mathfrak{g}) \otimes U(\mathfrak{g})$ by $J = 1 + \frac{hr}{2} \pmod{h^2}$. Show that in fact $r \in \mathfrak{g} \otimes \mathfrak{g}$ and that $\frac{r - r_{21}}{2}$ is a coboundary structure for $U\mathfrak{g}$. In other words, twisting will only produce quantizations of coboundary Lie bialgebras.

Lecture 10

Formal groups and h-formal groups

In the last lecture, we defined quantization of Lie bialgebras and quantized enveloping algebras. We now turn to the dual concept of formal groups and h-formal groups.

10.1 Definition

Let $\mathfrak{g}$ be a Lie algebra defined over a field k of characteristic zero and $U(\mathfrak{g})$ its enveloping algebra. The (full) dual space $U(\mathfrak{g})^*$ can be described in the following way. Let $x_1, x_2, \ldots$ be a basis of $\mathfrak{g}$. Recall the canonical (vector space) isomorphism $S(\mathfrak{g}) \simeq U(\mathfrak{g})$ given by the symmetrization map $x_1 \ldots x_r \mapsto \sum_{\sigma \in S_r} x_{\sigma(1)} \ldots x_{\sigma(r)}$. Thus, $U(\mathfrak{g})^* \simeq k[[x_1^*, x_2^*, \ldots]]$ (as a vector space). Furthermore, the usual coproduct on $U(\mathfrak{g})$ endows $U(\mathfrak{g})^*$ with an algebra structure, and it is clear that $U(\mathfrak{g})^* \simeq k[[x_1^*, x_2^*, \ldots]]$ (as an algebra). Finally, the algebra structure on $U(\mathfrak{g})$ gives rise to a map

$$k[[x_1^*, x_2^*, \ldots]] \to k[[x_1^*, x_2^*, \ldots]] \hat{\otimes} k[[y_1^*, y_2^*, \ldots]] := k[[x_1^*, x_2^*, \ldots y_1^*, y_2^*, \ldots]] \tag{10.1}$$

where we use x's and y's to distinguish the two components of the tensor product. Notice that $U(\mathfrak{g})^*$ is a local algebra and that $U(\mathfrak{g})^* = \varprojlim U(\mathfrak{g})^*/I^n$ where $I = \sum x_i^* k[[x_1^*, x_2^*, \ldots]]$ is the maximal ideal.

Geometrically, $U(\mathfrak{g})^*$ is realized in the following way (when we have $k = \mathbb{R}$ or $\mathbb{C}$ and $\dim \mathfrak{g} < \infty$): let G be the simply connected Lie group such that $\mathrm{Lie}(G) = \mathfrak{g}$. Identify $U(\mathfrak{g})$ with the algebra of left-invariant differential operators on G, and consider the pairing

$$\begin{aligned} U(\mathfrak{g}) \times C^\infty(G) &\to k, \\ (D, f) &\mapsto D.f(e). \end{aligned}$$

Let $J \subset C^\infty(G)$ be the ideal of functions vanishing at the identity e. Then $U(\mathfrak{g})^* = \varprojlim C^\infty(G)/J^n$.

This motivates the following two definitions:

Definition: A *formal group* is a (topological) Hopf algebra isomorphic to $U(\mathfrak{g})^*$ for some Lie algebra $\mathfrak{g}$. In particular, it is commutative, but not necessarily cocommutative. An *h-formal group* is a (topological) deformation Hopf algebra H such that H/hH is a formal group.

In the previous definition, the term "topological" should be understood in the sense of (10.1). If $\mathfrak{g}$ is the Lie algebra of a Lie group G, we will sometimes denote by $\overline{\mathcal{C}(G)}$ the formal group $U(\mathfrak{g})^*$.

Remark: The notion of a Poisson-Lie group extends to the case of formal groups: a *Poisson formal group* is a formal group equipped with a Poison bracket compatible with the group structure. A Lie bialgebra structure on $\mathfrak{g}$ induces a Poisson structure on $U(\mathfrak{g})^*$. Moreover, the following result holds (see [Drl]).

Proposition 10.1. *The category of Lie bialgebras is equivalent to the category of Poisson formal groups.*

The proof of this proposition is an easy exercise.

The notion of quantization of a Poisson formal group is the same as for Poisson-Lie groups.

10.2 Duality

Let H be a quantized enveloping algebra and $\mathfrak{g}$ its quasiclassical limit. Our goal in this section is to construct a quantized enveloping algebra H^* whose quasiclassical limit is $\mathfrak{g}^*$.

Our discussion in section 1 shows the following.

Proposition 10.2. *Let H be a quantized enveloping algebra. Then the dual Hopf algebra $H' = \mathrm{Hom}_K(H,K)$ is an h-formal group. Conversely, the dual Hopf algebra of an h-formal group is a quantized enveloping algebra.*

Here, by the dual of an h-formal group H' we mean the space of continuous K-linear maps $H' \to K$.

In particular, H' is not a quantized enveloping algebra and is not cocommutative up to first order. Nevertheless, using H', we now define the correct dual H^* of H, which is a QUE algebra.

For simplicity, we will only consider the finite-dimensional case, i.e H is a quantization of a finite-dimensional Lie bialgebra $(\mathfrak{g}, [\,,\,], \delta)$. Let J be the maximal ideal of H'. Set

$$H^* = \widehat{\sum_{n\geq 0} h^{-n} J^n}$$

where $\widehat{\Sigma}$ denotes the h-adic completion.

Proposition 10.3. *H^* is a quantized enveloping algebra and its quasiclassical limit is $\mathfrak{g}^*$.*

Proof: It is clear that H^* is torsion free, complete and separated. Thus it is a topologically free K-module. We claim that it is a deformation of $U(\mathfrak{g}^*)$. Indeed, consider the projection maps

$$\phi_n : J^n \to J^n/(hJ^n + J^{n+1}) \simeq U(\mathfrak{g}^*)_n$$

(where $U(\mathfrak{g}^*)_n$ is the image of the degree n elements in $S\mathfrak{g}^*$ under the symmetrization map $S\mathfrak{g}^* \to U\mathfrak{g}^*$)

$$\Phi : H^* \to U(\mathfrak{g}^*),$$
$$ch^{-i} \mapsto \phi_i(c)$$

where $c \in J^i$. Then Φ is surjective and $\operatorname{Ker}\Phi = hH^*$. Moreover, Φ is an algebra homomorphism; in order to see this, it is enough to check it on generators (i.e elements of degree 1):

For $x^* \in \mathfrak{g}^*$, let us set $\overline{x^*} = \frac{x^*}{h} \in h^{-1}J \subset H^*$.
Let $\theta \in \mathfrak{g}$, and write

$$\Delta_H(\theta) \equiv \theta \otimes 1 + 1 \otimes \theta + h\Delta_1(\theta) \pmod{h^2}$$

for the coproduct in H. Then, for $x^*, y^* \in \mathfrak{g}^*$, we have

$$\overline{x^*}.\overline{y^*}(\theta) \equiv \frac{1}{h^2}(x^* \otimes y^*)(\theta \otimes 1 + 1 \otimes \theta) + \frac{1}{h}(x^* \otimes y^*)(\Delta_1(\theta)) \pmod 1.$$

Thus,

$$\begin{aligned}\overline{x^*}.\overline{y^*} - \overline{y^*}.\overline{x^*} &\equiv \frac{1}{h}(x^* \otimes y^*)(\Delta_1(\theta) - \Delta_1^{op}(\theta)) \pmod 1 \\ &\equiv \frac{1}{h}(x^* \otimes y^*)(\delta(\theta)) \pmod 1 \\ &\equiv \frac{1}{h}[x^*, y^*](\theta) \pmod 1.\end{aligned}$$

Thus $\Phi([\overline{x^*}, \overline{y^*}]) = [x^*, y^*] = [\Phi(\overline{x^*}), \Phi(\overline{y^*})]$, and Φ is a morphism of algebras.
Furthermore, we have

$$\Delta_{H'}(\overline{x^*}) - (\overline{x^*} \otimes 1 + 1 \otimes \overline{x^*}) \in h(h^{-1}J \otimes h^{-1}J)$$

so that $\Phi(\Delta_{H'}(\overline{x^*})) = x^* \otimes 1 + 1 \otimes x^* = \Delta_{U\mathfrak{g}^*}(\Phi(\overline{x^*}))$. Finally, we have

$$(\Delta_{H'}(\overline{x^*}) - \Delta_{H'}^{op}(\overline{x^*}))(u \otimes v) \equiv \overline{x^*}([u, v]) \pmod 1$$

which shows that H^* is a quantization of $\mathfrak{g}^*$.

□

Conversely, any quantized enveloping algebra contains an h-formal group, constructed in the following way: let H be a quantized universal enveloping algebra and set

$$\begin{aligned}\delta_n : H &\to H^{\otimes n} \\ x &\mapsto (Id - \epsilon)^{\otimes n} \Delta_n(x)\end{aligned}$$

where $\Delta_n(x) = (\dots(\Delta \otimes 1 \otimes 1)(\Delta \otimes 1)\dots)$ is the iterated cocommutator. For instance we have $\delta_1(a) = a - \epsilon(a)$ and $\delta_2(a) = \Delta(a) - 1 \otimes a - a \otimes 1 + \epsilon(a)$.

Proposition 10.4. *The space* $L = \{a \in H \mid \delta_n(a) \in h^n H^{\otimes n}\ \forall n\}$ *is an h-formal group.*

Proof (sketch): From the relation

$$xy - \epsilon(xy) = (x - \epsilon(x))(y - \epsilon(y)) + \epsilon(x)(y - \epsilon(y)) + \epsilon(y)(x - \epsilon(x))$$

we deduce that

$$\delta_n(ab) = \sum_{i,j} \left((a_1^i - \epsilon(a_1^i))(b_1^j - \epsilon(b_1^j)) + \epsilon(a_1^i)(b_1^j - \epsilon(b_1^j)) + \epsilon(b_1^j)(a_1^i - \epsilon(a_1^i))\right) \otimes \dots$$

where $\Delta_n(a) = \sum_i a_1^i \otimes \dots \otimes a_n^i$ and $\Delta_n(b) = \sum_j b_1^j \otimes \dots \otimes b_n^j$. This implies that if $\delta_n(a) \equiv \delta_n(b) \equiv 0 \pmod{h^n}$ then $\delta_n(ab) \equiv 0 \pmod{h^n}$. Thus L is a subalgebra of H. A formula similar to the one above for the coproduct shows that L is stable under the coproduct. Finally, to show that L is commutative up to first order we have to show that $\frac{ab-ba}{h} \in L$ for all $a, b \in L$. This follows from the same computation as above using the formula

$$ab - ba = (a - \epsilon(a))(b - \epsilon(b)) - (b - \epsilon(b))(a - \epsilon(a)).$$

□

Exercise 10.1. Let A be the quantized enveloping algebra defined over k and generated by E,H with relations $[H, E] = 2E$ and

$$\begin{aligned}\Delta(E) &= E \otimes q^H + 1 \otimes E, & \Delta(H) &= H \otimes 1 + 1 \otimes H \\ S(E) &= -Eq^H, & S(H) &= -H, \\ \epsilon(E) &= 0, & \epsilon(H) &= 0.\end{aligned}$$

Show that $A \simeq A^*$.

10.3 R-matrices and R-forms

10.3.1 Comodules

To the concept of a module over a k-algebra A there corresponds the concept of *comodule* over a k-coalgebra C: a (right) comodule over C is a k-vector space equipped with a linear map $\rho : V \to C \otimes V$ making the following diagram commutative:

$$\begin{array}{ccc} V & \xrightarrow{\rho} & C \otimes V \\ \rho \downarrow & & \downarrow \Delta^{op} \otimes 1 \\ C \otimes V & \xrightarrow{1 \otimes \rho} & C \otimes C \otimes V \end{array}$$

and satisfying the (co)unit axiom $(\epsilon \otimes 1)\rho = 1$. The notion of a left comodule is defined in an analogous way, with Δ^{op} replaced by Δ. Similarly, a morphism between two comodules V_1 and V_2 is a linear map making the following diagram commutative

$$\begin{array}{ccc} V_1 & \xrightarrow{\rho} & C \otimes V_1 \\ f \downarrow & & \downarrow f \otimes 1 \\ V_2 & \xrightarrow{\rho} & C \otimes V_2 \end{array}$$

Let us denote by $\mathrm{Comod}_f(C)$ the category of finite dimensional (right) comodules over C.

Let A be a finite-dimensional k-algebra and A^* the dual coalgebra. In this case, the categories $\mathrm{Rep}_f\, A$ of finite dimensional left A-modules and the category $\mathrm{Comod}_f(A^*)$ of finite-dimensional right A^*-comodules are equivalent: to the A-module V with action

$$a : A \otimes V \to V$$

we assign the A^*-comodule V with coaction defined by the composition

$$a' : V \overset{c \otimes 1}{\to} A^* \otimes A \otimes V \overset{a \otimes 1}{\to} A^* \otimes V$$

where $c : k \to A^* \otimes A$ is the canonical element.

These two categories are antiequivalent to the categories of finite dimensional left A^*-comodules and right A-modules: to V with action $A \otimes V \to V$ we assign the vector space V^* with coaction $a^* : V^* \to V^* \otimes A^*$.

When A is infinite-dimensional, the coalgebra A^* is a topological coalgebra, and we have to replace the notion of comodules by "continuous" comodules.

10.3.2 Universal R-forms (coquasitriangular structures)

Let us now examine the concept dual to the notion of an R-matrix in a quantized enveloping algebra. Introduce the following notation: if $(B, m, \Delta, i, \epsilon, S)$ is a Hopf algebra and f, g are linear forms $B \to k$, we set $f * g : B \to k$, $x \mapsto (f \otimes g)\Delta(x)$.

Definition: Let $(A, m, \Delta, i, \epsilon, S)$ be an h-formal group. A bilinear map $R^* : A \otimes A \to k$ is said to be a *universal R-form* if

1. There exists $(R^*)^{-1} : A \otimes A \to k$ such that

$$(R^*) * (R^*)^{-1} = (R^*)^{-1} * (R^*) = \epsilon \otimes \epsilon,$$

2. $m^{op} = (R^*) * m * (R^*)^{-1}$,

3. $R^*(m \otimes 1) = (R^*)^{13} * (R^*)^{23}, \qquad R^*(1 \otimes m) = (R^*)^{13} * (R^*)^{12}.$

In other words, R^* measures the defect of commutativity of A. It is clear that if H is a quasitriangular quantized enveloping algebra with R-matrix R then R induces a universal R-form on the dual h-formal group $H' = \mathrm{Hom}_{k[[h]]}(H, k[[h]])$.

Recall that if (H, R) is a quasitriangular Hopf algebra then R induces an R-matrix R_V in $\mathrm{End}(V) \otimes \mathrm{End}(V)$ for every H-module V. In a similar way, a universal R-form of an h-formal group H' induces an R-matrix in the space $\mathrm{End}(V) \otimes \mathrm{End}(V)$ for every H'-comodule: consider the composition

$$\begin{aligned} V \otimes V &\overset{\rho \otimes \rho'}{\to} (H' \otimes V) \otimes (H' \otimes V) \\ &\overset{(R^*)^{13}}{\to} V \otimes V \end{aligned}$$

which we write shortly as $R^*(\rho \otimes \rho')$ this is an element of $\mathrm{End}(V) \otimes \mathrm{End}(V)$.

Proposition 10.5. *The element* $R_V = R^*(\rho \otimes \rho') \in \mathrm{End}(V) \otimes \mathrm{End}(V)$ *is an R-matrix, i.e R is invertible and*

$$R_{12} R_{13} R_{23} = R_{23} R_{13} R_{12}.$$

Proof: The proof is analogous to the proof of Proposition 9.3. The axioms for universal R-forms are dual to the axioms for R-matrices. In terms of pictures, this amounts to interchanging left and right, and it suffices to look at the pictures of Proposition 9.3 in a mirror to obtain a proof of Proposition 10.5.

□

Lecture 11

Infinite dimensional quantum groups

In this lecture, we define some quantizations of the infinite dimensional Lie bialgebras discussed previously, using the *RTT formalism*, a quantum analogue of the constructions of Lecture 6.

11.1 The RTT formalism and h-formal groups

We give in this section a presentation of the h-formal group $U_h(sl_n)'$ (where $U_h(sl_n)$ is the Drinfeld-Jimbo quantum group), which reflects more vividly the fact that it is a deformation of the algebra of formal functions on SL_n.

11.1.1 Formal groups revisited

Let us describe in some detail the Hopf algebra $\mathcal{C}(SL_n)$ of algebraic functions on SL_n and its completion $U(sl_n)'$, and give a presentation well suited for quantization. The Hopf algebra $\mathcal{C}(SL_n)$ is generated by e^*_{ij}, for $i, j = 1 \dots n$ (the usual coordinate functions). In order to write down the relations, let us arrange the generators in a matrix $L = (E_{ij} \otimes e^*_{ij})^n_{i,j=1} \in \mathrm{End}(\mathbb{C}^n) \otimes \mathcal{C}(SL_n)$ and consider

$$L^{13} \in \mathrm{End}(\mathbb{C}^n) \otimes \mathrm{End}(\mathbb{C}^n) \otimes \mathcal{C}(SL_n),$$
$$L^{23} \in \mathrm{End}(\mathbb{C}^n) \otimes \mathrm{End}(\mathbb{C}^n) \otimes \mathcal{C}(SL_n).$$

The relations can now be written as

$$L^{13}L^{23} = L^{23}L^{13}, \qquad Det(L) = 1,$$
$$\Delta(L) = L^{12}L^{13}, \qquad S(L) = L^{-1},$$
$$\epsilon(L) = 1.$$

Remark: One can think of L as a matrix whose entries (e^*_{ij}) are elements (and generators) of this algebra. Then $L^{12}L^{13}$ is a matrix whose entries are elements of $\mathcal{C}(SL_n)\otimes\mathcal{C}(SL_n)$. This matrix is simply given by usual matrix multiplication of L with itself. For instance, for $n=2$,

$$L^{12}L^{13} = \begin{pmatrix} e_{11}^*\otimes e_{11}^* + e_{12}^*\otimes e_{21}^* & e_{11}^*\otimes e_{12}^* + e_{12}^*\otimes e_{22}^* \\ e_{21}^*\otimes e_{11}^* + e_{22}^*\otimes e_{21}^* & e_{21}^*\otimes e_{12}^* + e_{22}^*\otimes e_{22}^* \end{pmatrix}.$$

Notice that L^{-1} is well defined since $Det(L)=1$. The relation $L_{13}L_{23}=L_{23}L_{13}$ expresses the commutativity of $\mathcal{C}(SL_n)$.

The reader is advised to write down these relations in components and check for himself that these indeed define the usual Hopf algebra structure on $\mathcal{C}(SL_n)$.

The topological completion of this algebra is defined in the following way. Let $J=\text{Ker}\ (\epsilon)$ be the maximal ideal of $\mathcal{C}(SL_n)$. It is generated by $e_{ij}^*-\delta_{ij}$. We have

$$U(sl_n)' = \varprojlim \mathcal{C}(SL_n)/J^n.$$

Remark: To obtain a similar presentation of GL_n, we replace the condition Det $L=1$ by the condition that L is invertible. In other terms, we formally adjoin a matrix L^{-1} and add the relation $LL^{-1}=1$ and extend the Hopf algebra structure in a unique compatible way. In this case, we have $\Delta(L^{-1}) = (L^{-1})^{12}(L^{-1})^{13}$, $S(L^{-1})=(L^{-1})^{-1}$, $\epsilon(L^{-1})=1$.

11.1.2 The RTT formalism

Let V be a finite dimensional topologically free K-module and let

$$R \in \text{End}(V)\otimes\text{End}(V)$$

be an R-matrix. We know how to obtain such a data from a comodule over an h-formal group with a universal R-form. Conversely, starting from V and R, Faddeev, Reshetikhin and Takthadjan gave in [FRT] a construction of an h-formal group $H(R)$ with universal R-form R^* and a $H(R)$-comodule structure $\rho: V\to V\otimes H(R)$ on V such that $R=R^*(\rho\otimes\rho)$.

The construction is based on the following observation. Let H be an h-formal group with universal R-form R^* and V a finite dimensional H-comodule. Let $R_V\in\text{End}(V)\otimes\text{End}(V)$ be the R-matrix induced by R^*. Using the definition of R_V, it is easy to check that the following two compositions are equal:

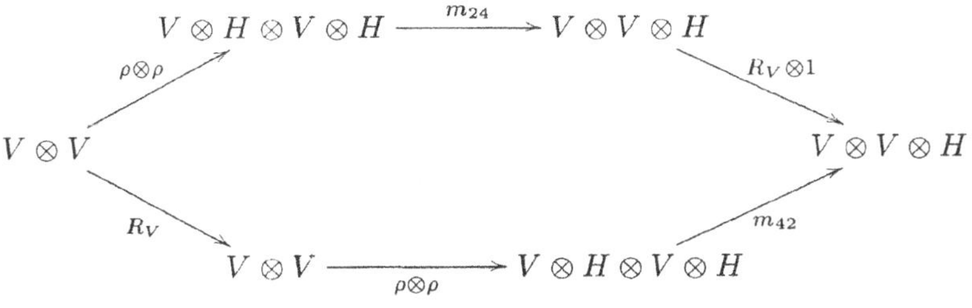

The idea of the FRT construction is to use these as defining relations for an h-formal group. Recall the notation σ for the usual permutation.

Theorem 11.1 (RTT formalism). *Let $W \simeq k^n$ be a finite dimensional vector space and $R \in \mathrm{End}(W) \otimes \mathrm{End}(W)[[h]]$ be an R-matrix such that $R \equiv 1 \pmod h$ which satisfies a "Hecke relation"*

$$(\sigma R - p)(\sigma R + q) = 0$$

where $p, q \equiv 1 \pmod h$. Define a $k[[h]]$-algebra $H(R)$ as follows: $H(R)$ is the topological completion of the algebra generated by elements t_{ij}, $i, j = 1 \ldots n$ arranged in a matrix $T = (t_{ij})_{i,j=1}^n \in \mathrm{End}(k^n) \otimes H(R)$ with relations

$$T \text{ is invertible,}$$

$$R^{12} T^{13} T^{23} = T^{23} T^{13} R^{12}.$$

Then the following define a Hopf algebra structure on $H(R)$:

$$\Delta(T) = T^{12} T^{13}, \qquad S(T) = T^{-1}, \qquad \epsilon(T) = 1.$$

Furthermore, $H(R)$ is an h-formal group, i.e it is a topologically free $k[[h]]$-module and $H(R)/hH(R) \simeq U(gl_n)'$ with its usual Hopf algebra structure. Moreover, there exists a universal R-form $R^* \in (H(R) \otimes H(R))^*$ inducing R on the $H(R)$-comodule $W[[h]]$.

Proof: See [FRT] for a proof. The "Hecke relation" condition is here to ensure that $H(R)$ is a flat deformation of $U(gl_n)'$.

Remarks: i) The word "topological completion" means the following : we first complete with respect to the maximal ideal $\mathrm{Ker}(\epsilon)$ and then take the h-adic completion.
ii)In this definition, the coproduct has a particularly nice form (it remains undeformed) but the relations defining the multiplication are more complex.
iii) This construction only defines deformations of $\mathcal{C}(GL_n)$. Analogues for other classical groups exist, but are slightly more complicated (see [CP]).

11.1.3 Examples

We give here RTT-presentations of the h-formal groups $U_h(gl_n)'$ and $U_h(sl_n)'$. Let $V = \mathbb{C}^n$ and $R \in \mathrm{End}(V) \otimes \mathrm{End}(V)[[h]]$ be the following R-matrix:

$$R = e^h \sum_{i=1}^{n} E_{ii} \otimes E_{ii} + \sum_{i \neq j} E_{ii} \otimes E_{jj} + (e^h - e^{-h}) \sum_{i<j} E_{ij} \otimes E_{ji}.$$

Then $H(R) = U_h(gl_n)'$. In order to obtain $U_h(sl_n)'$, we have to add one more relation: the *quantum determinant*.

Proposition 11.1. *The following element is in the center of* $U_h(gl_n)'$*:*

$$\mathrm{qdet} = \sum_{\gamma \in S_n} (-e^{-h})^{l(\gamma)} t_{\gamma(1)1} \dots t_{\gamma(n)n}$$

where $l(\gamma)$ *is the usual length of an element* $\gamma \in S_n$.

Proposition 11.2 ([CP]). *We have*

$$U_h(sl_n)' = U_h(gl_n)'/(qdet - 1).$$

Remarks: As the reader has probably guessed, the R-matrix above comes from the universal R-form on $U_h(sl_n)'$ or, equivalently the universal R-matrix on $U_h(sl_n)$. In this case, R is, up to a factor, the image of this universal R-matrix on the $U_h(sl_n)$-module playing the role of the standard n-dimensional representation of sl_n.

11.2 RTT formalism and infinite dimensional quantum groups

We will now use a variant of the RTT formalism to obtain quantizations of the infinite-dimensional Lie bialgebras defined in Lecture 6, when $\mathfrak{g} = gl_n$. Set $\Omega = \sum_{i,j} E_{ij} \otimes E_{ji} \in gl_n \otimes gl_n$.

Let $r \in (\mathfrak{g} \otimes \mathfrak{g})((z))$ be a unitary r-matrix with a spectral parameter such that $r = \frac{\Omega}{z} + \mathcal{O}(1)$. We associate to r a Manin triple $(\mathfrak{g}((t)), \mathfrak{g}(r), \mathfrak{g}[[t]])$ and a Lie bialgebra $\mathfrak{g}(r)$. The defining relations for $\mathfrak{g}(r)$ can be written as follows: consider the generating function $\tau(u) = r^{21}(t-u) \in \mathfrak{g} \otimes \mathfrak{g}(r)[[u]]$. Then

$$[\tau^{13}(u), \tau^{23}(v)] = [r^{21}(u-v), \tau^{13}(u) + \tau^{23}(v)], \tag{11.1}$$

$$\delta(\tau(u)) = [\tau^{12}(u), \tau^{13}(u)]. \tag{11.2}$$

Here is the principle of the construction of a quantization of $\mathfrak{g}(r)$. The Manin triple defines a (topological) Lie bialgebra structure on $\mathfrak{g}[[t]]$ which is dual $\mathfrak{g}(r)$. This in turn induces a Poisson structure on the (pro)algebraic group $GL_n[[t]]$. The Lie algebra of this Poisson-Lie group is the dual to $\mathfrak{g}(r)$. Thus, if we have a suitable R-matrix $R \in (gl_n \otimes gl_n)((t))[[h]]$, we may define by the RTT formalism an h-formal group $H(R)$ quantizing the algebra of functions on $GL_n[[t]]$. To obtain a quantization of $\mathfrak{g}(r)$, it remains to construct the quantized universal enveloping algebra dual to $H(R)$.

However, for the construction outlined above to yield a flat deformation of $U(gl_n[[t]])'$, we need to restrict ourselves to a particular class of R-matrices.

Definition: Let V be a finite dimensional vector space. The following equation for an element $R \in \mathrm{End}(V) \otimes \mathrm{End}(V)((u))[[h]]$:

$$R_{12}(u_1 - u_2)R_{13}(u_1 - u_3)R_{23}(u_2 - u_3) = R_{23}(u_2 - u_3)R_{13}(u_1 - u_3)R_{12}(u_1 - u_2)$$

is called the *quantum Yang-Baxter equation with spectral parameter* and its solutions are called *R-matrices with spectral parameter.* Furthermore, R is called unitary if

$$R(u)R_{21}(-u) = 1.$$

The simplest example of R-matrix with spectral parameter is given by Yang's R-matrix: $V = \mathbb{C}^n$ and $R_Y(u) = 1 - \frac{h\sigma}{nu}$. We will say that $R(u)$ is a *deformation of Yang's solution* if the following holds: expanding $R(u, h) \in k[[u, h, h/u]]$, we have $\lim_{s\to 0} R(su, sh) = 1 - \frac{\sigma h}{nu}$.

Proposition 11.3 (RTT formalism). *Let $R(u) \in \mathrm{End}(k^n)\otimes\mathrm{End}(k^n)((u))[[h]]$ be an R-matrix with spectral parameter which is a deformation of Yang's solution, satisfying $R \equiv 1 \pmod h$. Let $H(R)$ be the topological completion of the $k[[h]]$-algebra generated by elements $t_{kl}[i]$ for $k, l = 1 \ldots n$ and $i \geq 0$. To write down the relations, set $T_i = (t_{kl}[i])_{k,l=1}^n$ and*

$$T(u) = T_0 + T_1 u + T_2 u^2 + \ldots$$

The relations are

$$T \text{ is invertible},$$

$$R(u - v)^{12}T^{13}(u)T^{23}(v) = T^{23}(v)T^{13}(u)R^{12}(u - v) \tag{11.3}$$

where we expand $R(u - v)$ in the region $|u| > |v|$. Then the following define a Hopf algebra structure on $H(R)$:

$$\Delta(T(u)) = T^{12}(u)T^{13}(u), \qquad S(T(u)) = T^{-1}(u), \epsilon(T(u)) = 1. \tag{11.4}$$

Moreover, $H(R)$ is an h-formal group, i.e it is a topologically free $k[[h]]$-module and $H(R)/hH(R) \simeq U(gl_n[[t]])'$.

Proof: see [EK4].

Remarks: As before, the term topological means that we first complete with respect to the maximal ideal $J = \mathrm{Ker}(\epsilon)$ and then take the h-adic completion.

In fact, in this case, the R-matrix automatically satisfies some unitarity condition as is shown by the following result, whose proof can be found in [EK4]

Proposition 11.4. *Let $R(u)$ be as in Proposition 11.3. Then $R(u)R^{21}(-u)$ is a scalar operator of the form $f(u)\mathrm{Id}$ where $f(u) \in 1 + hk((u))[[h]]$ such that $f(u) = f(-u)$.*

By the above proposition, we can always rescale $R(u)$ (multiplying by some scalar in $k((u))[[h]]$ which doesn't change $H(R)$) so $R(u)R^{21}(-u) = 1$. Let us write $R(u) = 1 - hr(u) \pmod{h^2}$. It is easy to see by an argument analogous to Proposition 9.4 that $r(u)$ is a classical r-matrix with spectral parameter and that $r^{21}(u)+r(-u) = 0$. Suppose furthermore that $r(u) = \frac{\Omega}{u}+\mathcal{O}(1) = \frac{\sigma}{u}+\mathcal{O}(1)$. We can then define a Lie bialgebra $\mathfrak{g}(r)$ by the construction described in Lecture 6.

Now let us consider the quantized universal enveloping algebra $A(R)$ associated to the h-formal group $H(R)$. By definition $A(R) = \sum_n h^{-n}J^n$. We claim that $A(R)$ is a quantization of $\mathfrak{g}(r)$. To see this, let us write $T(u) \equiv 1 + ht(u) \pmod{h^2}$ with $t(u) \in \mathrm{End}\mathbb{C}^n \otimes A(R)[[u]]$. Then $A(R)$ is generated by the coefficients of $t(u)$ with commutation and cocommutation relations deduced from (11.3) and (11.4):

$$\begin{aligned}
[t_{13}(u), t_{23}(v)] &= [r^*_{12}(u-v), t_{13}(u) + t_{23}(v)] \qquad (11.5)\\
&\quad + h(r^*_{12}(u-v)t_{13}(u)t_{23}(v) - t_{23}(v)t_{13}(u)r^*_{12}(u-v)),\\
\Delta(t(u)) &= t_{12}(u) + t_{13}(u) + ht_{12}(u)t_{13}(u), \qquad (11.6)
\end{aligned}$$

where $r^* = \frac{1-R(u)}{h}$. Comparing (11.5) with (11.1) shows that $A(R)$ is a quantization of $\mathfrak{g}(r)$.

11.3 Examples

The RTT formalism is a powerful method for constructing quantized enveloping algebras from R-matrices. We will now apply it to the Yangian, dual Yangian and elliptic Lie bialgebras for $\mathfrak{g} = gl_n$.

11.3.1 The Yangian

We use here Yang's R-matrix $R_Y(u) = 1 - \frac{h\sigma}{nu}$. The RTT-formalism yields the following quantization of the Yangian Lie bialgebra:

Definition: The *Yangian* is the quantized universal enveloping algebra generated by $t_{ij}[m]$, $i, j = 1 \dots n$, $m \geq 0$ with the following relations.
Set $T_m = (t_{ij}[m])_{i,j=1}^n$ and $T = 1 + \sum_{m\geq 0} T_m u^{-m-1}$. The relations are

$$R_Y(u-v)T^{13}(u)T^{23}(v) = T^{13}(u)T^{23}(v)R_Y(u-v),$$

$$\Delta(T(u)) = T^{12}(u)T^{13}(u), \qquad S(T(u)) = T^{-1}(u), \qquad \epsilon(T(u)) = 1.$$

11.3.2 The dual Yangian

In the same way, we can define a quantization of the dual Yangian:

Definition: The *dual Yangian* is the quantized universal enveloping algebra generated by $t_{ij}[m]$, $i,j = 1\dots n$, $m \leq -1$ with the following relations.
Set $T_m = (t_{ij}[m])_{i,j=1}^n$ and $T = \sum_{m \geq 0} T_{-m-1}u^m$. The relations are

$$R_Y(u-v)T^{13}(u)T^{23}(v) = T^{13}(u)T^{23}(v)R_Y(u-v),$$

$$\Delta(T(u)) = T^{12}(u)T^{13}(u), \qquad S(T(u)) = T^{-1}(u), \qquad \epsilon(T(u)) = 1.$$

11.3.3 Quantum elliptic algebra

We give here a quantization of the elliptic algebra $gl_n(r_{ell})$ defined by the elliptic r-matrix $r_{ell}(u)$ of lecture 6. For this, we need a quantum elliptic R-matrix:

Proposition 11.5. *The element*

$$R_{ell}(u) = n\sigma - h[r_{ell}(u - \frac{h}{n^2}) - r_{ell}(-\frac{h}{n^2})] \in gl_n \otimes gl_n((u))[[h]]$$

where σ is the permutation is an R-matrix. Furthermore, $R_{ell}(u)R_{ell}^{-1}(-u) \in \mathbb{C}((u))[[h]]\mathrm{Id}$ (is a scalar).

Proof: see [Che1], [Bel].

Since rescaling R doesn't affect the algebra $H(R)$, we can apply the RTT formalism:

Definition: The *quantum elliptic algebra* is the quantized enveloping algebra generated by $t_{ij}[m]$, $i,j = 1\dots n$, $m \leq -1$ with the following relations.
Set $T_m = (t_{ij}[m])_{i,j=1}^n$ and $T = \sum_{m \geq 0} T_{-m-1}u^m$. The relations are

$$R_{ell}(u-v)T^{13}(u)T^{23}(u) = T^{13}(u)T^{23}(v)R_{ell}(u-v),$$

$$\Delta(T(u)) = T^{12}(u)T^{13}(u), \qquad S(T(u)) = T^{-1}(u), \qquad \epsilon(T(u)) = 1.$$

It is a quantization of $gl_n(r_{ell})$.

11.3.4 Quantized affine Lie algebra

We have already defined a quantization of the affine Lie bialgebra in the context of Kac-Moody algebras. When $\mathfrak{g} = gl_n$, there is also an equivalent realization of this quantization in terms of the RTT formalism. However, as we saw, the Manin triple defining the affine Lie bialgebra involves two points on $\mathbb{P}^1$ rather than one and the construction is more involved. We will just state it here, and refer the reader to [DF] and [CP] for more details.

Proposition 11.6. *The following element*

$$R(u) = \sum_{i=1}^{n} E_{ii} \otimes E_{ii} + \frac{e^{-u} - 1}{q - q^{-1}} \sum_{i \neq j} E_{ii} \otimes E_{jj}$$
$$+ \frac{e^{-u}(q - q^{-1})}{e^{-u}q - q^{-1}} \sum_{i<j} E_{ij} \otimes E_{ji} + \frac{q - q^{-1}}{e^{-u}q - q^{-1}} \sum_{i>j} E_{ij} \otimes E_{ji}$$

is an R-matrix (as usual, $q = e^h$).

Definition: Let $U(R)$ be the Hopf algebra defined as follows: $U(R)$ is generated by elements $l_{ij}^{\pm}[\mp m]$, $i, j = 1 \ldots n$, $m \geq 0$ where $l_{ij}^{+}[0], l_{ji}^{-}[0] = 0$ for $j > i$. Let $L_m^{\pm} = (l_{ij}^{\pm}[\mp m])_{i,j=1}^{n}$ and set $L^{\pm}(u) = L_0^{\pm} + L_1^{\pm} u^{\pm 1} + L_2^{\pm} u^{\pm 2} + \ldots$. The defining relations are

$$l_{ii}^{+}[0] l_{ii}^{-}[0] = l_{ii}^{-}[0] l_{ii}^{+}[0] = 1,$$
$$R(u - v)^{12} (L^{\pm})^{13}(u) (L^{\pm})^{23}(v) = (L^{\pm})^{23}(v) (L^{\pm})^{13}(u) R(u - v)^{12},$$
$$R(u - v)^{12} (L^{+})^{13}(u) (L^{-})^{23}(v) = (L^{-})^{23}(v) (L^{+})^{13}(u) R(u - v)^{12},$$

$$\Delta(L^{\pm}(u)) = L_{12}^{\pm}(u) L_{13}^{\pm}(u), \qquad S(L^{\pm}(u)) = L^{\pm}(u)^{-1}, \qquad \epsilon(L^{\pm}(u)) = 1.$$

Proposition 11.7. *The quantized affine Lie algebra $U_h(\hat{gl}_n)$ is isomorphic to $U(R)$.*

Proof: see [DF] and [Be].

Remark: i) There is a similar construction of the full quantized affine Lie algebra (including the central extension)- see [RS].
ii) There is yet another presentation of the quantized affine Lie algebra (and of the Yangian), due to Drinfeld, which is more suitable for studying finite dimensional representations (*Drinfeld's new realization*-see [Dr4] and [Be]).

Lecture 12

The quantum double

We have seen in Lecture 4 how to embed any Lie bialgebra in a quasitriangular one-its Drinfeld double. In this lecture we give the quantum analogue of this important construction.

12.1 The quantum double for finite dimensional Hopf algebras

Let us briefly recall the main properties of the Drinfeld double of a (finite dimensional) Lie bialgebra $\mathfrak{g}$. By definition, $D\mathfrak{g} = \mathfrak{g} \oplus \mathfrak{g}^*$ as a vector space, $\mathfrak{g}$ and $\mathfrak{g}^*$ are Lie subalgebras, and the cocommutator is $\delta_{D\mathfrak{g}} = \delta_{\mathfrak{g}} - \delta_{\mathfrak{g}^*}$. Finally the commutator can be characterized by the following condition: the canonical element $r \in \mathfrak{g} \otimes \mathfrak{g}^* \subset D\mathfrak{g} \otimes D\mathfrak{g}^*$ is a quasitriangular structure for $D\mathfrak{g}$. This motivates the following proposition:

Proposition 12.1. *Let H be a finite dimensional Hopf algebra and H^{*op} its dual Hopf algebra with opposite comultiplication Δ^{op}. Let $R \in H \otimes H^{*op}$ be the canonical element. Then there exists a unique Hopf algebra structure on the vector space $D(H) = H \otimes H^{*op}$ such that the following hold:*

1. *H and H^{*op} are Hopf subalgebras of $D(H)$ (via the canonical embeddings),*
2. *The multiplication map $H \otimes H^{*op} \to D(H)$, $h \otimes h' \mapsto h.h'$ is an isomorphism of vector spaces,*
3. *$R \in D(H) \otimes D(H)$ is a quasitriangular structure for $D(H)$.*

Proof: By 2., every $d \in D(H)$ can be uniquely written as a product $d = ab$, $a \in H$, $b \in H^{*op}$. Condition 1. thus determines the coproduct on $D(H)$. Hence, the only remaining step is to compute products of the form ba, $a \in H$, $b \in H^{*op}$. We will derive a formula for these products using 3. and then show that this indeed induces a Hopf algebra structure on $D(H)$.

We will use the pictorial technique. Let a_i be a basis of H and a_i^* the dual basis of H^{*op}. By definition, $R = \sum_i a_i \otimes a_i^*$. Viewing R as a map $k \to D(H) \otimes D(H)$

and omitting the symbol for the ground field, we will represent R as follows:

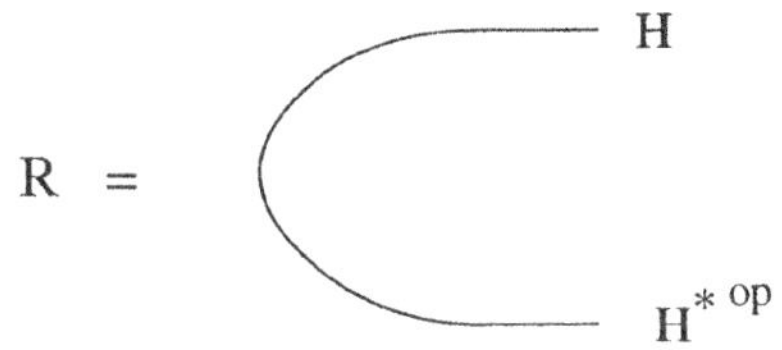

Likewise, the natural pairing $H \otimes H^{*op} \to k$ will be denoted by the picture

Notice that it follows from the definition of R that $(1 \otimes \Delta)R = R_{13}R_{12}$, $(\Delta \otimes 1)R = R_{13}R_{23}$ (these do not involve products ab, $a \in H$, $b \in H^{*op}$). In pictures, this can be drawn as:

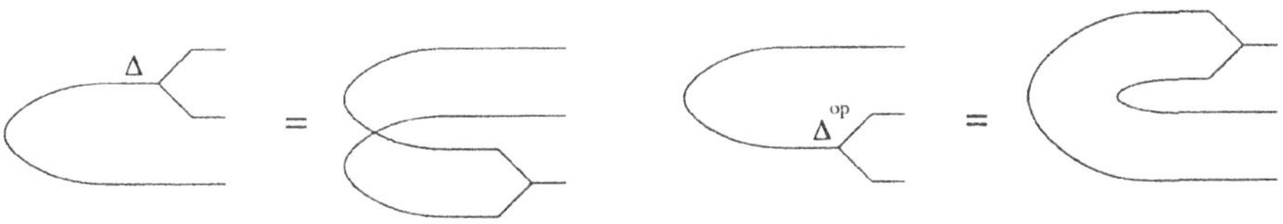

(12.1)

Here we write Δ for the coproduct on H^* and Δ^{op} for the coproduct on H^{*op}. Intuitively, the pictures above show that we can "turn around" the coproduct. We will also need the following identities $(S \otimes S)R = R, (1 \otimes S^{-1})R = (S \otimes 1)R = R^{-1}$ (the first identity is clear; to prove $(S \otimes 1)R = R^{-1}$, notice that $R(S \otimes 1)R = ((S \otimes S)R)(S \otimes 1)R = m_{23}(S \otimes S \otimes 1)R_{13}R_{12} = (S \otimes 1)m_{23}(1 \otimes \Delta)R = 1$; the second relation is proved in the same fashion). In pictures, this means:

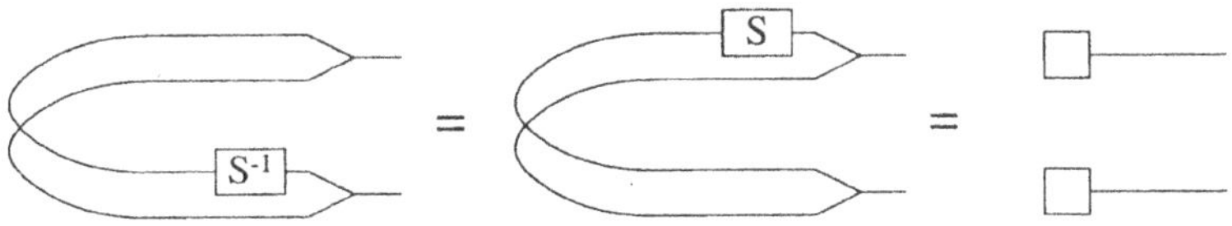

(12.2)

The following rules are obviously true:

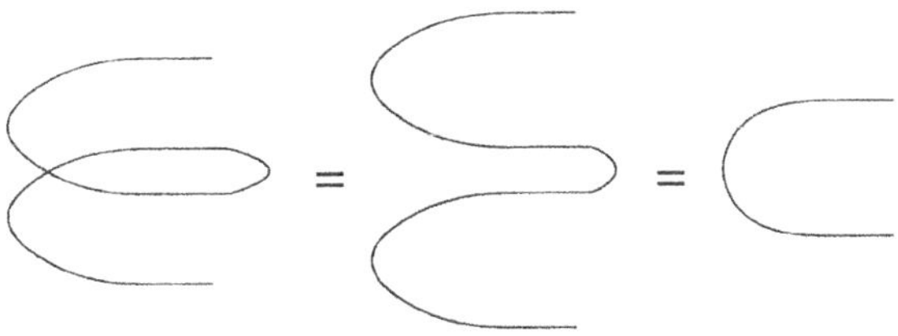

(12.3)

Now let us suppose that R is a quasitriangular structure for $D(H)$. In particular, R is a solution of the QYBE, i.e

$$\sum_{i,j,k} a_i a_j \otimes a_i^* a_k \otimes a_j^* a_k^* = \sum_{i,j,k} a_j a_i \otimes a_k a_i^* \otimes a_k^* a_j^* \tag{12.4}$$

or, in pictures:

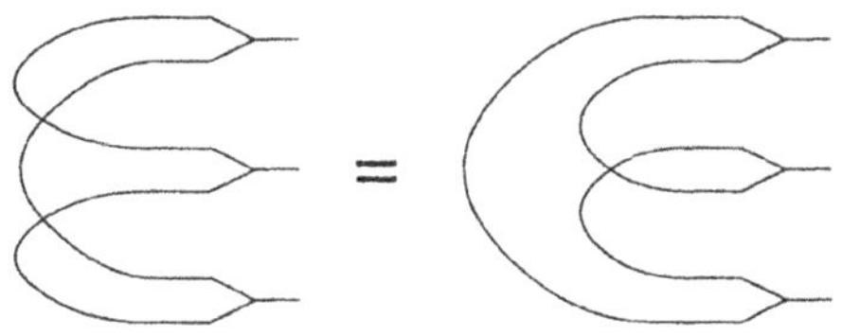

(12.5)

The first term contains products of the form $a_i^* a_k$. We can therefore view (12.4) as a system of linear equations in the unknowns $a_i^* a_k$.

Lemma 12.1. *The vectors* $v_{ik} = \sum_j a_i a_j \otimes a_j^* a_k^*$ *form a basis of* $H \otimes H^*$.

Proof: Consider the Hopf algebra H^{op} (with opposite comultiplication). Its dual is $(H^{op})^* = (H^*)_{op}$ (with opposite multiplication). We have

$$v_{ik} = a_i \otimes a_k^* \sum_j a_j \otimes a_j^* = a_i \otimes a_k^* R$$

in $H^{op} \otimes (H^{op})^*$. One can show, by arguments similar to those given for (12.2), that $R(1 \otimes S)(R) = (1 \otimes S)(R)R = 1$. The lemma follows.

□

Hence, the products $a_i^* a_k$ are uniquely determined by the QYBE.
Claim: the product is given by the following picture:

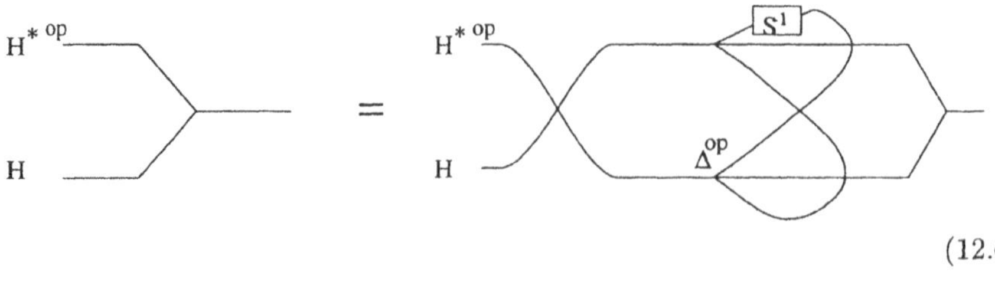

(12.6)

To prove this, substitute (12.6) in the QYBE (12.5):

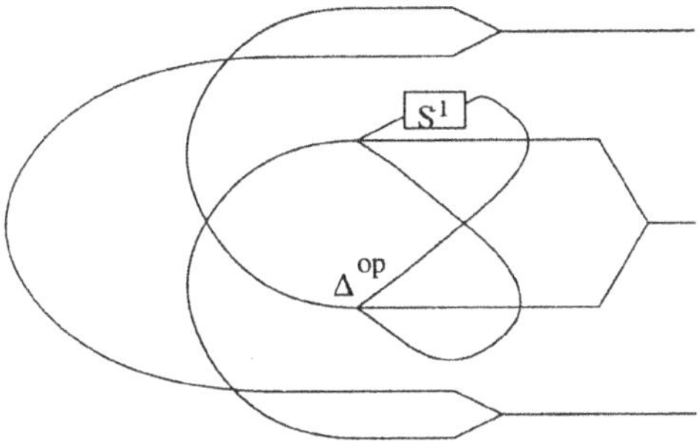

(12.7)

Using (12.1) twice as shown by the arrows below, we can rewrite it as:

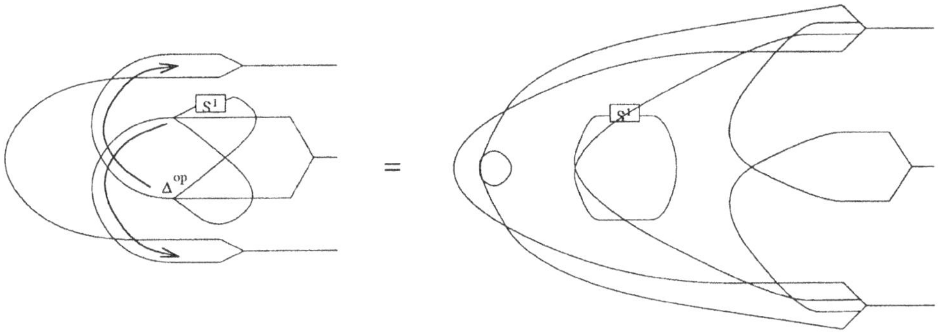

Now, removing the loops by (12.3) and moving the antipode around by (12.2), we have:

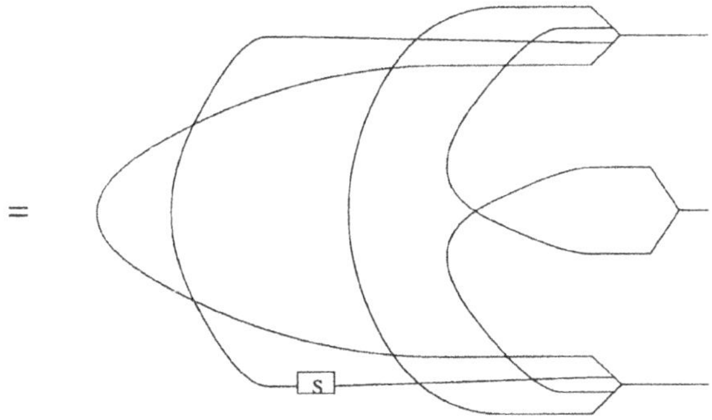

Finally, using (12.2), we get:

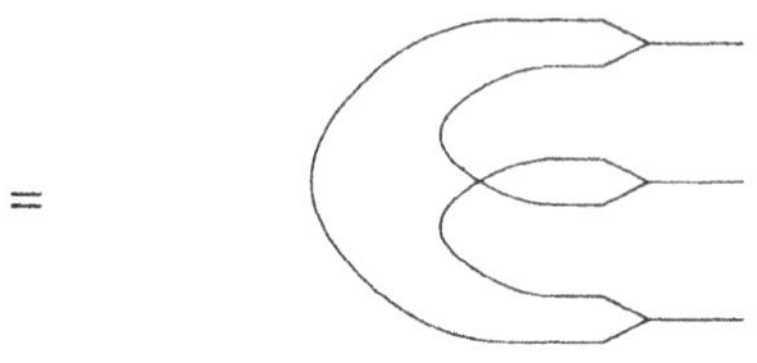

Similar pictures show that (12.6) indeed defines an algebra structure on $D(H)$ (i.e we have $a(b_1b_2) = (ab_1)b_2$ $(a_1a_2)b = a_1(a_2b)$) and that Δ, Δ^{op} extend to an algebra homomorphism denoted $\Delta : D(H) \to D(H)$.

Now let us check that R is a quasitriangular structure for $D(H)$. It is enough to prove that $\Delta^{op}(x)R = R\Delta(x)$ for $x \in H, H^{*op}$. We do this for $x \in H$ and leave the other, similar case to the reader. We have to show:

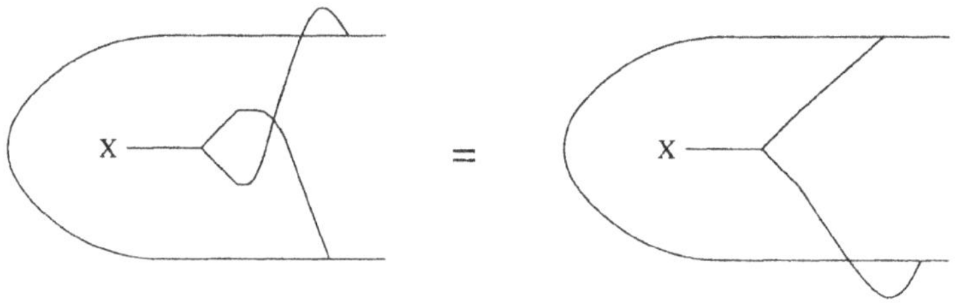

(12.8)

Using (12.6), we see that the r.h.s of (12.8) is:

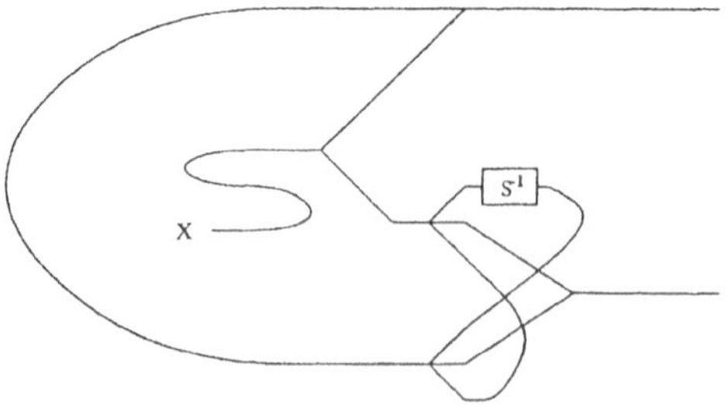

(12.9)

We now use (12.6) as shown below to obtain:

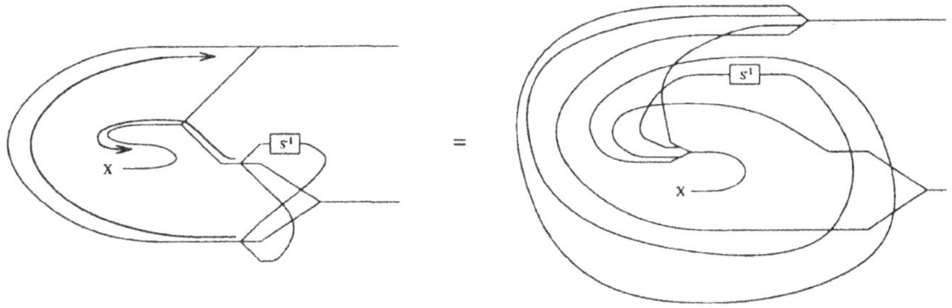

(12.10)

where we applied (12.1)according to the arrows. Removing the loop and applying (12.1) yields that the right-hand side of (12.10) is:

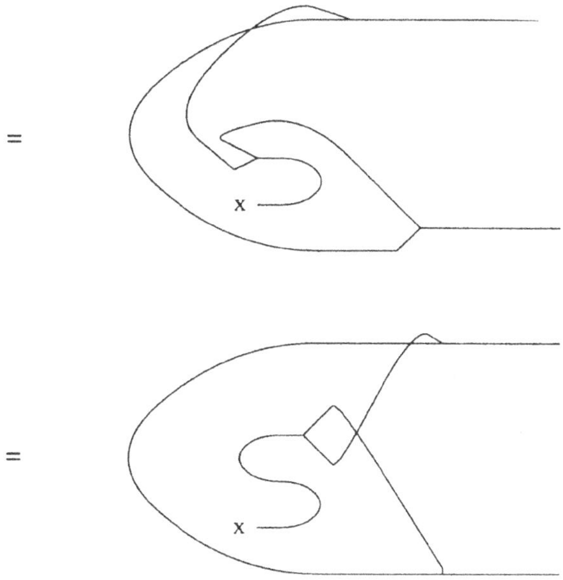

(12.11)

This concludes the proof of Proposition 12.1.

□

Definition: The quasitriangular Hopf algebra $D(H)$ is called the *quantum double* of H.

Example: Let G be a finite group and let $H = \mathbb{C}[G]$. Then $H^{*op} = \mathcal{F}(G)^{op}$, i.e it is the commutative algebra of functions on G with comultiplication given by $\Delta(\phi)(x \otimes y) = \phi(yx)$. Then the quantum double is the Hopf algebra $D(H) = \mathbb{C}[G] \otimes \mathcal{F}(G)^{op}$ in which $\mathbb{C}[G]$ and $\mathcal{F}(G)^{op}$ are Hopf subalgebras, with the product determined by the relation

$$\phi(y).g = g.\phi(g^{-1}yg)$$

as can be seen from (12.6). Thus, as an algebra, $D(H)$ is the semidirect product $\mathbb{C}[G] \ltimes \mathcal{F}(G)$. This is the simplest example of an interesting noncommutative and noncocommutative Hopf algebra.

12.2 The quantum double for quantized universal enveloping algebras

The above construction easily extends to the case of quantized enveloping algebras:

Proposition 12.2. *Let H be a quantized enveloping algebra and H^{*op} its dual quantized enveloping algebra with the opposite comultiplication. The following conditions define a unique structure of quantized enveloping algebra on the K-module $D(H) = H \otimes H^{*op}$:*

1. *H, H^{*op} are Hopf subalgebras of $D(H)$,*

2. *The map $H \otimes H^{*op} \to D(H)$, $h \otimes h' \mapsto h.h'$ is an isomorphism of vector spaces,*

3. *the canonical element $R \in H \otimes H^{*op} \subset D(H) \otimes D(H)$ is a quasitriangular structure for $D(H)$.*

Moreover, the product between H^{*op} and H is determined by:

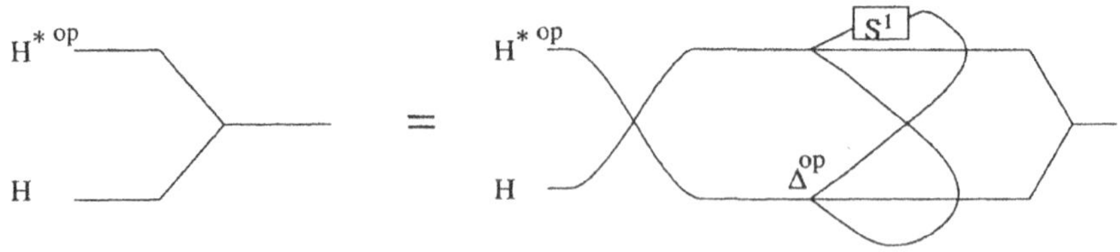

(12.12)

Proof: It follows word for word, picture for picture the proof of Proposition 12.1.

□

The following result shows that the quantum double is really the quantum analogue of the classical double:

Proposition 12.3. *Let H be a quantized enveloping algebra and let $\mathfrak{g}$ be the quasiclassical limit of H. Then $D(H)$ is a quantization of $D\mathfrak{g}$.*

Proof: It is clear that $U := D(H)/hD(H) \simeq U(D\mathfrak{g})$ as a vector space. Since H and H^{*op} are Hopf subalgebras of $D(H)$, $U(\mathfrak{g})$ and $U(\mathfrak{g}^{*op})$ are Hopf subalgebras of U. Furthermore, the classical limit r of R is the canonical element of $\mathfrak{g} \otimes \mathfrak{g}^*$ and satisfies the $CYBE$. As in (12.1), this completely determines the algebra structure of U and $U = U(D\mathfrak{g})$ as a Hopf algebra. This statement and the fact that H and H^{*op} are quantizations of $\mathfrak{g}$ and $\mathfrak{g}^{*op}$ respectively, imply the proposition.

□

Definition: The quantized enveloping algebra $D(H)$ is called the *quantum double* of H.

12.3 Quasitriangular structure on $U_h(\mathfrak{g})$

Let $\mathfrak{g}$ be a simple Lie algebra, $\mathfrak{n}_- \oplus \mathfrak{h} \oplus \mathfrak{n}_+$ a polarization of $\mathfrak{g}$ and $< , >_{\mathfrak{g}}$ an invariant nondegenerate bilinear form. Recall that the standard Lie bialgebra structure on $\mathfrak{g}$ is obtained by considering $\mathfrak{g}$ as "almost" the double of $\mathfrak{n}_- \oplus \mathfrak{h}$. In particular, it is quasitriangular. We will now apply the same method to produce a quasitriangular structure for the Drinfeld-Jimbo quantum groups $U_h(\mathfrak{g})$.

Definition: Let A, B be two bialgebras and $< , >$ a bilinear pairing $A \otimes B \to \mathbb{C}$. We say that $< >$ is a *duality pairing of bialgebras* if the induced maps

$$< a, . >: A \to B^*,$$
$$< ., b >: B \to A^*,$$

are algebra homomorphisms. In other terms, we have the following invariance properties:

$$< a.a', b > = < a \otimes a', \Delta(b) >,$$
$$< a, b.b' > = < \Delta(a), b \otimes b' >,$$
$$< 1, b > = \epsilon(b),$$
$$< a, 1 > = \epsilon(a).$$

If, furthermore, A, B are Hopf algebras, we say that $< , >$ is a *duality pairing of Hopf algebras* if

$$< S(a), b > = < a, S(b) > .$$

It is easy to check that if A, B are bialgebras (resp. Hopf algebras) and $< , >$ is a duality pairing, then the maps $< a, . >$ and $< ., b >$ are bialgebra (resp. Hopf algebra) homomorphisms.

Now recall the notations from section 9.3: $U_h(\mathfrak{g})$ is generated by e_i, f_i, h_i, $i = 1, \ldots n$ with relations (9.2-9.5), and $U_h(\mathfrak{b}_+)$ (resp. $U_h(\mathfrak{b}_-)$) is the Hopf subalgebra generated by e_i, h_i (resp. f_i, h_i). Similarly, let $U_h(\mathfrak{n}_+)$ (resp. $U_h(\mathfrak{n}_-)$) be the subalgebra generated by e_i (resp. f_i).

Lemma 12.2 (Triangular decomposition of $U_h(\mathfrak{g})$). *The multiplication map*

$$U_h(\mathfrak{n}_-)\otimes U_h(\mathfrak{h})\otimes U_h(\mathfrak{n}_+) \to U_h(\mathfrak{g})$$

is an isomorphism of $\mathbb{C}[[h]]$*-modules.*

Proof: This is true because the statement holds modulo h.

We will use the following important fact, and refer the reader to [J] and [Lu] for a proof:

Drinfeld pairing: There exists a unique duality pairing

$$<,>: U_h(\mathfrak{b}_+)\hat{\otimes} U_h(\mathfrak{b}_-)^{op} \to \mathbb{C}((h))$$

defined on generators by

$$< H_i, H_j > = \frac{1}{h} < h_i, h_j >_{\mathfrak{g}}, \quad < E_i, F_j > = \frac{\delta_{ij}}{e^{d_i h} - e^{-d_i h}},$$

$$< H_i, F_j > = < E_i, H_j > = 0,$$

$$< 1, 1 > = 1, \qquad < E_i, 1 > = < 1, F_i > = < 1, H_i > = < H_i, 1 > = 0.$$

Moreover, $<,>$ induces a Hopf algebra isomorphism

$$U_h(\mathfrak{b}_\pm) \tilde{\to} U_h(\mathfrak{b}_\mp)^{*op}$$

in the quantized enveloping algebra sense. The form $<,>$ is called the *Drinfeld pairing.*

Exercise 12.1. Show that we have

$$< e^{hH_i}, e^{hH_j} > = e^{h<h_i,h_j>}.$$

This shows that $D(U_h(\mathfrak{b}_+)) \simeq U_h(\mathfrak{b}_+) \otimes U_h(\mathfrak{b}_-)$ (with the algebra structure given by (12.12)). Furthermore, it is easy to check that the map

$$E_i \otimes 1 \to E_i,$$
$$1 \otimes F_i \to F_i,$$
$$H_i \otimes 1 \to H_i,$$
$$1 \otimes H_i \to H_i,$$

extends to a surjective Hopf algebra map

$$\pi : U_h(\mathfrak{b}_+) \otimes U_h(\mathfrak{b}_-) \twoheadrightarrow U_h(\mathfrak{g}).$$

In particular, the quasitriangular structure on $D(U_h(\mathfrak{b}_+))$ induces a quasitriangular structure on $U_h(\mathfrak{g})$, which can be explicitly written as

$$R = q^{\sum_p x_p \otimes x_p} \sum_i a_i \otimes a^i \tag{12.13}$$

where (x_i) is an orthonormal basis of $\mathfrak{h}$, (a_i), $i \in I$ a basis of $U_h(\mathfrak{n}_+)$ and (a^i), $i \in I$ the dual basis of $U_h(\mathfrak{n}_-)$.

Remark: We slightly abuse terminology in the last statement: (a_i) and (a^i) are dual basis of $U_h(\mathfrak{n}_+) \otimes \mathbb{C}((h))$ and $U_h(\mathfrak{n}_-) \otimes \mathbb{C}((h))$.

Exercise 12.2. Show that $\mathrm{Ker}(\pi)$ is the two-sided ideal generated by $H_i \otimes 1 - 1 \otimes H_i$.

Problem 12.1. Show that formula (12.13) coincides with the formula (9.9) when $\mathfrak{g} = sl_2$.

Problem 12.2 (Radford's theorem). Let $(H, m, \Delta, S, i, \epsilon)$ be a finite dimensional Hopf algebra. Show that there exists $m \geq 1$ such that $S^m = 1$ (hint: consider the quantum double $D(H)$ of H and use Problem 9.1)

Lecture 13

Tensor categories and quasi-Hopf algebras

We now develop the formalism of tensor categories which is the categorical structure underlying Hopf algebras and quantum groups. As usual, we will work over $\mathbb{C}$, but results of this lecture are valid for any field of characteristic zero.

13.1 Semigroup categories

13.1.1 Definition

Let $\mathcal{C}$ be a category. Let $\otimes : \mathcal{C} \times \mathcal{C} \to \mathcal{C}$ be a bifunctor. We want to define the notion of associativity for $\otimes$. As usual in category theory, it is unrealistic to require that $(X \otimes Y) \otimes Z = X \otimes (Y \otimes Z)$. On the other hand, it is *not enough* to just require $(X \otimes Y) \otimes Z$ and $X \otimes (Y \otimes Z)$ to be isomorphic. Thus, we need something intermediate between these two extremes.

Recall that a morphism Ψ between two functors $F, G : \mathcal{C} \to \mathcal{C}'$ is a collection of morphisms

$$\psi_X : F(X) \to G(X) \qquad (X \in \mathcal{O}b(\mathcal{C}))$$

such that for every $f \in \mathrm{Hom}(X, Y)$ the diagram

$$\begin{array}{ccc} F(X) & \xrightarrow{F(f)} & F(Y) \\ \downarrow{\scriptstyle \psi_X} & & \downarrow{\scriptstyle \psi_Y} \\ G(X) & \xrightarrow[G(f)]{} & G(Y) \end{array} \tag{13.1}$$

is commutative. We say that Ψ is an isomorphism if all the maps ψ_X are isomorphisms.

The correct notion of associativity is the following: we have a (bi)functor

$$\begin{aligned} \cdot \otimes \cdot : \ \mathcal{C} \times \mathcal{C} &\to \mathcal{C}, \\ (X, Y) &\to X \otimes Y \end{aligned}$$

and an *associativity constraint* is an isomorphism of (tri)functors

$$\Phi : (\cdot \otimes \cdot) \otimes \cdot \xrightarrow{\sim} \cdot \otimes (\cdot \otimes \cdot). \tag{13.2}$$

Now consider tensor products of more than three factors. In this case, we have more than two ways of putting brackets on the product $V_1 \otimes \ldots \otimes V_n$. Ideally, we would like to have a canonical isomorphism between any two bracketings like

$$(V_1 \otimes (V_2 \otimes \ldots (\ldots))) \otimes V_n.$$

For example, let us consider $W, X, Y, Z \in \mathcal{O}b(\mathcal{C})$, and consider the two bracketings $((W \otimes X) \otimes Y) \otimes Z$ and $W \otimes (X \otimes (Y \otimes Z))$. There are two ways of passing from one to another using associativity isomorphisms, namely

$$\begin{array}{ccc}
 & ((W \otimes X) \otimes Y) \otimes Z & \\
 \swarrow^{\Phi_{W,X,Y} \otimes 1_Z} & & \searrow^{\Phi_{W \otimes X,Y,Z}} \\
(W \otimes (X \otimes Y)) \otimes Z & & (W \otimes X) \otimes (Y \otimes Z) \\
\downarrow \Phi_{W,X \otimes Y,Z} & & \Phi_{W,X,Y \otimes Z} \downarrow \\
W \otimes ((X \otimes Y) \otimes Z) & \xrightarrow[1_W \otimes \Phi_{X,Y,Z}]{} & W \otimes (X \otimes (Y \otimes Z))
\end{array} \tag{13.3}$$

It is therefore natural to require the above diagram to be commutative for every quadruple $W, X, Y, Z \in \mathcal{O}b(\mathcal{C})$. This requirement is known as the *pentagon* axiom, or pentagon relation.

Definition: A *semigroup* category is an abelian category equipped with a bifunctor

$$\cdot \otimes \cdot : \mathcal{C} \times \mathcal{C} \to \mathcal{C}$$

and an isomorphism of (tri)functors:

$$\Phi : (\cdot \otimes \cdot) \otimes \cdot \xrightarrow{\sim} \cdot \otimes (\cdot \otimes \cdot),$$
$$\Phi_{X,Y,Z} : (X \otimes Y) \otimes Z \xrightarrow{\sim} X \otimes (Y \otimes Z)$$

satisfying the pentagon relation (13.3) for every quadruple $W, X, Y, Z \in \mathcal{O}b(\mathcal{C})$.

It turns out that this is a sufficient condition for the existence of only one natural way of identifying any two bracketings. In other terms, all the higher analogs of the pentagon relation (for more than four objects) follow from the pentagon relation (13.3).

Theorem 13.1 (MacLane-Stasheff coherence theorem). *Let $\mathcal{C}$ be a semigroup category, $V_1, \ldots V_n \in \mathcal{O}b(\mathcal{C})$ and A_1, A_2 two bracketings of the expression $V_1 \otimes \ldots \otimes V_n$. Then any two isomorphisms $A_1 \xrightarrow{\sim} A_2$ composed of associativity isomorphisms Φ's and their inverses, are equal.*

We reformulate the above theorem in topological terms, and refer the reader to [Mac] for a proof. A topological proof can be found in the book [ShSt] (with a picture of the proof on the cover).

Let $n \in \mathbb{N}$ and let A_n be the following 2-dimensional CW complex. The 0-dimensional cells are labeled by all possible bracketings of the expression $V_1 \otimes \ldots \otimes V_n$. Attach a 1-dimensional cell between any two adjacent bracketings (i.e such that we can pass from one to the other via *one* associativity isomorphism). Finally, glue pentagonal 2-dimensional cells for each diagram of the form (13.3), and a square 2-dimensional cell for each diagram of the form (13.1) corresponding to functoriality of the associativity isomorphisms.

Theorem 13.2 (MacLane-Stasheff coherence theorem). $\pi_1(A_n) = \{1\}$.

13.1.2 Examples

Example 13.1 (Vector spaces). Let $\mathcal{V}ec$ be the category of k-vector spaces, and let

$$\otimes : \mathcal{V}ec \times \mathcal{V}ec \to \mathcal{V}ec$$

be the usual tensor product. Then $\mathcal{V}ec$ is a semigroup category when equipped with the usual identification:

$$\begin{aligned} \Phi_{X,Y,Z} : (X \otimes Y) \otimes Z &\to X \otimes (Y \otimes Z), \\ (x \otimes y) \otimes z &\to x \otimes (y \otimes z) \end{aligned}$$

From now on, we will always identify the vector spaces $(X \otimes Y) \otimes Z$ and $X \otimes (Y \otimes Z)$ via $\Phi_{X,Y,Z}$ and we write $\Phi_{X,Y,Z} = 1$.

Notice that we can specialize this example to many others: the category $\operatorname{Rep} G$ of left (right) modules over a group G, the category $\operatorname{Rep} \mathfrak{g}$ of modules over a Lie algebra $\mathfrak{g}$ (with usual tensor products), etc., are semigroup categories.

Example 13.2 (Modules over a bialgebra). The following example is fundamental. Let $(A, m, \Delta, i, \epsilon)$ be a bialgebra, and let $\operatorname{Rep} A$ be the category of left A-modules. We now define a notion of tensor product $V \otimes W$ of two A-modules V, W:

As a vector space, $V \otimes W$ is the usual tensor product, and the action is given by

$$\pi_{V \otimes W}(a) = (\pi_V \otimes \pi_W)(\Delta(a)). \tag{13.4}$$

The associativity isomorphisms are trivial: $\Phi_{V,W,Z} = 1$.

Exercise 13.1. Check that $\operatorname{Rep} A$ with $\otimes$ and Φ is indeed a semigroup category.

This gives an interpretation for the axioms defining a bialgebra. In particular, the coassociativity of Δ makes it possible to choose $\Phi \equiv 1$.

Exercise 13.2. Suppose that A is a $\mathbb{C}$-Hopf algebra (i.e it has an antipode S). Show that S allows us to define a structure of an A-module on $\mathrm{Hom}_{\mathbb{C}}(X, Y)$ for any two A-modules X, and Y, by

$$a.f(x) = \sum_i a_i'.f(S(a_i'').x)$$

where $\Delta(a) = \sum_i a_i' \otimes a_i''$. Check that this coincides with the usual A-action on $\mathrm{Hom}_{\mathbb{C}}(X, Y)$ when A is a group algebra or a universal enveloping algebra.

Notice that in the two preceding examples, we have $\Phi \equiv 1$.
Definition: A semigroup category is called *strict* if

$$(X \otimes Y) \otimes Z = X \otimes (Y \otimes Z) \qquad \forall\, X, Y, Z$$

and the associativity isomorphisms are trivial:

$$(X \otimes Y) \otimes Z = X \otimes (Y \otimes Z), \qquad \Phi_{X,Y,Z} = 1 \qquad \forall\, X, Y, Z \in \mathcal{O}b(\mathcal{C}).$$

Example 13.3 (Finite groups). We now give an example of a semigroup category which is not strict. Let G be a finite group and let $\mathcal{F}(G)$ be the algebra of functions on G. As we have seen, $\mathcal{F}(G)$ is a bialgebra, and we can define the tensor product $V \otimes W$ of two $\mathcal{F}(G)$-modules by (13.4). Let us restrict ourselves to the category $\mathcal{C}$ of one dimensional $\mathcal{F}(G)$-modules, and take one representative for each equivalence class. Thus, as a set, $\mathcal{C} \simeq G$ and with this identification, $g \otimes h \simeq gh$.

We will now describe all semigroup structures on $\mathcal{C}$, i.e collections of maps

$$\Phi_{X,Y,Z} : (X \otimes Y) \otimes Z \to X \otimes (Y \otimes Z)$$

($\Phi_{X,Y,Z} \in \mathrm{Aut}\,(X \otimes Y \otimes Z) \simeq \mathbb{C}^*$) satisfying the pentagon relations. If we view Φ as a function $\Phi : G \times G \times G \to \mathbb{C}^*$, the pentagon relations become

$$\Phi(W, X, Y)\Phi(W, XY, Z)\Phi(X, Y, Z)\Phi(W, X, YZ)^{-1}\Phi(WX, Y, Z)^{-1} = 1.$$

This is just the condition that Φ be a 3-cocycle on G with coefficients in $\mathbb{C}^*$: $\Phi \in Z^3(G, \mathbb{C}^*)$. In particular, if $\Phi \neq 1$, we obtain a non-strict semigroup category.

13.1.3 Tensor functors

As usual, having defined semigroup categories, we need to define the notion of a functor between semigroup categories which "respects" the semigroup structure:

Definition: Let $\mathcal{C}, \mathcal{C}'$ be two semigroup categories and $F : \mathcal{C} \to \mathcal{C}'$ a functor. An isomorphism of (bi)functors

$$J : F(\cdot) \otimes F(\cdot) \to F(\cdot \otimes \cdot)$$

is called a tensor structure on F if the following diagram is commutative:

$$\begin{array}{ccccc}
(F(X) \otimes F(Y)) \otimes F(Z) & \xrightarrow{J_{X,Y} \otimes 1} & F(X \otimes Y) \otimes F(Z) & \xrightarrow{J_{X \otimes Y, Z}} & F((X \otimes Y) \otimes Z) \\
\downarrow{\scriptstyle \Phi^{\mathcal{C}'}} & & & & \downarrow{\scriptstyle F(\Phi^{\mathcal{C}})} \\
F(X) \otimes (F(Y) \otimes F(Z)) & \xrightarrow{1 \otimes J_{Y,Z}} & F(X) \otimes F(Y \otimes Z) & \xrightarrow{J_{X, Y \otimes Z}} & F(X \otimes (Y \otimes Z))
\end{array} \tag{13.5}$$

A functor F equipped with a tensor structure is called a tensor functor.

Recall that a functor $L : A \to B$ is an *equivalence* of categories if there exists a functor $L' : B \to A$ such that $L'L \simeq \mathrm{Id} : A \to A$ and $LL' \simeq \mathrm{Id} : B \to B$. One can show (c.f. [Mac]) that L is an equivalence if and only if it is *essentially surjective* (i.e $\forall\, b \in \mathcal{O}b(B)$, there exists $a \in \mathcal{O}b(A)$ such that $L(a) \simeq b$), and *fully faithful* (i.e $L : \mathrm{Mor}_A(X,Y) \xrightarrow{\sim} \mathrm{Mor}_B(L(X), L(Y))$).

Definition: A tensor functor $F : \mathcal{C} \to \mathcal{C}'$ is said to be an *equivalence of semigroup categories* if F is an equivalence of categories.

Example 13.4 (Forgetful functor). Let A be a bialgebra and Rep $_A$ the (semigroup) category of left A-modules. Let $F_A : \mathcal{C}_A \to \mathcal{V}ec$ be the forgetful functor. Then F is a tensor functor with $J = 1$.

Example 13.5 (Finite groups). Let G be a finite group, $\mathcal{F}(G)$ the algebra of functions on G and $\mathcal{C}$ the category of one-dimensional $\mathcal{F}(G)$-modules. To each $\Phi \in Z^3(G, \mathbb{C}^*)$ we associate a semigroup structure on $\mathcal{C}$ denoted by $\mathcal{C}_\Phi$. Let us determine which $\mathcal{C}_\Phi$ are tensor equivalent:

Let us first see for which $\Phi, \Phi' \in Z^3(G, \mathbb{C}^*)$ the trivial (identity) functor $F : \mathcal{C}_\Phi \to \mathcal{C}_{\Phi'}$ has a tensor structure. By definition, this is the case if and only if there exists

$$J_{X,Y} : F(X) \otimes F(Y) \to F(X \otimes Y)$$

satisfying (13.5), or in other terms $J : G \times G \to \mathbb{C}^*$ such that

$$\Phi(X,Y,Z)J(Y,Z)J(X,YZ) = J(X,Y)J(XY,Z)\Phi'(X,Y,Z)$$

which we rewrite as

$$\begin{aligned}
\Phi/\Phi'(X,Y,Z) &= J(X,Y)J(X,YZ)^{-1}J(XY,Z)J(Y,Z)^{-1} \\
&= dJ
\end{aligned}$$

where d is the differential in the cochain complex of G.

Hence F has a tensor structure if and only if $\Phi/\Phi' \in B^3(G, \mathbb{C}^*)$ (coboundaries) and the set of all such structures is identified with the set of $J : G\times G \to \mathbb{C}^*$ such that $dJ = \Phi/\Phi'$. It is clear that F is then an equivalence of tensor categories.

Now let (F, J) be an equivalence of tensor categories $\mathcal{C}_\Phi \to \mathcal{C}_{\Phi'}$. It is clear that $F(X \otimes Y) = F(X) \otimes F(Y)$, i.e the action of F on objects is given by an automorphism $F \in \text{Aut}\,(G)$. Composing (F, J) with $(Id, J^{-1}) : \mathcal{C}_\Phi \tilde{\to} \mathcal{C}_{\Phi dJ}$, we can assume that $J = 1$. But then $\Phi' = F.\Phi$. As a conclusion, we see that $\mathcal{C}_\Phi$ and $\mathcal{C}_{\Phi'}$ are tensor equivalent if and only if $\Phi = \Phi'$ in $H^3(G, \mathbb{C}^*)/\text{Aut}\,(G)$.

13.2 Monoidal categories

13.2.1 Units in semigroup categories

Definition: Let $\mathcal{C}$ be a semigroup category. An object $c \in \mathcal{O}b(\mathcal{C})$ is called a *unit* if it is equipped with an isomorphism

$$u : c \tilde{\to} c \otimes c$$

making the following diagram commutative

$$\begin{array}{ccc} c & \xrightarrow{u} & c\otimes c \\ {\scriptstyle u}\downarrow & & \downarrow{\scriptstyle 1\otimes u} \\ c\otimes c & \xrightarrow{\Phi\circ(u\otimes 1)} & c\otimes(c\otimes c) \end{array} \tag{13.6}$$

and if the the functors $\mathcal{C} \to \mathcal{C}$

$$\begin{aligned} \cdot \otimes c : X &\to X \otimes c, \\ c \otimes \cdot : X &\to c \otimes X, \end{aligned}$$

are (auto)equivalences of categories.

Lemma 13.1. *Let $\mathcal{C}$ be a semigroup category and c a unit of $\mathcal{C}$. Then there exist unique, functorial isomorphisms*

$$\begin{aligned} r_X &: X \tilde{\to} X \otimes c, \\ l_X &: X \tilde{\to} c \otimes X, \end{aligned}$$

such that $r_c = l_c = u$ and

$$\Phi_{X,c,c}(r_X \otimes 1_c) = 1_X \otimes u, \tag{13.7}$$

$$\Phi^{-1}_{c,c,X}(1_c \otimes l_X) = u \otimes 1_X. \tag{13.8}$$

Proof: Uniqueness is clear from (13.7). To prove existence, set

$$\begin{aligned}\tilde{r}_{X\otimes c} &= \Phi^{-1}_{X,c,c}(1_X \otimes u),\\ \tilde{l}_{c\otimes X} &= \Phi_{c,c,X}(u \otimes 1_X),\end{aligned}$$

and

$$\begin{aligned}r_X &= (\cdot \otimes c)^{-1}(\tilde{r}_{X\otimes c}),\\ l_X &= (c \otimes \cdot)^{-1}(\tilde{l}_{c\otimes X}).\end{aligned}$$

Using (13.6), it is easy to check (13.7).

□

Corollary 13.1. *For any $X, Y \in \mathcal{O}b(\mathcal{C})$, the following diagram is commutative:*

$$\begin{array}{ccc} (X \otimes c) \otimes Y & \xrightarrow{\Phi} & X \otimes (c \otimes Y) \\ & {\scriptstyle r_X^{-1}\otimes 1} \searrow \quad \swarrow {\scriptstyle 1\otimes l_Y^{-1}} & \\ & X \otimes Y & \end{array} \tag{13.9}$$

Definition: A semigroup category is called a *monoidal category* if there exists a unit object $c \in \mathcal{O}b(\mathcal{C})$.

As usual, the unit is (essentially) unique.

Lemma 13.2. *Let $\mathcal{C}$ be a monoidal category. Then the unit object c is unique, up to a unique isomorphism.*

Proof: If c, c' are both units, there is a canonical isomorphism $\tau : c \xrightarrow{\sim} c'$ given by the composition

$$c \xrightarrow{r'_c} c \otimes c' \xrightarrow{l_{c'}} c'$$

where we have used obvious notations. This is the unique isomorphism compatible with u, u', i.e making the following diagram commutative:

$$\begin{array}{ccc} c & \xrightarrow{\tau} & c' \\ {\scriptstyle u}\downarrow & & \downarrow{\scriptstyle u'} \\ c \otimes c & \xrightarrow[\tau\otimes\tau]{} & c' \otimes c' \end{array}$$

□

From now on, we will now denote by c the unit object of $\mathcal{C}$, and identify $c \simeq c \otimes c$ via u, and $X \otimes c \simeq X \simeq c \otimes X$ via r_X and l_X.

We extend the notion of a tensor functor to monoidal categories by requiring, in addition to (13.5), that

$$F(c_{\mathcal{C}}) = c_{\mathcal{C}'}, \tag{13.10}$$

$$J_{X,c} = J_{c,X} = 1_{F(X)}. \tag{13.11}$$

Examples: All semigroup categories discussed in the preceding section are monoidal. For example, if A is a bialgebra over k, then the unit object of Rep A is k with action $\pi_k(a) = \epsilon(a)$.

13.2.2 MacLane's theorem

As we have seen, and as more examples will show, it is very important to allow our monoidal categories to be non-strict. Hence we should a priori take the associativity morphisms into account in all of our constructions, and this can become very cumbersome. However, we have the following result:

Theorem 13.3 (MacLane). *Every monoidal category is tensor equivalent to a strict one.*

We refer the reader to [Mac] or [Kass] for a proof. What this theorem means in practice is that we can forget about brackets for all our general categorical constructions. However, in order to make the category strict, it may be necessary to add new objects to the category (of course, isomorphic to already existing objects). For instance, as we saw in example 13.3, this is necessary for $\mathcal{C}_\Phi$ if $[\Phi] \in H^3(G, \mathbb{C}^*)$ is nonzero.

13.3 Quasi-bialgebras and quasi-Hopf algebras

The category of modules over a bialgebra is a strict monoidal category. We now define a more general algebraic structure corresponding to arbitrary (non strict) monoidal categories.

13.3.1 Definition

Definition: Let A be an associative algebra, and let $\Delta : A \to A \otimes A$, $\epsilon : A \to k$ be morphisms of algebra such that

$$(\epsilon \otimes 1)\Delta = (1 \otimes \epsilon)\Delta = 1.$$

Let $\Phi \in A \otimes A \otimes A$ be an invertible element. Then A is called a *quasi-bialgebra* if

1. Φ satisfies the pentagon relation:

$$\Phi_{1,2,34}\Phi_{12,3,4} = \Phi_{2,3,4}\Phi_{1,23,4}\Phi_{1,2,3}, \tag{13.12}$$

2. Δ is quasi-coassociative

$$\Phi(\Delta \otimes 1)\Delta(x)\Phi^{-1} = (1 \otimes \Delta)\Delta(x), \tag{13.13}$$

3. Φ satisfies the counit axiom

$$(1 \otimes \epsilon \otimes 1)(\Phi) = 1. \tag{13.14}$$

The element $\Phi \in A \otimes A \otimes A$ is called an associator.

Proposition 13.1. *Let $(A, \Delta, \epsilon, \Phi)$ be a quasi-bialgebra. Then the category* Rep A *of A-modules equipped with the tensor product*

$$\pi_{V \otimes W}(a) = (\pi_V \otimes \pi_W)(\Delta(a)), \tag{13.15}$$

the unit object $c = k$ with $\pi_k(a) = \epsilon(a)$, and the associativity isomorphisms

$$\Phi_{X,Y,Z} : (X \otimes Y) \otimes Z \to X \otimes (Y \otimes Z), \tag{13.16}$$

$$x \otimes y \otimes z \to \Phi . x \otimes y \otimes z \tag{13.17}$$

is a monoidal category.

Proof: The pentagon relation (13.3) follows from (13.12); the fact that Φ is A-linear follows from (13.13) and the unit axioms follow from $(1 \otimes \epsilon \otimes 1)\Phi = 1$. Details are left to the reader.

□

A quasi-Hopf algebra is a quasi-bialgebra with an antipode. More precisely,

Definition: A *quasi-Hopf algebra* is a quasi-bialgebra $(A, \Delta, \epsilon, \Phi)$ equipped with an antihomomorphism $S : A \to A$ and elements α, $\beta \in A$ such that

$$m(S \otimes \alpha)\Delta(x) = \epsilon(x)\alpha, \qquad m(1 \otimes \beta S)\Delta(x) = \epsilon(x)\beta, \qquad \forall x \in A,$$

$$m(S \otimes \alpha \otimes \beta S)\Phi = 1, \qquad m(1 \otimes \beta S \otimes \alpha)(\Phi^{-1}) = 1.$$

Proposition 13.1 admits the following converse.

Proposition 13.2. *Let A be an associative algebra and let $\Delta : A \to A \otimes A$, $\epsilon : A \to k$ be morphisms of algebras satisfying $(\epsilon \otimes 1)\Delta = (1 \otimes \epsilon)\Delta = 1$. Consider the category* Rep A *of A-modules with tensor product and unit as in Proposition 13.1. Any monoidal structure on* Rep A *defines a quasi-bialgebra structure on A, via*

$$\Phi = \tilde{\Phi}_{A,A,A}(1 \otimes 1 \otimes 1)$$

where $\tilde{\Phi}$ denotes the associativity isomorphisms of Rep A.

Proof: Clear.

□

13.3.2 Equivalence of quasi-bialgebras and twists

To the notion of (tensor) equivalence of monoidal categories there corresponds the following notion of equivalence of quasi-bialgebras:

Definition: An equivalence of quasi-bialgebras between two quasi-bialgebras $(A, \Delta, \epsilon, \Phi)$ and $(A', \Delta', \epsilon', \Phi')$ is a pair (θ, J) where $\theta : A \xrightarrow{\sim} A'$ is an algebra isomorphism and $J \in A' \otimes A'$ is an invertible element such that

$$\Delta' = J^{-1}((\theta \otimes \theta)\Delta)J, \tag{13.18}$$

$$\Phi' = J_{2,3}^{-1} J_{1,23}^{-1} \Phi J_{12,3} J_{1,2}, \tag{13.19}$$

$$(\epsilon \otimes 1)J = (1 \otimes \epsilon)J = 1. \tag{13.20}$$

The element J is called a *twist.*

Proposition 13.3. *If two quasi-bialgebras are equivalent then the associated monoidal categories are tensor equivalent.*

Proof: Let $(A, \Delta, \epsilon, \Phi)$ and $(A', \Delta', \epsilon', \Phi')$ be as above and (θ, J) an equivalence of quasi-bialgebras. Then θ induces an equivalence of categories

$$F : \operatorname{Rep} A \xrightarrow{\sim} \operatorname{Rep} A'$$

and

$$J_{X,Y} : \begin{array}{rl} F(X) \otimes F(Y) & \to F(X \otimes Y), \\ x \otimes y & \to J.x \otimes y \end{array} \tag{13.21}$$

is an isomorphism of functors. Moreover, condition (13.5) is equivalent to (13.19) and (13.11) follows from (13.20). Thus $\operatorname{Rep} A$ and $\operatorname{Rep} A'$ are equivalent as monoidal categories.

□

Remark: This gives us an interpretation of the twisting procedure described in Lecture 9. In particular, twisting is much more natural in the framework of quasi-bialgebras rather than bialgebras. Notice that condition (3) is required in Lecture 9 to stay within the category of bialgebras (i.e $\Phi = 1$).

13.3.3 "Nonabelian cohomology"

Let A be an associative k-algebra, $\Delta : A \to A \otimes A$ and $\epsilon : A \to k$ be algebra homomorphisms satisfying $(\epsilon \otimes 1)\Delta = (1 \otimes \epsilon)\Delta = 1$.

Introduce the group $\overline{\Gamma}_A$ of invertible elements J of $A \otimes A$ satisfying:

$$(\epsilon \otimes 1)J = (1 \otimes \epsilon)J = 1, \qquad \Delta(x)J = J\Delta(x),\ x \in A.$$

Then we have a right action of $\overline{\Gamma}_A$ by (13.18), (13.19) on the set of quasi-bialgebra structures on A with counit ϵ and comultiplication Δ, and equivalence classes are $\overline{\Gamma}_A$-orbits (we only consider equivalences which are trivial at the level of the algebra, i.e $\theta = 1$ in the notations of section 13.3.2.)

In particular, let us introduce the following "nonabelian cohomology group": the "cycles" are

$$\begin{aligned} Z^3_{nonab} = \{\Phi \in A^{\otimes 3} \mid\ & \Phi \text{ is invertible, } (1 \otimes \epsilon \otimes 1)\Phi = 1, \\ & \Phi \text{ satisfies the pentagon relation,} \\ & \Phi(\Delta \otimes 1)\Delta(x) = (1 \otimes \Delta)\Delta(x)\Phi\} \end{aligned}$$

and two cycles Φ, Φ' are cohomologous (denoted $\Phi \sim \Phi'$) if there exists $J \in \overline{\Gamma}_A$ with

$$\Phi' = J^{-1}_{2,3} J^{-1}_{1,23} \Phi J_{12,3} J_{1,2}.$$

We set $H^3_{nonab}(A, \Delta) = Z^3_{nonab}/\sim$. In particular, if Δ is coassociative, we denote by $0 \in H^3_{nonab}(A)$ the class of $\Phi = 1$. With this definition, a quasi-bialgebra $(A, \Delta, \epsilon, \Phi)$ is equivalent to a bialgebra (with $\Phi = 1$) if and only if $[\Phi] = 0 \in H^3_{nonab}(A, \Delta)$.

Remark: For $A = \mathcal{F}(G)$, we have $H^3_{nonab}(A, \Delta) = H^3(G, \mathbb{C}^*)$, as we have already seen.

More generally, one can consider the set of all quasi-bialgebra structures $(A, \Delta, \epsilon, \Phi)$ on an associative algebra equipped with a homomorphism $\epsilon : A \to k$. In this case, the action is enlarged to the group Γ_A of invertible elements of $A \otimes A$ satisfying

$$(\epsilon \otimes 1)J = (1 \otimes \epsilon)J = 1.$$

The corresponding notion of "nonabelian cohomology" is the following: cycles Z^3_{nonab} are now pairs (Φ, Δ) satisfying

$$(\epsilon \otimes 1)\Delta = (1 \otimes \epsilon)\Delta = 1, \qquad (1 \otimes \epsilon \otimes 1)(\Phi) = 1,$$

$$\begin{aligned} \Phi_{1,2,34}\Phi_{12,3,4} &= \Phi_{2,3,4}\Phi_{1,23,4}\Phi_{1,2,3}, \\ \Phi(\Delta \otimes 1)\Delta(x)\Phi^{-1} &= (1 \otimes \Delta)\Delta(x). \end{aligned}$$

Two cycles (Φ_1, Δ_1) and (Φ_2, Δ_2) are cohomologous if there exists $J \in \Gamma_A$ such that

$$\Phi_2 = J^{-1}_{2,3} J^{-1}_{1,23} \Phi_1 J_{12,3} J_{1,2},$$

$$\Delta_2 = J^{-1}\Delta_1 J.$$

We set $H^3_{nonab}(A) = Z^3_{nonab}/\sim$, and equivalence classes of quasi-bialgebras structures on (A, ϵ) are now identified with elements of $H^3_{nonab}(A)$.

We can now give an illustration of MacLane's theorem: let $\varphi : \tilde{G} \to G$ be a surjective group homomorphism between two groups, and denote by $\varphi^* : \mathcal{F}(G) \hookrightarrow \mathcal{F}(\tilde{G})$ the induced homomorphism between function algebras. Let $\tilde{\mathcal{C}}$ be the category of 1-dimensional $\mathcal{F}(\tilde{G})$-modules with $\mathrm{Hom}_{\tilde{\mathcal{C}}}(X,Y) = \mathrm{Hom}_{\mathcal{F}(G)}(X,Y)$ where we consider X, Y as $\mathcal{F}(G)$-modules via φ^*. Then $\tilde{\mathcal{C}} = \tilde{G}$ but $x, y \in \tilde{G}$ are isomorphic if $\varphi(x) = \varphi(y)$.

Let us now choose a function $f : \tilde{G} \times \tilde{G} \to \mathbb{C}^*$ and define a tensor product on $\tilde{\mathcal{C}}$ by defining it on objects to be the multiplication in $\tilde{G}$ and on morphisms by $a \otimes b = abf(g,h)f(g',h')^{-1}$, for $a : g \to g'$ and $b : h \to h'$. Denote the obtained category with tensor product by $\tilde{G}_f$.

Now let $\Phi \in Z^3(G, \mathbb{C}^*)$ and set $\tilde{\Phi} = \varphi^*(\Phi) \in Z^3(\tilde{G}, \mathbb{C}^*)$. Assume that $\tilde{\Phi} = df$. Then the identity map 1 satisfies the requirements for an associativity map, and the functor $\varphi : \tilde{G} \to G$ with tensor structure f^{-1} is a tensor equivalence between $(\tilde{G}_f, 1)$ and (G, Φ).

If $[\Phi] \neq 0$ in $H^3(G, \mathbb{C}^*)$ then (G, Φ) cannot be twisted into a strict structure. However, $(\tilde{G}_f, 1)$ is a strict category. Thus, Maclane's theorem in this case follows from the fact that for any element $\Phi \in H^3(G, \mathbb{C}^*)$ there exists $\tilde{G}$ such that the pullback of Φ to $\tilde{G}$ is trivial. (For example, one may take $\tilde{G}$ to be a free non-abelian group).

Lecture 14

Braided tensor categories and quasitriangular quasi-Hopf algebras

14.1 Braided monoidal categories

14.1.1 Motivation

The monoidal categories $\mathcal{V}ec$, $\operatorname{Rep} G$ and $\operatorname{Rep} \mathfrak{g}$ (for a group G and a Lie algebra $\mathfrak{g}$) all have an additional structure: there is a functorial isomorphism

$$\beta_{X,Y} : X \otimes Y \tilde{\to} Y \otimes X \tag{14.1}$$

(the trivial map $\beta_{X,Y}(x \otimes y) = y \otimes x$). Notice that we have $\beta_{X,Y}\beta_{Y,X} = 1$.

Now consider the category $\operatorname{Rep} A$ where A is a bialgebra. Let X, Y be two A-modules. By definition, the action of A on $X \otimes Y$ and $Y \otimes X$ is respectively given by

$$\pi_{X \otimes Y}(a) = (\pi_X \otimes \pi_Y)(\Delta(a)),$$
$$\pi_{Y \otimes X}(a) = \sigma \circ (\pi_X \otimes \pi_Y)(\Delta^{op}(a)) \circ \sigma,$$

where we use σ to denote the usual permutation of factors. Hence, to define a functorial A-linear map $\beta_{X,Y} : X \otimes Y \to Y \otimes X$, we need an element $R \in A \otimes A$ satisfying

$$R\Delta(a) = \Delta^{op}(a)R, \qquad \forall\, a \in A.$$

In particular, $\beta_{X,Y}\beta_{Y,X} = 1$ if and only if $RR_{21} = 1$.

We want to define the notion of a tensor category including such functorial commutativity isomorphisms. Recall that in monoidal categories, we have a canonical isomorphism between any two brackettings of a word $V_1 \otimes \ldots \otimes V_n$. When considering commutativity isomorphisms (14.1) one could also require a canonical isomorphism between any two brackettings and reordering of a word $V_1 \otimes \ldots \otimes V_n$:

$$(V_{i_1} \otimes (V_{i_2} \otimes \ldots) \ldots) \otimes V_{i_n}). \tag{14.2}$$

However, this condition in particular implies (and is in fact equivalent to) the commutativity of the diagram

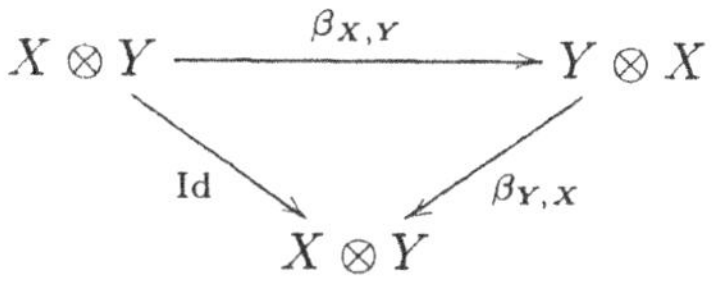

(14.3)

(i.e $\beta_{X,Y}\beta_{X,Y} = 1$) which, as we will see, is too restrictive.
In the next section, we introduce the correct notion.

14.1.2 The braid group

Let $z_1, \ldots z_n$ be n distinct points in $\mathbb{C}$ and set $x_i = (z_i, 0) \in \mathbb{C} \times \mathbb{R}$, $y_i = (z_i, 1) \in \mathbb{C} \times \mathbb{R}$.

Definition: A *braid* in *n strands* is an isotopy class of a union of n disjoint smooth curves (strands) in $\mathbb{C} \times \mathbb{R}$ with endpoints $\{x_1, \ldots x_n, y_1, \ldots y_n\}$ such that each curve intersects any horizontal plane $\mathbb{C} \times \{c\}$ at at most one point:

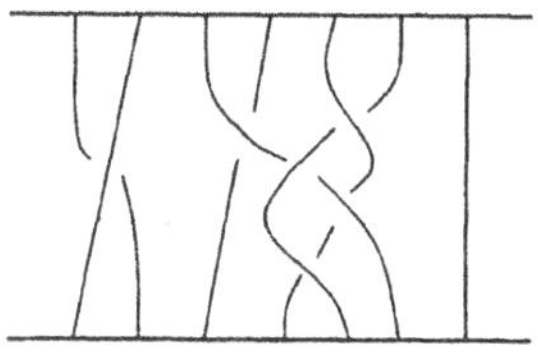

In other words, we require that the curves are "monotonous". The collection of n-braids forms an (infinite) group B_n: multiplication is defined by concatenation of strands:

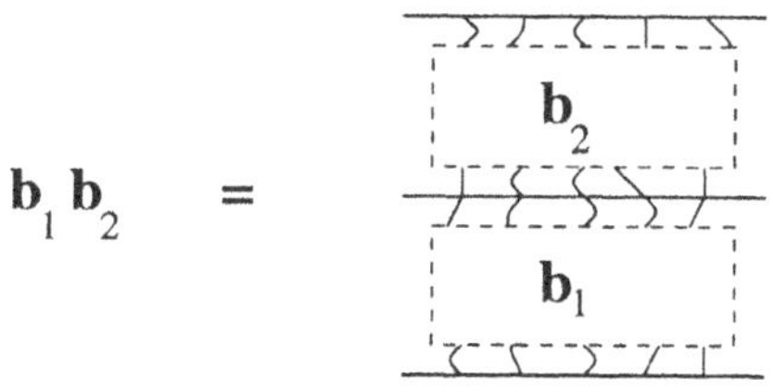

Notice that the curves go from up to down. For example, the unit braid is just

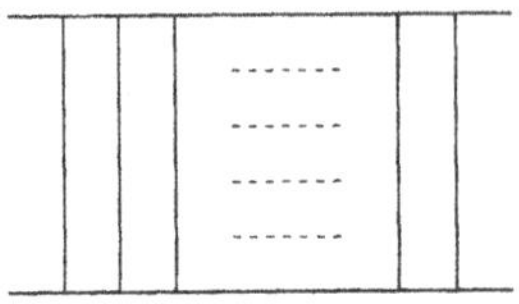

and the inverse of a braid is the symmetric braid with respect to a horizontal plane. It is clear that this group is independent of the choice of $z_1, \dots z_n$ as any two marked planes $(\mathbb{C}, z_1, \dots z_n)$ and $(\mathbb{C}, z'_1, \dots z'_n)$ are homeomorphic. The group B_n is called the *Braid group*.

The following is a well-known result, due to E. Artin:

Theorem 14.1 (Artin). *The Braid group has generators*

$\mathbf{b}_i =$ 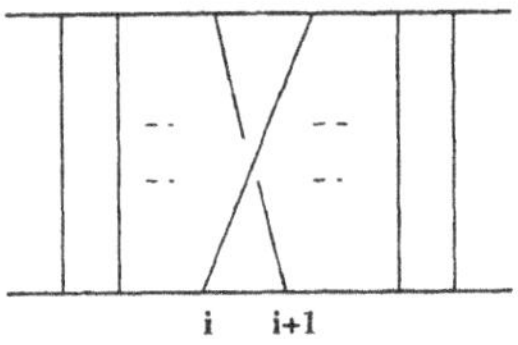

and (braid) relations

$$b_i b_j = b_j b_i \qquad (|i-j| > 1),$$
$$b_i b_{i+1} b_i = b_{i+1} b_i b_{i+1}.$$

Notice the analogy with the symmetric group: S_n is generated by the transpositions $\tau_i : i \leftrightarrow i+1$ with relations

$$\tau_i^2 = 1,$$
$$\tau_i \tau_j = \tau_j \tau_i \qquad (|i-j| > 1),$$
$$\tau_i \tau_{i+1} \tau_i = \tau_{i+1} \tau_i \tau_{i+1}.$$

In particular, there is a projection $\pi : B_n \twoheadrightarrow S_n$, $b_i \mapsto \tau_i$.

Definition: The normal subgroup $PB_n = \operatorname{Ker} \pi$ is called the *pure Braid group*.

The Braid group is a fundamental object in geometry for the following reason: consider the "configuration space" $X = \mathbb{C}^n \backslash \{z_i = z_j\}$ of n distinct points on $\mathbb{C}$. The symmetric group S_n acts naturally on X by permutation of the components.

Proposition 14.1. *We have*

$$\pi_1(X/S_n) = B_n.$$

Proof: Let $x = (z_1, \dots z_n) \in X/S_n$. A loop with base point x can be represented as a braid in n strands connecting the set of points $\{y_i = (z_i, 1)\}$ to the set of points $\{x_i = (z_i, 0)\}$ in $\mathbb{C} \times \mathbb{R}$. Moreover, two such loops are homotopic if and only if the corresponding braids are isotopic.

□

Exercise 14.1. Show that $\pi_1(X) = PB_n$.

Now let C be a monoidal category equipped with an isomorphism

$$\beta : \cdot \otimes \cdot \to \cdot \otimes^{op} \cdot,$$
$$\beta_{X,Y} : X \otimes Y \tilde{\to} Y \otimes X.$$

Let $V_1, \dots V_n \in Ob(C)$ and consider all expressions obtained by bracketing and reordering $V_1 \otimes \dots \otimes V_n$ as in (14.2). We assign to every isomorphism composed of Φ's, β's and their inverses an element of B_n in the following way: to an associativity morphism Φ we assign 1; to $\beta_{V_{i_k}, V_{i_{k+1}}}$ we assign b_k; more generally, to

$$\beta_{A,B} : \dots \otimes (V_{i_a} \dots V_{i_b}) \otimes (V_{i_{b+1}} \dots V_{i_c}) \otimes \dots \tilde{\to} \dots \otimes (V_{i_{b+1}} \dots V_{i_c}) \otimes (V_{i_a} \dots V_{i_b}) \otimes \dots$$

we assign the braid

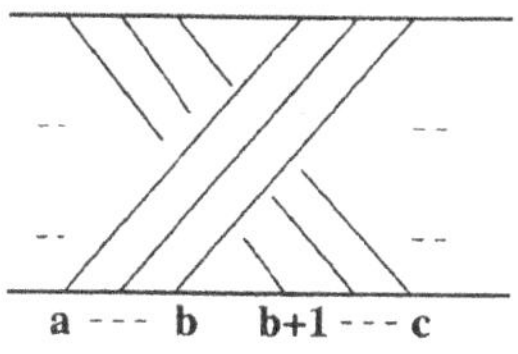

Finally, to the composition of morphisms we associate the product of the corresponding braids.

Here is the condition that we require:
Consistency condition. For any two expressions A and B of the form (14.2), a morphism $\varphi : A \tilde{\to} B$ composed of Φ's and β's only depends on its image in B_n.

For instance, the above condition implies that the following diagram is commutative:

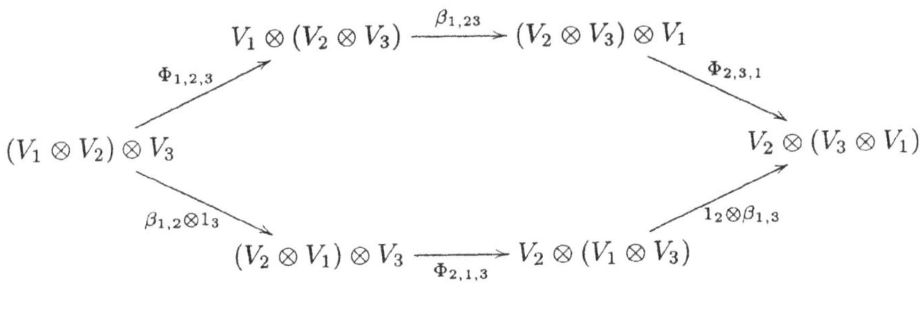

as both compositions correspond to the braid

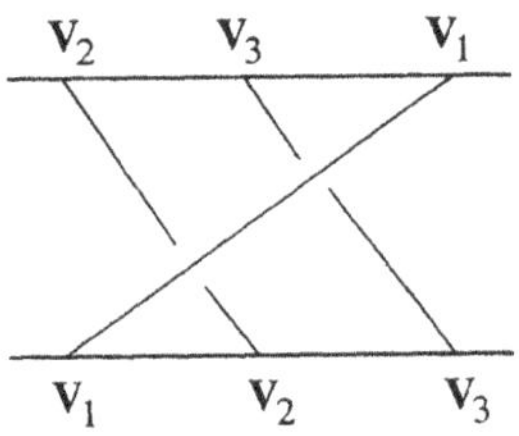

(14.4)

This is called the *hexagon* axiom.

Conversely, it turns out that (14.4) (and its mirror image) for every triple $V_1, V_2, V_3 \in \mathcal{O}b(\mathcal{C})$ is enough to guarantee the consistency condition. We thus make the following definition:

Definition: A *braided* monoidal category is a monoidal category $\mathcal{C}$ equipped with an isomorphism of (bi)functors

$$\beta : \cdot \otimes \cdot \to \cdot \otimes^{op} \cdot,$$
$$\beta_{X,Y} : X \otimes Y \xrightarrow{\sim} Y \otimes X$$

such that

1. The hexagon axiom (14.4) is satisfied,
2. The hexagon axiom with β^{-1} instead of β is satisfied,
3. $\beta_{c,X} = \beta_{X,c} = 1$.

The functorial isomorphism $\beta_{X,Y}$ is called the *commutativity isomorphism* or the *braiding*.

Theorem 14.2 (Braided analog of MacLane coherence theorem). *The consistency condition is satisfied in a braided monoidal category.*

14.1.3 Braided tensor functors

A *braided tensor functor* $(F, J) : \mathcal{C} \to \mathcal{C}'$ between two braided monoidal categories is a tensor functor between the underlying monoidal categories making the following diagram commutative:

$$\begin{array}{ccc} F(X \otimes Y) & \xrightarrow{J^{-1}_{X,Y}} & F(X) \otimes F(Y) \\ {\scriptstyle F(\beta_{X,Y})}\downarrow & & \downarrow{\scriptstyle \beta'_{F(X),F(Y)}} \\ F(Y \otimes X) & \xrightarrow{J^{-1}_{Y,X}} & F(Y) \otimes F(X) \end{array} \tag{14.5}$$

where β, β' are the braidings in $\mathcal{C}, \mathcal{C}'$.

14.1.4 Braid group representations

Our previous discussion shows that for any braided monoidal category $\mathcal{C}$ and any $V \in \mathcal{O}b(\mathcal{C})$, we have a representation of $B_n \to \mathrm{Aut}\,(V^{\otimes n})$. More generally, for any $V_1, \ldots V_n \in \mathcal{O}b(\mathcal{C})$ we have a representation of B_n in

$$\bigoplus_{\gamma \in S_n} V_{\gamma(1)} \otimes \ldots \otimes V_{\gamma(n)}. \tag{14.6}$$

14.1.5 Symmetric categories

Definition: A braided monoidal category is called symmetric if $\beta_{X,Y}\beta_{Y,X} = 1$ for all $X, Y \in \mathcal{O}b(\mathcal{C})$.

In this case, it is easy to see that the representations (14.6) of the Braid group factor through S_n.

Examples: The following monoidal categories are symmetric: $\mathcal{V}ec$, $\mathrm{Rep}\, G$, $\mathrm{Rep}\, \mathfrak{g}$. In general, if A is a cocommutative Hopf algebra then $\mathrm{Rep}\, A$ is symmetric.

14.2 Quasitriangular Quasi-Hopf algebras

We can now give the categorical interpretation of (quasi)triangular structures for Hopf algebras:

Proposition 14.2. *Let H be a Hopf algebra and* $\mathrm{Rep}\, H$ *the monoidal category of H-modules. Then* $\mathrm{Rep}\, H$ *is braided if and only if H admits a quasitriangular structure. Moreover,* $\mathrm{Rep}\, H$ *is a symmetric category if and only if H admits a triangular structure.*

Proof: Let R be a quasitriangular structure of H. Set

$$\beta_{X,Y} : X \otimes Y \stackrel{\sim}{\to} Y \otimes X, \qquad (14.7)$$
$$x \otimes y \to \sigma \circ R.(x \otimes y).$$

Then $\beta_{X,Y}$ is H-linear by $R\Delta = \Delta^{op}R$. Furthermore, the hexagon axioms (14.4) follow from the hexagon relations

$$(1 \otimes \Delta)R = R_{13}R_{12}, \qquad (14.8)$$
$$(\Delta \otimes 1)R = R_{13}R_{23}. \qquad (14.9)$$

Applying $1 \otimes \epsilon \otimes 1$ to (14.8) yields $(\epsilon \otimes 1)R = (1 \otimes \epsilon)R = 1$, i.e $\beta_{c,X} = \beta_{X,c} = 1$. Thus $\operatorname{Rep} H$ is a braided monoidal category.

The converse is proved along the same lines (set $R = \sigma \circ \beta_{H,H}(1 \otimes 1)$). Finally, $\operatorname{Rep} H$ is symmetric if and only $RR_{21} = 1$, i.e if and only if R is a triangular structure.

□

The following definition now becomes quite natural:

Definition: A *quasitriangular* quasi-Hopf algebra is a pair (H, R), where H is a quasi-Hopf algebra and $R \in H \otimes H$ is invertible and such that

$$R\Delta = \Delta^{op}R, \qquad (14.10)$$

$$(1 \otimes \Delta)R = \Phi_{231}^{-1}R_{13}\Phi_{213}R_{12}\Phi_{123}^{-1},$$
$$(\Delta \otimes 1)R = \Phi_{312}R_{13}\Phi_{132}^{-1}R_{23}\Phi_{123},$$

where Φ is the associator of H. We say that R is a quasitriangular structure, or R-matrix for H. We say that R is *triangular* if $RR_{21} = 1$.

Of course, forgetting the antipode, we can define in a similar way the notion of a *quasitriangular quasi-bialgebra*.

Proposition 14.3. *Let (H, R) be a quasitriangular quasi-bialgebra. The monoidal category* $\operatorname{Rep} H$ *equipped with the commutativity isomorphism (14.7) is braided.*

Proof: left to the reader.

□

Problem 14.1. Prove that the R-matrix of a quasitriangular quasi-Hopf algebra satisfies the following version of the quantum Yang-Baxter equation:

$$R_{12}\Phi_{312}R_{13}\Phi_{132}^{-1}R_{23}\Phi_{123} = \Phi_{321}R_{23}\Phi_{231}^{-1}R_{13}\Phi_{213}R_{12}.$$

14.2.1 Equivalence of quasitriangular quasi-Hopf algebras

The notion of a twist naturally extends to the quasitriangular case:

Proposition 14.4. *Let* $(A, \Delta, \epsilon, \Phi, R)$ *be a quasitriangular quasi-bialgebra and let* $J \in \Gamma_A$ *be a twist. Set*

$$\Delta' = J^{-1}\Delta J, \tag{14.11}$$

$$\Phi' = J_{2,3}^{-1} J_{1,23}^{-1} \Phi J_{12,3} J_{1,2}, \tag{14.12}$$

$$R' = (J_{21})^{-1} R J. \tag{14.13}$$

Then $(A, \Delta', \epsilon, \Phi', R')$ *is a quasitriangular quasi-bialgebra.*

Proof: Equation (14.13) comes from (14.5).

Definition: The quasitriangular quasi-bialgebra $(A, \Delta', \epsilon, \Phi', R')$ is called the *twist* of $(A, \Delta, \epsilon, \Phi, R)$ by J.

Lecture 15

Knizhnik-Zamolodchikov equations and the Drinfeld category

In this lecture we give the fundamental example, due to Drinfeld, of a non-trivial quasitriangular quasi-Hopf algebra (i.e which is not a Hopf algebra), defined via monodromy and asymptotics of the *Knizhnik-Zamolodchikov* system of differential equations. We take $k = \mathbb{C}$ here.

15.1 The KZ equations:

15.1.1 Definition

Let $\mathfrak{g}$ be a finite dimensional Lie algebra, and $\Omega \in S^2\mathfrak{g} \subset U(\mathfrak{g}) \otimes U(\mathfrak{g})$ an invariant symmetric element. We will consider multivalued functions $\mathbb{C}^n \to U(\mathfrak{g})^{\otimes n}[[h]]$ which may have singularities and which take values in a finite dimensional subspace of $U(\mathfrak{g})^{\otimes n}$ modulo every power of h. For $n \in \mathbb{N}$, consider the following system of differential equations with regular singularities for a function $u : \mathbb{C}^n \to U(\mathfrak{g})^{\otimes n}[[h]]$

$$\frac{\partial u}{\partial z_i} = \hbar \sum_{i \neq j} \frac{\Omega_{ij}}{z_i - z_j}.u \qquad (i = 1 \ldots n) \qquad (\mathrm{KZ}_n)$$

where we set $\hbar = \frac{h}{2i\pi}$ and, as usual, the notation Ω_{ij} means that Ω acts in the ith and j^{th} components of the tensor product $U(\mathfrak{g})^{\otimes n}[[h]]$. These equations are called the *Knizhnik-Zamolodchikov equations.*

Proposition 15.1. *The system* (KZ_n) *is consistent and* $\mathfrak{g}$*-invariant, i.e we have*

$$[\frac{\partial}{\partial z_i} - \hbar \sum_{i \neq i} \frac{\Omega_{ij}}{z_i - z_j}, \frac{\partial}{\partial z_k} - \hbar \sum_{l \neq k} \frac{\Omega_{kl}}{z_k - z_l}] = 0$$

and if u is a solution and $g \in \mathfrak{g}$, then $\Delta^n(g).u$ is also a solution.

Proof: Straightforward computation, using the $\mathfrak{g}$-invariance of Ω and the relation

$$[\Omega_{ij}, \Omega_{ik} + \Omega_{jk}] = 0.$$

□

In other words, the system (KZ_n) defines a flat connection on the trivial bundle $U(\mathfrak{g})^{\otimes n}[[h]]$ over the *configuration space* $X_n = \mathbb{C}^n \backslash \{z_i = z_j\}$, with regular singularities along the hyperplanes $z_i = z_j$. It is clear that this system is invariant under the action of S_n by permutation of the components, and therefore induces a flat connection on the trivial bundle $U(\mathfrak{g})^{\otimes n}[[h]]$ over the space X_n/S_n.

Alternatively, we can consider (KZ_n) as a system of differential equations for a function

$$u : \mathbb{C}^n \to V_1 \otimes \ldots \otimes V_n[[h]]$$

for any $U(\mathfrak{g})$-modules $V_1, \ldots V_n$, and we obtain a flat connection on the trivial bundle $V_1 \otimes \ldots \otimes V_n[[h]]$ over X_n.

The Knizhnik-Zamolodchikov equations play a fundamental role in many areas of mathematics, such as Conformal Field theory and the Wess-Zumino-Witten model or representation theory of affine Lie algebras.

15.1.2 Link with the CYBE

The observant reader may have recognized Yang's r-matrix $r(z) = \frac{\Omega}{z}$ in the definition of the (KZ_n) equations. This is more than a coincidence: following Cherednik [Che2], consider systems of differential equations of the form

$$\frac{\partial u}{\partial z_i} = \sum_{i<j} r_{ij}(z_i - z_j).u - \sum_{i>j} r_{ji}(z_j - z_i).u \qquad (i = 1 \ldots n) \qquad (KZ_n^r)$$

where $r : \mathbb{C} \to \mathfrak{g} \otimes \mathfrak{g}$ is a meromorphic function.

Proposition 15.2. *The system* KZ_n^r *is consistent if and only if* $r(z)$ *satisfies the CYBE with spectral parameters:*

$$[r_{12}(z_1-z_2), r_{13}(z_1-z_3)]+[r_{12}(z_1-z_2), r_{23}(z_2-z_3)]+[r_{13}(z_1-z_3), r_{23}(z_2-z_3)] = 0.$$

Proof: Straightforward computation (left to the reader).

In particular, we can consider the trigonometric KZ equations defined by the r-matrix (6.6), or the elliptic KZ equations defined by the r-matrix (6.7) (for $\mathfrak{g} = sl_n$). See [EFK] and [E] for more details.

15.2 Monodromy of the KZ equations

The aim of this section is to explain how to use the KZ equations to build a nontrivial quasitriangular quasi-Hopf algebra structure on $U(\mathfrak{g})[[h]]$.

15.2.1 The KZ associator

To turn $U(\mathfrak{g})[[h]]$ (with its usual Hopf algebra structure) into a quasi-Hopf algebra, we must construct an associator $\Phi \in U(\mathfrak{g})^{\otimes 3}[[h]]$ satisfying relations (13.13-13.14) and the pentagon relation (13.12).

Consider the system (KZ_3) and let us set $z_1 = 0$, $z_2 = z$, $z_3 = 1$. We are left with

$$\frac{1}{\hbar}\frac{du}{dz} = (\frac{\Omega_{12}}{z} + \frac{\Omega_{23}}{z-1})u. \tag{KZ'$_3$}$$

We let Γ be the set of solutions of (KZ'_3) over $I = (0, 1)$.

Proposition 15.3. *There exists a unique solution of* (KZ'_3) *in* Γ *of the form*

$$F_0(z) = (1 + \sum_{n\geq 1} z^n F_0^{(n)})z^{\hbar\Omega_{12}}. \tag{15.1}$$

where $F_0^{(n)} \in U(\mathfrak{g})^{\otimes 3}$ *and where* $z^{\hbar\Omega_{12}}$ *stands for the series* $\exp(\hbar\Omega_{12}\log(z))$.

Proof: Substituting (15.1) into (KZ'_3) yields

$$\begin{aligned}&\frac{1}{\hbar}(\sum_{n=1}^{\infty} nz^{n-1}F_0^{(n)})z^{\hbar\Omega_{12}} + (1+\sum_{n=1}^{\infty} z^n F_0^{(n)})\Omega_{12}\frac{z^{\hbar\Omega_{12}}}{z}\\ =&(\frac{\Omega_{12}}{z} + \frac{\Omega_{23}}{z-1})(1+\sum_{n\geq 1} z^n F_0^{(n)})z^{\hbar\Omega_{12}}.\end{aligned}$$

Equating powers of z, we get the recurrence relation:

$$nF_0^{(n)} - \hbar[\Omega_{12}, F_0^{(n)}] = -\hbar\Omega_{23}(1 + F_0^{(1)} + \ldots + F_0^{(n-1)}). \tag{15.2}$$

Since $n\mathrm{Id} - \hbar\mathrm{ad}(\Omega_{12})$ is invertible, this uniquely defines the coefficients $F_0^{(n)}$, $n \geq 1$. The convergence of $1 + \sum_n F_0^{(n)} z^n$ follows from the general principle that a formal solution to a differential equation with regular singularities is in fact analytic (see [CL]).□

Similarly, there exists a unique solution of (KZ'_3) of the form

$$F_1(z) = (1 + \sum_{n\geq 1}(1-z)^n F_1^{(n)})(1-z)^{\hbar\Omega_{23}}. \tag{15.3}$$

Definition: The *KZ associator* $\Phi_{KZ} \in U(\mathfrak{g})^{\otimes 3}[[h]]$ is defined by the formula

$$F_0(z) = F_1(z)\Phi_{KZ}. \tag{15.4}$$

It is clear that Φ_{KZ} is constant (since $F_0(z)$, $F_1(z)$ are both solutions of (KZ'_3)) and $\mathfrak{g}$-invariant (since (KZ'_3) is $\mathfrak{g}$-invariant).

Remark: Notice that the KZ_n system also makes sense for a function $u : \mathbb{C}^n \to V_1 \otimes \ldots V_n[[h]]$ where $V_1, \ldots V_n$ are (finite-dimensional) $\mathfrak{g}$-modules. In particular, for $n = 3$, and when Ω acts semisimply on $V_1 \otimes V_2 \otimes V_3$ (in particular if $\mathfrak{g}$ is simple), we can reformulate the above definition as follows:

Definition: Let $v \in V_1 \otimes V_2 \otimes V_3[[h]]$ be an eigenvector of Ω_{12}, i.e $\Omega_{12}.v = kv$, $k \in \mathbb{C}[[h]]$. We say that $\phi \in \Gamma$ has *asymptotics* v around 0 if

$$\phi(z) = z^{\hbar k}(v + r(z))$$

for some vector valued function $r(z)$ analytic and vanishing at 0.

In a similar fashion, we define the notion of asymptotics around 1 of a function $\phi \in \Gamma$, replacing the above equation by

$$\phi(z) = (1 - z)^{\hbar k}(v + r(1 - z)) \tag{15.5}$$

for $v \in V_1 \otimes V_2 \otimes V_3[[h]]$ such that $\Omega_{23}.v = kv$.

Corollary 15.1. *For any eigenvector $v \in V_1 \otimes V_2 \otimes V_3[[h]]$ of Ω_{12}, there exists a unique function $\phi \in \Gamma$ whose asymptotics around 0 is v. Extending this correspondence by linearity, we obtain an isomorphism*

$$\Psi_0 : V_1 \otimes V_2 \otimes V_3[[h]] \tilde{\to} \Gamma.$$

Similarly, considering asymptotics around 1, we have an isomorphism

$$\Psi_1 : V_1 \otimes V_2 \otimes V_3[[h]] \tilde{\to} \Gamma.$$

Composing the two, we obtain an isomorphism

$$\Psi_{KZ} = \Psi_1^{-1}\Psi_0 : V_1 \otimes V_2 \otimes V_3[h]] \tilde{\to} V_1 \otimes V_2 \otimes V_3[[h]].$$

It is given by $\Psi_{KZ} = \Phi_{KZ|V_1 \otimes V_2 \otimes V_3}$.

Proof: It suffices to set $\Psi_0(v) = F_0 v$ and $\Psi_1(v) = F_1 v$.

□

Geometrically, Φ_{KZ} gives the monodromy of KZ_3' from the asymptotic region $0 < z \ll 1$ to the region $0 \ll z < 1$:

One should think of each point as representing an object V_i, $i = 1, 2, 3$: the monodromy thus goes from a position where V_1 and V_2 are very close to each other to a position where V_2 and V_3 very close to each other. We can think of "very close" as "being in the same bracket", and Φ_{KZ} intuitively plays the role of an associator. We spend the next section proving that Φ_{KZ} is indeed an associator.

Problem 15.1. Show that $\Phi_{KZ}^{-1} = \Phi_{KZ}^{321}$.

15.2.2 Quasi-Hopf structure

To check that $\Phi = \Phi_{KZ}$ indeed defines a quasi-Hopf algebra structure on $U(\mathfrak{g})[[h]]$, we have to check the following relations:

$$(1 \otimes \Delta)\Delta = \Phi(\Delta \otimes 1)\Delta\Phi^{-1}, \tag{15.6}$$

$$(1 \otimes \epsilon \otimes 1)\Phi = 1. \tag{15.7}$$

The pentagon relation:

$$\Phi_{1,2,34}\Phi_{12,3,4} = \Phi_{2,3,4}\Phi_{1,23,4}\Phi_{1,2,3}. \tag{15.8}$$

Equation (15.6) is nothing but the $\mathfrak{g}$-invariance of Φ. To see (15.7), note that applying $(1 \otimes \epsilon \otimes 1)$ to (KZ'_3) gives the trivial system

$$\begin{aligned}\frac{1}{\hbar}(1 \otimes \epsilon \otimes 1)\frac{dF}{dz} &= \left(\frac{(1 \otimes \epsilon \otimes 1)\Omega_{12}}{z} + \frac{(1 \otimes \epsilon \otimes 1)\Omega_{23}}{z-1}\right)(1 \otimes \epsilon \otimes 1)F \\ &= 0\end{aligned}$$

since $(1 \otimes \epsilon \otimes 1)\Omega_{12} = (1 \otimes \epsilon \otimes 1)\Omega_{23} = 0$. In particular, $(1 \otimes \epsilon \otimes 1)F_0(z) = (1 \otimes \epsilon \otimes 1)F_1(z) = 1$.

To prove the pentagon relation, consider the system (KZ'_4) obtained from (KZ_4) by setting $z_1 = 0$, $z_4 = 1$. We are left with

$$\begin{aligned}\frac{1}{\hbar}\frac{\partial F}{\partial z} &= \left(\frac{\Omega_{12}}{z} + \frac{\Omega_{23}}{z-u} + \frac{\Omega_{24}}{z-1}\right)F, \\ \frac{1}{\hbar}\frac{\partial F}{\partial u} &= \left(\frac{\Omega_{13}}{u} + \frac{\Omega_{23}}{u-z} + \frac{\Omega_{34}}{u-1}\right)F,\end{aligned} \tag{KZ'$_4$}$$

for a function $F : I^2 \to U(\mathfrak{g})^{\otimes 4}[[h]]$. This system is coherent because (KZ_4) is coherent.

Consider the following asymptotic zones of the simply connected domain $0 < z < u < 1 \subset I^2$:

1. $O_1 : z \ll u \ll 1$

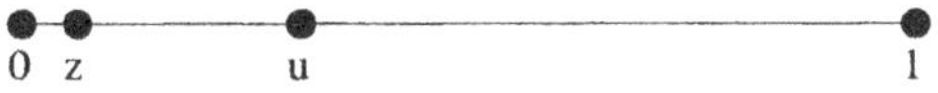

2. $O_2 : u - z \ll u \ll 1$

3. $O_3 : u - z \ll 1 - z \ll 1$

4. $O_4 : 1 - u \ll 1 - z \ll 1$

5. $O_5 : z \ll 1,\ 1 - u \ll 1$

As for the ($\mathrm{KZ'}_3$) equations, we can define the notion of asymptotics of a solution in one of the above zones, and in particular, there are solutions of ($\mathrm{KZ'}_4$) with the following asymptotic behavior:

$$\begin{aligned}
W_1 &\sim z^{\hbar\Omega_{12}} u^{\hbar(\Omega_{13}+\Omega_{23})} && \text{in } O_1,\\
W_2 &\sim (u-z)^{\hbar\Omega_{23}} u^{\hbar(\Omega_{12}+\Omega_{13})} && \text{in } O_2,\\
W_3 &\sim (u-z)^{\hbar\Omega_{23}} (1-z)^{\hbar(\Omega_{24}+\Omega_{34})} && \text{in } O_3,\\
W_4 &\sim (1-u)^{\hbar\Omega_{34}} (1-z)^{\hbar(\Omega_{23}+\Omega_{24})} && \text{in } O_4,\\
W_5 &\sim z^{\hbar\Omega_{12}} (1-u)^{\hbar\Omega_{34}} && \text{in } O_5.
\end{aligned}$$

Lemma 15.1. *We have the following equalities:*

$$W_1(z,u) = W_2(z,u)\Phi_{1,2,3} \tag{15.9}$$

$$W_2(z,u) = W_3(z,u)\Phi_{1,23,4} \tag{15.10}$$

$$W_3(z,u) = W_4(z,u)\Phi_{2,3,4} \tag{15.11}$$

$$W_4(z,u) = W_5(z,u)\Phi_{1,2,34}^{-1} \tag{15.12}$$

$$W_5(z,u) = W_1(z,u)\Phi_{12,3,4}^{-1} \tag{15.13}$$

Proof: Consider the following cycle connecting the asymptotic zones $O_1, \ldots O_5$:

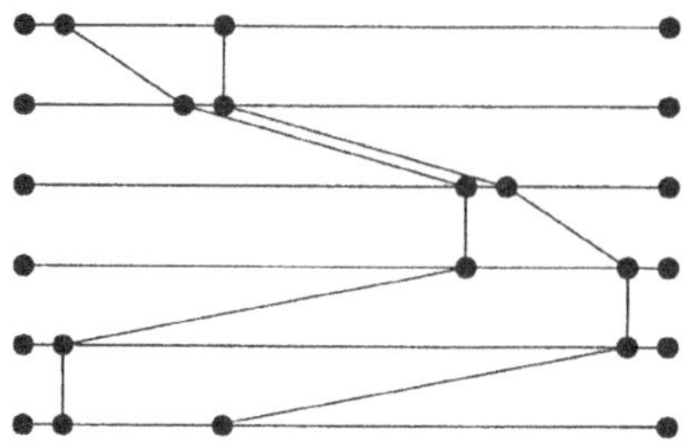

The monodromy $M_{O_1 \to O_2} \in U(\mathfrak{g})^{\otimes 3}[[h]]$ is defined by

$$W_1(z,u) = W_2(z,u) M_{O_1 \to O_2}.$$

In these zones, we have $z, u \ll 1$ so that we can approximate ($\mathrm{KZ'}_4$) by

$$\frac{1}{\hbar}\frac{\partial F}{\partial z} = \left(\frac{\Omega_{12}}{z} + \frac{\Omega_{23}}{z-u}\right)F,$$
$$\frac{1}{\hbar}\frac{\partial F}{\partial u} = \left(\frac{\Omega_{13}}{u} + \frac{\Omega_{23}}{u-z}\right)F.$$

Freezing the variable u and setting $\tilde{z} = \frac{z}{u}$, we obtain a system proportional to ($\mathrm{KZ'}_3$). In particular, the monodromies are the same: $M_{O_1 \to O_2} = \Phi_{1,2,3}$.

Let us now justify equation (15.10): set $x = z$, $y = z - u$ and define $\tilde{F}$ by $\tilde{F}(x,y) = F(z,u)$. Then F is a solution of ($\mathrm{KZ'}_4$) if and only if $\tilde{F}$ satisfies

$$\frac{1}{\hbar}\frac{\partial \tilde{F}}{\partial x} = \left(\frac{\Omega_{12}}{x} + \frac{\Omega_{24}}{x-1} + \frac{\Omega_{13}}{x-y} + \frac{\Omega_{34}}{x-y-1}\right)\tilde{F}, \qquad (*)$$
$$\frac{1}{\hbar}\frac{\partial \tilde{F}}{\partial y} = \left(-\frac{\Omega_{13}}{x-y} + \frac{\Omega_{23}}{y} - \frac{\Omega_{34}}{x-y-1}\right)\tilde{F},$$

Freezing the variable $y = z - u$, and considering $(*)$ in the asymptotic zones O_2, O_3, we can approximate the above system by

$$\frac{1}{\hbar}\frac{\partial \tilde{F}}{\partial x} = \left(\frac{\Omega_{12}+\Omega_{13}}{x} + \frac{\Omega_{24}+\Omega_{34}}{x-1}\right)\tilde{F}. \qquad (**)$$

But from the relations

$$(1 \otimes \Delta \otimes 1)\Omega_{12} = \Omega_{12} + \Omega_{13},$$
$$(1 \otimes \Delta \otimes 1)\Omega_{23} = \Omega_{24} + \Omega_{34},$$

it follows that $(**)$ is obtained by applying $1 \otimes \Delta \otimes 1$ to ($\mathrm{KZ'}_3$). In particular, the monodromy is given by

$$M_{O_2 \to O_3} = (1 \otimes \Delta \otimes 1)\Phi_{1,2,3} = \Phi_{1,23,4}.$$

The other equalities (15.11-15.13) are proved in the same way.

□

Corollary 15.2. *The element Φ_{KZ} satisfies the pentagon relation.*

Proof: It is enough to observe that the cycle $O_1 \to O_2 \to O_3 \to O_4 \to O_5 \to O_1$ is homotopically trivial, and hence that monodromy around it is trivial.

Thus we have proved:

Theorem 15.1 (Drinfeld). *$(U(\mathfrak{g})[[h]], \Delta, \epsilon, \Phi_{KZ})$ is a quasi-bialgebra over $\mathbb{C}[[h]]$.*

As in the case of bialgebras, one can show that any quasi-bialgebra which is a deformation of a quasi-Hopf algebra automatically has a structure of a quasi-Hopf algebra. Hence there exists $S : U(\mathfrak{g})[[h]] \to U(\mathfrak{g})[[h]]$, $\alpha, \beta \in U(\mathfrak{g})[[h]]$, such that $(U(\mathfrak{g})[[h]], \Delta, \epsilon, S, \Phi_{KZ}, \alpha, \beta)$ is a quasi-Hopf algebra.

15.2.3 Braided (quasitriangular) structure

We now show that the quasi-Hopf algebra $(U(\mathfrak{g})[[h]], \Delta, \epsilon, \Phi_{KZ})$ admits a quasitriangular structure. According to the general principle of using monodromies of the KZ equations, we make the following definition:

Consider the system KZ_2:

$$\frac{1}{\hbar}\frac{\partial F}{\partial z_1} = \frac{\Omega}{z_1 - z_2} F,$$
$$\frac{1}{\hbar}\frac{\partial F}{\partial z_2} = \frac{\Omega}{z_2 - z_1} F.$$

The solutions are given by $F(z_1, z_2) = e^{\frac{\hbar}{2i\pi}(z_1 - z_2)\Omega} C$, for some constant C and the monodromy around the following half circle is equal to $e^{\frac{\hbar\Omega}{2}}$:

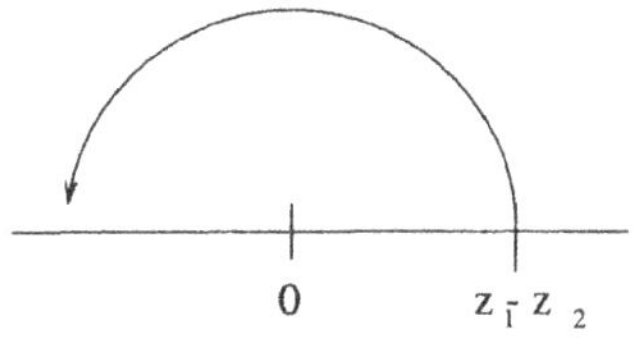

Definition: We set

$$R = e^{\frac{\hbar\Omega}{2}} \in U(\mathfrak{g})^{\otimes 2}[[h]]. \tag{15.14}$$

Proposition 15.4. *$(U(\mathfrak{g})[[h]], \Delta, \epsilon, \Phi_{KZ}, R)$ is a quasitriangular quasi-bialgebra over $\mathbb{C}[[h]]$.*

Proof: Notice that $\Delta^{op} R = R\Delta$ since R is $\mathfrak{g}$-invariant. We now prove the hexagon relations.

Consider the following simply connected domain

$$K' = \{z \in \mathbb{C},\ \mathrm{Im}\ z \geq 0,\ z \neq 0, 1\} \subset \mathbb{P}^1(\mathbb{C}).$$

Set $\phi : \mathbb{C}^3 \to \mathbb{C}$ to be the rational function

$$(z_1, z_2, z_3) \mapsto \frac{z_2 - z_1}{z_3 - z_1}$$

and let $K = \phi^{-1}(K')$. It is clear that K is also simply connected.

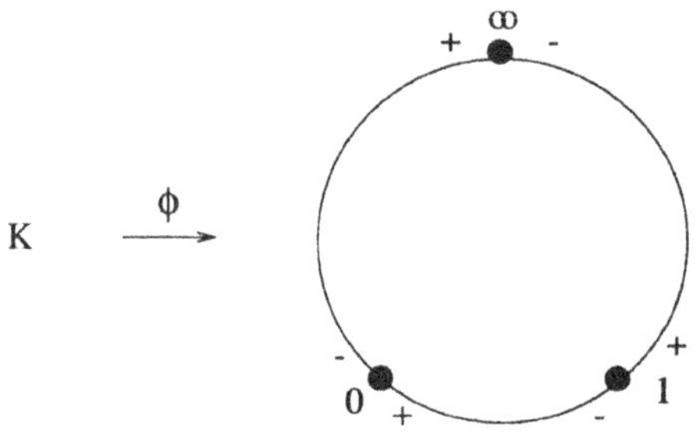

Now consider the following six (real) asymptotic zones of K corresponding to $0^{\pm}$, $1^{\pm}$ and $\infty^{\pm}$ of K':

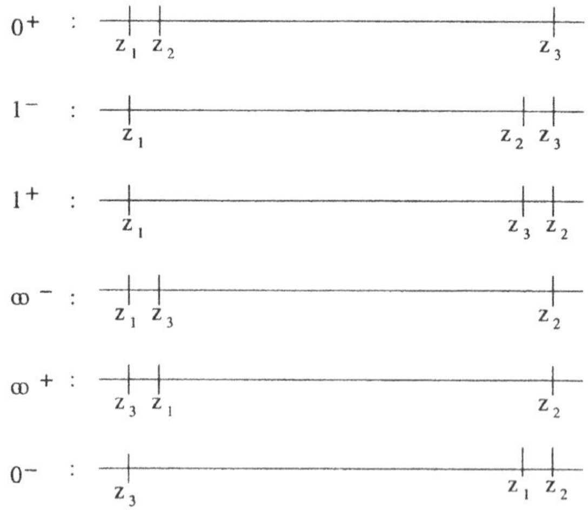

Let Γ_K be the set of solutions of (KZ$_3$) in K. We define the notions of asymptotics of a function $\varphi \in \Gamma_K$ in each of these regions and the notion of monodromy in a manner similar to (15.1) and (15.4).

Let us compute the monodromies along the following two curves γ_1 and γ_2 in K:

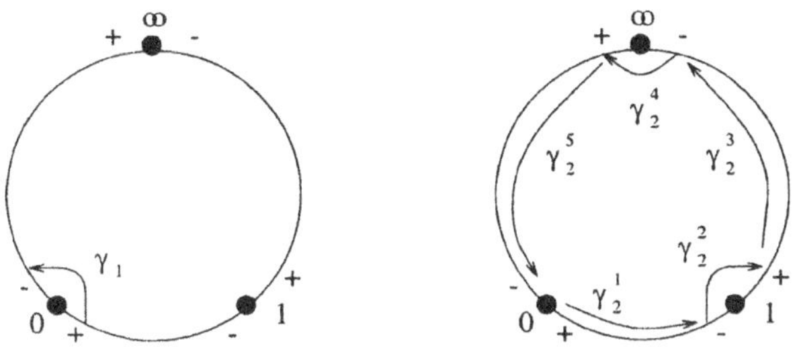

By definition, monodromy along the first arc $\gamma_2^{(1)}$ is Φ. Next, notice that the monodromy along $\gamma_2^{(3)}$, i.e along

is the same as along $-\gamma_2^{(1)}$, exchanging z_2 and z_3, or equivalently Ω_{12} and Ω_{13}. Thus $M_{\gamma_2^{(3)}} = \Phi_{132}^{-1}$. Similarly, $M_{\gamma_2^{(5)}} = \Phi_{312}$.

Around the point 1, we have $|z_2 - z_3| \ll |z_1 - z_2|, |z_1 - z_3|$, so that (KZ$_3$) can be approximated by the (KZ$_2$) system

$$\frac{1}{\hbar}\frac{\partial F}{\partial z_2} = -\frac{1}{\hbar}\frac{\partial F}{\partial z_3} = \frac{\Omega_{23}}{z_2 - z_3}F.$$

In particular, monodromy around $\gamma_2^{(2)}$ is $M_{\gamma_2^{(2)}} = R_{23}$. Similarly, $M_{\gamma_2^{(4)}} = R_{13}$.

Finally, to compute monodromy along γ_1, i.e along

let us make the change of variables $x = \frac{z_1 - z_2}{2}$, $y = \frac{z_1 + z_2}{2}$, $z = z_3$ and set $\tilde{F}(x, y, z) = F(z_1, z_2, z_3)$. After freezing $z_1 - z_2$, we can rewrite (KZ$_3$) as

$$\frac{1}{\hbar}\frac{\partial \tilde{F}}{\partial y} = \left(\frac{\Omega_{13}}{z_1 - z_3} + \frac{\Omega_{23}}{z_2 - z_3}\right)F,$$
$$\frac{1}{\hbar}\frac{\partial \tilde{F}}{\partial z} = \left(\frac{\Omega_{13}}{z_3 - z_1} + \frac{\Omega_{23}}{z_3 - z_2}\right)F.$$

In the zone around 0, this can be approximated by

$$\frac{1}{\hbar}\frac{\partial \tilde{F}}{\partial y} = \frac{\Omega_{13}+\Omega_{23}}{y-z}\tilde{F} = \frac{(\Delta\otimes 1)\Omega_{12}}{y-z}\tilde{F},$$
$$\frac{1}{\hbar}\frac{\partial \tilde{F}}{\partial z} = \frac{\Omega_{13}+\Omega_{23}}{z-y}\tilde{F} = \frac{(\Delta\otimes 1)\Omega_{23}}{z-y}\tilde{F},$$

which is just $(\Delta \otimes 1)$ applied to (KZ_2). Hence $M_{\gamma_1} = (\Delta \otimes 1)R$.

Since γ_1, γ_2 are homotopic, monodromies along γ_1 and γ_2 agree and we conclude that

$$(\Delta \otimes 1)R = \Phi_{321}R_{13}\Phi_{132}^{-1}R_{23}\Phi_{123}. \tag{15.15}$$

The second hexagon relation is proved in a similar fashion.

□

15.3 The Drinfeld category

Definition: The quasitriangular quasi-bialgebra

$$A = (U(\mathfrak{g})[[h]], \Delta, \epsilon, \Phi_{KZ}, R)$$

is called the *KZ quasi-bialgebra.*

It is clear that A is a deformation of the Hopf algebra $U(\mathfrak{g})$. Furthermore, A can be equipped with an antipode S, making it into a quasi-Hopf algebra (the *KZ quasi-Hopf algebra*).

Consider the following category $\mathcal{D}$. Objects of $\mathcal{D}$ are modules over $U(\mathfrak{g})$ and for any $V, W \in \mathcal{O}b(\mathcal{D})$, we set

$$\mathrm{Hom}_{\mathcal{D}}(V,W) = \mathrm{Hom}_{U(\mathfrak{g})}(V,W) \otimes_{\mathbb{C}} \mathbb{C}[[h]].$$

The category $\mathcal{D}$ is called the *Drinfeld category.* It is a full subcategory of Rep A which is stabe under tensor products, and it is thus a braided monoidal category.

Remark: Let $\mathfrak{g}_h$ be a deformation Lie algebra of $\mathfrak{g}$ and $\Omega_h \in (S^2\mathfrak{g}_h)^{\mathfrak{g}_h}$ an invariant symmetric element. The above construction extends and we obtain in this way a quasi-bialgebra $A_{\mathfrak{g}_h} = (U(\mathfrak{g}_h), \Delta, \epsilon, \Phi_{KZ}, R)$.

15.4 Braid group representation

Recall that there is a canonical action of the Braid group B_n on $V^{\otimes n}$ for any object V in a braided monoidal category. In particular, the braided monoidal structure of $\mathcal{D}$ gives rise to an action $\rho_{\mathcal{D}} : B_n \to \mathrm{Aut}(V^{\otimes n}[[h]])$ for any finite-dimensional $\mathfrak{g}$-module V. On the other hand, the KZ_n connection on the trivial bundle $V^{\otimes n}[[h]] \times X_n/S_n$ gives rise to a representation $\rho_{KZ} : \pi_1(X_n/S_n) \simeq B_n \to \mathrm{Aut}(V^{\otimes n}[[h]])$ (defined up to conjugation). Since $\mathcal{D}$ is defined by means of the KZ equations, it is not surprising that

Proposition 15.5. *The representations $\rho_{\mathcal{D}}$ and ρ_{KZ} are isomorphic.*

Proof: Let us prove this for $n = 3$. In this case, the Braid group B_3 is generated by b_1, b_2. In $\pi_1(X_3/S_3)$, the element b_1 corresponds to the loop $\sigma_1 : (0, 1, \infty) \to (1, 0, \infty)$ while b_2 is represented by the loop $\sigma_2 : (0, 1, \infty) \to (-\infty, 0, 1) \to (-\infty, 1, 0) \to (0, \infty, 1)$.

The representation $\rho_{\mathcal{D}}$ is defined by $\rho_{\mathcal{D}}(b_1) = R_{12} = e^{\frac{h\Omega_{12}}{2}}$ and $\rho_{\mathcal{D}}(b_2) = \Phi_{KZ}^{-1} R^{23} \Phi_{KZ}$, which correspond to the monodromy of KZ_3 along the loops σ_1 and σ_2 respectively. This is the same as ρ_{KZ}.

□

Remark: The assumption $k = \mathbb{C}$ (or $k \supset \mathbb{C}$) is essential: the associator Φ_{KZ} is defined by analytic methods. However, we will see that when $\mathfrak{g}$ is defined over $\mathbb{Q}$, there exists an associator defined over $\mathbb{Q}$, and hence over any field $k \supset \mathbb{Q}$.

Lecture 16

Quasi-Hopf quantized enveloping algebras

We now return to quantum groups and define the "quasi" version of quantized enveloping algebras and Lie bialgebras. As we will see, this is a much more flexible framework. Most definitions and results here are due to Drinfeld-see [Dr4]

16.1 Quasi-Hopf quantized enveloping algebras

16.1.1 Definition

Let k be a field of characteristic zero, and as usual, set $K = k[[h]]$. We will use the following abbreviation: QUE algebra stands for "quantized universal enveloping algebra".

Definition: A *deformation quasi-bialgebra* is a topologically free quasi-bialgebra $(A, \Delta, \epsilon, \Phi)$ over K. We say that $(A, \Delta_A, \epsilon_A, \Phi_A)$ is a deformation of a quasi-bialgebra $(A_0, \Delta_{A_0}, \epsilon_{A_0}, \Phi_{A_0})$ defined over k if

$$(A/hA, \Delta_A \bmod h, \epsilon_A \bmod h, \Phi_A \bmod h) = (A_0, \Delta_{A_0}, \epsilon_{A_0}, \Phi_{A_0}).$$

The notion of deformation quasi-Hopf algebra is defined analogously.

In a similar fashion, we extend the notion of QUE algebra:

Definition: A *quasi-Hopf QUE algebra* is a deformation quasi-Hopf algebra $(H, \Delta, \epsilon, S, \Phi)$ of $(U(\mathfrak{g}), \Delta_0, \epsilon_0, S, 1)$ for some Lie algebra $\mathfrak{g}$, such that

$$\mathrm{Alt}(\Phi) \equiv 0 \pmod{h^2} \tag{16.1}$$

where

$$\mathrm{Alt}(x \otimes y \otimes z) = \frac{1}{6}(x \otimes y \otimes z + y \otimes z \otimes x + z \otimes x \otimes y - y \otimes x \otimes z - z \otimes y \otimes x - x \otimes z \otimes y).$$

Of course, in the above, Δ_0 and ϵ_0 are the usual Hopf algebra structures on $U(\mathfrak{g})$.

Remark: Condition (16.1) is of technical importance; see [ShSt] for more details.

16.1.2 Examples

Example 16.1 (Hopf QUE algebras). Any QUE algebra is in particular a quasi-Hopf QUE algebra, with $\Phi = 1$ (like the Drinfeld-Jimbo quantum groups $U_h(\mathfrak{g})$).

Example 16.2 (KZ quasi-Hopf algebra). $\mathcal{D} = (U(\mathfrak{g})[[h]], \Delta_0, \epsilon_0, S, \Phi_{KZ})$ is a quasi-Hopf QUE algebra as we saw in the last lecture.

In the same way, we define the notion of a quasitriangular quasi-Hopf QUE algebra and a triangular quasi-Hopf QUE algebra $(H, \Delta, \epsilon, S, \Phi, R)$. For instance, the quasi-Hopf QUE algebras of examples 16.1 and 16.2 are quasitriangular with universal R-matrices given by (12.13) and (15.14) respectively.

Exercise 16.1. Let $(H, \Delta, \epsilon, S, \Phi)$ be a quasi-Hopf QUE algebra. If R is a quasitriangular structure on it, show that $R \equiv 1 \pmod{h}$.

16.1.3 Twists

We can apply the general twisting procedure as in Lecture 14 to the particular case of quasi-Hopf QUE algebras.

Exercise 16.2. Let $(A, \Delta, \epsilon, S, \Phi)$ be a quasi-Hopf QUE algebra, and let $J \in \Gamma_A$. Show that $J \equiv 1 \pmod{h}$.

This implies that condition (16.1) is preserved by twisting.
The following important results are due to Drinfeld:

Proposition 16.1 (Drinfeld). *Any quasi-Hopf QUE algebra is twist equivalent to one where $\Phi \equiv 1 \pmod{h^2}$. Furthermore, if R is a quasitriangular structure, we can twist it into a symmetric quasitriangular structure $\tilde{R} = \tilde{R}_{21}$, preserving $\Phi \equiv 1 \pmod{h^2}$.*

Proof: see [Dr4] for the first assertion.

Problem 16.1. Prove the second assertion.

Exercise 16.3. Show that twisting transforms a triangular quasi-Hopf QUE algebra into a triangular quasi-Hopf QUE algebra.

16.2 Lie quasibialgebras

We now examine the structure induced on $U(\mathfrak{g})$ by any quasi-Hopf QUE algebra H with $H/hH = U(\mathfrak{g})$.

16.2.1 Definition

Let $(H, \Delta, \epsilon, S, \Phi)$ be a quasi-Hopf QUE algebra, which is a deformation of $U(\mathfrak{g})$, and let us suppose that $\Phi \equiv 1 \pmod{h^2}$ (see prop.16.1). As in the QUE algebra case, let us set

$$\delta(x) = \frac{\Delta(\tilde{x}) - \Delta^{op}(\tilde{x})}{h} \pmod{h} \qquad x \in \mathfrak{g},\ \tilde{x} \equiv x \pmod{h}. \tag{16.2}$$

We also set

$$\varphi = \frac{\text{Alt } \Phi}{h^2} \pmod{h}. \tag{16.3}$$

Lemma 16.1. *We have:*

1. $\delta : \mathfrak{g} \to \Lambda^2\mathfrak{g}$ *is a 1-cocycle,*
2. $\varphi \in \Lambda^3\mathfrak{g}$,
3. $\text{Jac}_\delta(x) = [x \otimes 1 \otimes 1 + 1 \otimes x \otimes 1 + 1 \otimes 1 \otimes x, \varphi]$,
4. $\text{Alt}((\delta \otimes 1 \otimes 1)\varphi) = 0$.

Proof: See [Dr4]. Statement 1. is proved as for Hopf algebras (i.e from $\Delta(ab) = \Delta(a)\Delta(b)$). Statement 2. is proved by a cohomological argument using the fact that Φ satisfies the pentagon relation. Statement 3. comes from coassociativity of Δ and 4. follows from the pentagon relation.

□

This motivates the following

Definition: A *Lie quasibialgebra* over k is a triple $(\mathfrak{g}, \delta, \varphi)$ where $\mathfrak{g}$ is a Lie algebra, $\delta : \mathfrak{g} \to \Lambda^2\mathfrak{g}$ a 1-cocycle and $\varphi \in \Lambda^3\mathfrak{g}$ satisfies conditions 3. and 4. above.

It is clear that any Lie bialgebra can be regarded as a Lie quasibialgebra by setting $\varphi = 0$.

16.2.2 Quantization of Lie quasibialgebras

Definition: Let $(H, \Delta, \epsilon, S, \Phi)$ be a quasi-Hopf QUE algebra and $H/hH = U(\mathfrak{g})$. The Lie quasibialgebra $(\mathfrak{g}, \delta, \varphi)$ with δ, φ defined by (16.2, 16.3) is called the quasiclassical limit of $(H, \Delta, \epsilon, S, \Phi)$. Conversely, we say that $(H, \Delta, \epsilon, S, \Phi)$ is a quantization of $(\mathfrak{g}, \delta, \varphi)$.

Remark: The problem of quantization of Lie quasibialgebras is open.

16.2.3 Quasitriangular Lie quasibialgebras

Let us now examine the quasiclassical limit of a quasitriangular structure:

Proposition 16.2. *Let $(H, \Delta, \epsilon, S, \Phi, R)$ be a quasitriangular quasi-Hopf QUE algebra with $\Phi \equiv 1 \pmod{h^2}$ and $H/hH = U(\mathfrak{g})$. Define $\Omega \in U(\mathfrak{g}) \otimes U(\mathfrak{g})$ by $\Omega = \frac{R_{21}R-1}{h} \pmod{h}$. Then $\Omega \in (S^2\mathfrak{g})^{\mathfrak{g}} \subset (U(\mathfrak{g})^{\otimes 2})^{\mathfrak{g}}$.*

Proof: It is obvious that Ω is symmetric. Furthermore, applying (14.10) twice, we see that

$$\begin{aligned}(\Delta \otimes 1)(R_{21}R) &= ((1 \otimes \Delta)(R))_{312}(\Delta \otimes 1)R \\ &= \Phi_{123}^{-1} R_{32} \Phi_{132} R_{31} \Phi_{312}^{-1} \Phi_{312} R_{13} \Phi_{132}^{-1} R_{23} \Phi_{123}.\end{aligned}$$

Reducing modulo h^2 (using $\Phi \equiv 1 \pmod{h^2}$), we see that $(\Delta \otimes 1)\Omega = \Omega_{13} + \Omega_{23}$. Hence the first component of Ω is primitive, and $\Omega \in \mathfrak{g} \otimes \mathfrak{g}$. The invariance of Ω follows from $\Delta R_{21} R = R_{21} R \Delta$.

□

Exercise 16.4. Show that Ω remains unchanged under twisting. Conclude that, after a suitable twist (where R becomes symmetric), we obtain the familiar formula

$$R \equiv 1 + \frac{h\Omega}{2} \pmod{h^2}.$$

Problem 16.2. Let $(H, \Delta, \epsilon, \Phi, S, R)$ and Ω be as above (i.e R is symmetric). Show that the quasiclassical limit of $(H, \Delta, \epsilon, \Phi, S, R)$ is $(\mathfrak{g}, 0, \frac{1}{4}[\Omega_{12}, \Omega_{23}])$. (Hint: use the quantum Yang-Baxter equation.)

In view of the results of the last problem, we say that the quasiclassical limit of the quasitriangular quasi-Hopf QUE algebra $(H, \Delta, \epsilon, \Phi, S, R)$ is the pair $(\mathfrak{g}, \Omega)$ (and $(H, \Delta, \epsilon, \Phi, S, R)$ is a quantization of $(\mathfrak{g}, \Omega)$). This notion is invariant under twisting.

Remark: Notice the difference with the previous definition: the quasiclassical limit of $(H, \Delta, \epsilon, S, \Phi, R)$ as a quasi-Hopf QUE algebra is $(\mathfrak{g}, \delta, \varphi)$ with $\delta(a) = [a \otimes 1 + 1 \otimes a, \rho]$ where $\rho = \frac{R-1}{h} \pmod{h}$ and $\rho + \rho_{21} = \Omega$.

In particular, if R is a triangular structure, then $\Omega = 0$.

Example 16.3. Let $\mathfrak{g}_h$ be a deformation Lie algebra over $\mathbb{C}[[h]]$ and $\Omega \in (S^2\mathfrak{g}_h)^{\mathfrak{g}_h}$. Then the KZ quasi-Hopf algebra $(U(\mathfrak{g}_h), \Delta, \epsilon, S, \Phi_{KZ}, e^{\frac{h\Omega}{2}})$ is a quantization of $(\mathfrak{g}, \Omega_0)$ where $\Omega \equiv \Omega_0 \pmod{h}$. We will see that in fact all quantizations of $(\mathfrak{g}, \Omega_0)$ can be twisted into one of the above type.

16.2.4 Twists

The notion of a twist at the quantum level induces the following notion of a twist at the quasiclassical level:

Definition: Let $(\mathfrak{g}, \delta, \varphi)$ be a Lie quasibialgebra, and let $f \in \Lambda^2\mathfrak{g}$. Set

$$\delta'(x) = \delta(x) + [x \otimes 1 + 1 \otimes x, f], \tag{16.4}$$

$$\varphi' = \varphi + \frac{1}{2}\mathrm{Alt}(\delta \otimes 1)f - \mathrm{CYB}(f). \tag{16.5}$$

Proposition 16.3. *$(\mathfrak{g}, \delta', \varphi')$ is a Lie quasibialgebra (the twist of $(\mathfrak{g}, \delta, \varphi)$ by f). Moreover, twisting by f_1 and then by f_2 is the same as twisting by $f_1 + f_2$.*

Proof (sketch): We have to check conditions 1.-4. It is clear that $\delta' : \mathfrak{g} \to \Lambda^2\mathfrak{g}$ is a 1-cocycle, and that $\varphi' \in \Lambda^3\mathfrak{g}$. Furthermore, from

$$\mathrm{Jac}(\delta)(f) + [x, \mathrm{CYB}(f)] = 0,$$

(see (3.1)), we see that the terms added in φ' insure condition 3. Finally, 4. follows from a direct computation. The second assertion of the proposition is straightforward.

□

The link between quasiclassical twists and quantum twists (section 16.1.3) is given by the following

Proposition 16.4. *Let $(H, \Delta, \epsilon, S, \Phi)$ be a quasi-Hopf QUE algebra, $J \in \overline{\Gamma}_H$ and let $(H, \Delta', \epsilon, S', \Phi')$ be the twist of H by J. Suppose further that $\Phi \equiv \Phi' \equiv 1 \pmod{h^2}$. Set $f \equiv \frac{J - J^{21}}{h} \pmod{h}$. Then the quasiclassical limit of $(H, \Delta', \epsilon, S', \Phi')$ is the twist by f of the quasiclassical limit of $(H, \Delta, \epsilon, S, \Phi)$.*

Proof: See [Dr4] for a detailed proof. Equation (16.4) comes from (13.18) and equation (16.5) comes from the pentagon relation (13.19).

□

Remark: The group of twists at the quantum level is in general nonabelian, while it is the abelian group $\Lambda^2\mathfrak{g}$ at the quasiclassical level.

16.3 Associators

Let us now return to the KZ quasi-Hopf algebra $(U(\mathfrak{g})[[h]], \Delta, \epsilon, S, \Phi_{KZ}, e^{\frac{h\Omega}{2}})$, and study the set of possible associators for it.

16.3.1 Definition

Definition: Let $\mathfrak{g}$ be a Lie algebra defined over k and $\Omega \in (S^2\mathfrak{g})^{\mathfrak{g}}$. An invertible element $\Phi \in \left(U(\mathfrak{g})^{\otimes 3}[[h]]\right)^{\mathfrak{g}}$ will be called an *associator* for $(\mathfrak{g}, \Omega)$ if

1. $(1 \otimes \epsilon \otimes 1)\Phi = 1$,

2. Φ satisfies the pentagon relation,

3. Φ satisfies the two hexagon axioms with $R = e^{\frac{h\Omega}{2}}$.

In other words, Φ is an associator if and only if $(U(\mathfrak{g})[[h]], \Delta, \epsilon, S, \Phi, e^{\frac{h\Omega}{2}})$ is a quasi-Hopf algebra.

Exercise 16.5. Show that $\Phi \equiv 1 \pmod h$.

We let $\mathrm{Ass}(\mathfrak{g}, \Omega)$ denote the set of all associators. So far, Φ_{KZ} is the only explicit example of an associator we have (note that $1 \notin \mathrm{Ass}(\mathfrak{g}, \Omega)$ if $\Omega \neq 0$).

16.3.2 Action of twists on $\mathrm{Ass}(\mathfrak{g}, \Omega)$

The following subgroup of $\Gamma_{U(\mathfrak{g})[[h]]}$ acts on $\mathrm{Ass}(\mathfrak{g}, \Omega)$:

$$\Gamma^0_{U(\mathfrak{g})[[h]]} = \Gamma_{U(\mathfrak{g})[[h]]} \cap (S^2U(\mathfrak{g}))^{\mathfrak{g}}[[h]]$$

by the usual formula:

$$J.\Phi = J_{2,3}^{-1} J_{1,23}^{-1} \Phi J_{12,3} J_{1,2}.$$

Indeed, the only thing to check is the equality

$$(J_{21})^{-1} e^{\frac{h\Omega}{2}} J = e^{\frac{h\Omega}{2}}.$$

This follows from the fact that $\Omega = \frac{1}{2}(\Delta(C) - C \otimes 1 - 1 \otimes C)$ where $C \in (U(\mathfrak{g})[[h]])^{\mathfrak{g}}$ is the quadratic "Casimir" $m(\Omega)$.

The following theorem gives a complete answer to the question raised at the beginning of this section:

Theorem 16.1 (Drinfeld). *The action of* $\Gamma^0_{U(\mathfrak{g})[[h]]}$ *on* $\mathrm{Ass}(\mathfrak{g}, \Omega)$ *is transitive. Moreover, for* $\Phi \in \mathrm{Ass}(\mathfrak{g}, \Omega)$,

$$\mathrm{Stab}(\Phi) = \{\Delta(u)(u^{-1} \otimes u^{-1}),\ u \in U(\mathfrak{g})^{\mathfrak{g}}[[h]],\ \epsilon(u) = 1\}.$$

Proof: See [Dr4]. This theorem is proved by cohomological methods.

Remark: Notice that $\mathrm{Stab}(\Phi)$ is independent of Φ. Thus, $\mathrm{Ass}(\mathfrak{g}, \Omega)$ is a principal homogeneous space of a group. However, there is no distinguished element and therefore no canonical group structure on it.

16.4 Classification of quasitriangular quasi-Hopf QUE algebras

The following important theorem settles the question of classification of quantization of quasitriangular Lie quasibialgebras.

Theorem 16.2 (Drinfeld). *Any quasitriangular quasi-Hopf QUE algebra is twist equivalent to one of the form*

$$(U(\mathfrak{g}_h), \Delta, \epsilon, S, \Phi, e^{\frac{h\Omega}{2}}) \tag{16.6}$$

where $\mathfrak{g}_h$ is a deformation Lie algebra, Δ and ϵ the usual coproduct and counit, $\Omega \in (S^2\mathfrak{g}_h)^{\mathfrak{g}_h}$ and Φ an associator. Moreover, $\mathfrak{g}_h$ and Ω are unique up to isomorphism.

Proof: see [Dr4].

As we have seen, the choice of the associator in (16.6) is irrelevant, up to twisting (and this is also true for deformation Lie algebras). Thus quasitriangular quasi-Hopf QUE algebra are classified by pairs $(\mathfrak{g}, \Omega)$ as above. In particular, this shows that any quasitriangular quasi-Hopf QUE algebra $(A, \Delta, \epsilon, S, \Phi, R)$, regarded as an associative algebra, is not only a deformation of a universal enveloping algebra, but a universal enveloping algebra itself.

16.5 The Drinfeld-Kohno theorem

Let us examine an important consequence of Theorem 16.2. Suppose $\mathfrak{g}$ is a (semi)simple complex Lie algebra. The Drinfeld-Jimbo quantum group $U_h(\mathfrak{g})$ is a quasitriangular QUE algebra with R-matrix given by (12.13). In particular it is a quasitriangular quasi-Hopf QUE algebra, with $\Phi = 1$. By Drinfeld's theorem, it is twist-equivalent to a quasi-Hopf algebra

$$(U(\mathfrak{g}_h), \Delta, \epsilon, S, \Phi, e^{\frac{h\Omega}{2}}) \tag{16.7}$$

where $\mathfrak{g}_h$ is a deformation of $\mathfrak{g}$ and $\Omega \in (S^2\mathfrak{g}_h)^{\mathfrak{g}_h}$. But it is well known that any deformation of a (semi)simple Lie algebra is trivial since $H^2(\mathfrak{g}, \mathfrak{g}) = 0$. Hence $\mathfrak{g}_h \simeq \mathfrak{g}[[h]]$. In particular, there is an isomorphism of *algebras* $s : U_h(\mathfrak{g}) \tilde{\to} U(\mathfrak{g})[[h]]$. Also, any invariant element of $\mathfrak{g} \otimes \mathfrak{g}[[h]]$ is a multiple of Ω_0, the Casimir element for the Killing form (normalized so that $< \alpha, \alpha >= 2$ for long roots).

One can show that in fact $\Omega_0 = \Omega$ by computing the eigenvalues of the braiding operator $R_{21}R$ acting on some highest weight modules (using the explicit form of the R-matrix (12.13)). Finally, we can take $\Phi = \Phi_{KZ}$ by Theorem 16.1. Alltogether we have the following result.

Theorem 16.3. *The braided monoidal category of modules over $U_h(\mathfrak{g})$ is equivalent to the Drinfeld category for $(\mathfrak{g}, \Omega_0)$.*

Remark: The isomorphism s is not known explicitly (but see [CP] for the case of sl_2).

Recall that if $\mathcal{C}$ is a braided monoidal category and $V \in \mathcal{O}b(\mathcal{C})$ then there is a canonical action of B_n in $V^{\otimes n}$ for $n \geq 1$ (when $\mathcal{C}$ is the category of modules over a quasitriangular Hopf algebra, this action is given in terms of the R-matrix). It is clear that a tensor equivalence $F : \mathcal{C}_1 \to \mathcal{C}_2$ between two braided monoidal categories induces an isomorphism of representations of B_n in $V^{\otimes n}$ and $F(V)^{\otimes n}$ for any $V \in \mathcal{O}b(\mathcal{C}_1)$.

Theorem 16.3 has the following important consequence: let V be a finite-dimensional $\mathfrak{g}$-module. Since $U_h(\mathfrak{g}) \simeq U(\mathfrak{g})[[h]]$ as an algebra, $V[[h]]$ has a structure of a $U_h(\mathfrak{g})$-module, i.e $V \in \operatorname{Rep} U_h(\mathfrak{g})$. Let $\rho : B_n \to \operatorname{Aut}(V^{\otimes n}[[h]])$ be the corresponding action of B_n on $V^{\otimes n}[[h]]$. Denote by $\rho_{KZ} : B_n \to \operatorname{Aut}(V^{\otimes n}[[h]])$ the monodromy representation of the KZ connection on the trivial bundle $V^{\otimes n} \times X/S_n$ (see Lecture 15).

Theorem 16.4 (Drinfeld-Kohno). *The representations ρ and ρ_{KZ} are isomorphic.*

Proof: By Theorem 16.3, the representation ρ is isomorphic to the representation of B_n on $V^{\otimes n}[[h]]$ when V is considered as an object in the Drinfeld category for $(\mathfrak{g}, \Omega)$. But by Proposition 15.5, this representation is given by the monodromy of the KZ connection.

□

This theorem is as beautiful as it is intriguing: the representation of B_n arising from $U_h(\mathfrak{g})$ is of purely algebraic origin (relations (9.2-9.5)), while the representation of B_n arising from monodromy of the KZ equations is a priori transcendental. Another proof of this result, which relies on an explicit computation of the monodromy of the KZ equations, can be found in [Var].

16.6 Geometric interpretation of Lie quasibialgebras

Recall the 1-1 correspondence between Lie bialgebras and Manin triples of Lecture 4. We now define an analogous construction for Lie quasibialgebras.

Definition: A *Manin pair* is a pair $(\mathfrak{g}, \mathfrak{g}_+)$ where $\mathfrak{g}$ is a finite dimensional Lie algebra equipped with a nondegenerate symmetric, invariant bilinear form $<,>$ and $\mathfrak{g}_+$ is a Lagrangian subalgebra (i.e a maximal isotropic subspace, which is a Lie subalgebra).

Notice the difference with Manin triples: the Lie subalgebra $\mathfrak{g}_+$ may not always have a Lagrangian complement which is a Lie subalgebra; but we can always find a Lagrangian complement, which is not necessarily a Lie subalgebra.

Definition: A *Manin pair with a Lagrangian complement* is a triple $(\mathfrak{g}, \mathfrak{g}_+, \mathfrak{g}_-)$ where $(\mathfrak{g}, \mathfrak{g}_+)$ is a Manin pair and where $\mathfrak{g}_-$ is a Lagrangian complement of $\mathfrak{g}_+$.

These notions can be extended to the infinite dimensional setting in the same manner as for usual Manin triples. Namely, one should add the condition that the form $<,>$ induces an isomorphism $\mathfrak{g}_- \xrightarrow{\sim} \mathfrak{g}_+^*$ (the full dual).

Example: Geometric Manin pairs

Let $\mathfrak{g}$ be a complex Lie algebra with a nondegenerate invariant form $(\,,\,)$. Let us consider the following general situation. Let X be a smooth projective complex curve, and ω a meromorphic nonzero one-form on X. Let $\{x_i\}$ be the set of zeros and poles of ω and let d_i be the order of the pole or zero at x_i. Let R be the algebra of functions on X, regular outside of $\{x_i\}$, and let $\mathcal{O}_{x_i}$ (resp. k_{x_i}) be the local ring (resp. the local field) at x_i and I_{x_i} the maximal ideal of $\mathcal{O}_{x_i}$. We have a natural embedding

$$\mathfrak{g}(R) \hookrightarrow \mathfrak{g} \otimes \big(\bigoplus_i k_{x_i}\big) =: \mathfrak{g}\big(\bigoplus_i k_{x_i}\big)$$

where we let $\mathfrak{g}(A) = \mathfrak{g} \otimes A$ for any commutative $\mathbb{C}$-algebra A. Endow $\mathfrak{g}\big(\bigoplus_i k_{x_i}\big)$ with the nondegenerate invariant form

$$< f, g >= \sum_i \mathrm{Res}_{x_i}(f, g)\omega.$$

By the residue theorem, $\mathfrak{g}(R)$ is isotropic. In fact, one can show that it is a *maximal* isotropic subalgebra. However, there are in general no Lagrangian subalgebra complementary to $\mathfrak{g}(R)$, and the pair $\big(\mathfrak{g}\big(\bigoplus_i k_{x_i}\big), \mathfrak{g}(R)\big)$ cannot be extended to a triple.

The link with Lie quasibialgebras is given by the following result which appears in [Dr4].

Theorem 16.5 (Drinfeld). *There is a* 1-1 *correspondence between Manin pairs with Lagrangian complement and Lie quasibialgebras.*

There is a 1-1 *correspondence between Manin pairs and equivalence classes (under twisting) of Lie quasibialgebras.*

Proof: We will restrict ourselves to the finite dimensional case. Let $(\mathfrak{g}, \mathfrak{g}_+, \mathfrak{g}_-)$ be a Manin pair with a complement. Restricting the commutator to $\mathfrak{g}_-$ gives two maps

$$[\,,\,]_+ : \Lambda^2\mathfrak{g}_- \to \mathfrak{g}_+, \tag{16.8}$$

$$[\,,\,]_- : \Lambda^2\mathfrak{g}_- \to \mathfrak{g}_-, \tag{16.9}$$

such that $[\,,\,] = [\,,\,]_+ + [\,,\,]_-$. Notice that $(\mathfrak{g}, \mathfrak{g}_+, \mathfrak{g}_-)$ is a Manin triple if and only if $[\,,\,]_+ = 0$. Now set

$$\delta = [\,,\,]_-^* : \mathfrak{g}_+ \simeq \mathfrak{g}_-^* \to \Lambda^2 \mathfrak{g}_-^* \simeq \Lambda^2 \mathfrak{g}_+.$$

We can view $[\,,\,]_+^*$ as a map $\mathfrak{g}_+^* \to \Lambda^2 \mathfrak{g}_+$, or as an element $\varphi = [\,,\,]_+^* \in \mathfrak{g}_+^{\otimes 3}$.

Lemma 16.2. *The triple $(\mathfrak{g}_+, \delta, \varphi)$ is a Lie quasibialgebra.*

Proof (sketch): The fact that δ is a 1-cocycle is proved in the same way as in Lecture 4. The fact that $\varphi \in \Lambda^3 \mathfrak{g}_+$ follows from the invariance of $<\,,\,>$. The two last conditions come from the Jacobi identity of $\mathfrak{g}$.

□

The inverse construction is a (quasi)analogue of the Drinfeld double construction:

If $(\mathfrak{g}_+, \delta, \varphi)$ is a Lie quasibialgebra, we set $\mathfrak{g} = \mathfrak{g}_+ \oplus \mathfrak{g}_-$ with $\mathfrak{g}_- = \mathfrak{g}_+^*$. The commutator $[\mathfrak{g}_+^*, \mathfrak{g}_-]$ is the sum of the dual of δ and φ as in (16.8-16.9), and the commutator $[\mathfrak{g}_+, \mathfrak{g}_-]$ is uniquely determined by the condition that $<\,,\,>$ be symmetric and invariant.

It remains to show that changing the Lagrangian complement amounts to twisting the corresponding Lie quasibialgebras.

The abelian group $\mathrm{Hom}_{\mathsf{skew}}(\mathfrak{g}_-, \mathfrak{g}_+)$ acts freely and transitively on the set of all Lagrangian complements $\tilde{V}$ to $\mathfrak{g}_+$ by:

$$\rho.\tilde{V} \to \begin{pmatrix} 1 & \rho \\ 0 & 1 \end{pmatrix} \tilde{V}.$$

We can view $\rho \in \mathrm{Hom}_{\mathsf{skew}}(\mathfrak{g}_-, \mathfrak{g}_+)$ as an element $\tilde{\rho} \in \Lambda^2 \mathfrak{g}$.

Problem 16.3. Prove this statement.

Lemma 16.3. *The Lie quasibialgebra $(\mathfrak{g}_+, \tilde{\delta}, \tilde{\varphi})$ associated to $(\mathfrak{g}, \mathfrak{g}_+, \rho.\mathfrak{g}_-)$ is the twist of the Lie quasibialgebra associated to $(\mathfrak{g}, \mathfrak{g}_+, \mathfrak{g}_-)$ by $\tilde{\rho}$.*

Proof: We will just show how δ transforms and leave the (more involved) case of φ to the intrepid reader. By definition, we have

$$< x, [y, y'] > = < \delta(x), y \otimes y' >, \qquad \forall\, x \in \mathfrak{g}_+,\ y, y' \in \mathfrak{g}_-$$

and

$$\begin{aligned} < \tilde{\delta}(x), y \otimes y' > &= < x, [y + \rho(y), y' + \rho(y')] > \\ &= < x, [y, y'] > + < x, [y, \rho(y')] > + < x, [\rho(y), y'] > \\ &= < x, [y, y'] > + < x, [(y \otimes 1)\tilde{\rho}, y'] > + < x, [y, (y' \otimes 1)\tilde{\rho}] > \\ &= < x, [y, y'] > + < [(1 \otimes x), \tilde{\rho}], y \otimes y' > + < [(x \otimes 1), \tilde{\rho}], y \otimes y' > \end{aligned}$$

where we used skew symmetry of $\tilde{\rho}$. From this it follows $\tilde{\delta} = \delta + \partial \tilde{\rho}$.

□

Lecture 17

Lie associators and the Grothendieck-Teichmuller group

We now define a special class of associators, which we will need in the construction of quantization of Lie bialgebras. Most results here are due to Drinfeld [Dr4].

17.1 Lie associators

These are associators given in some sense by a "universal" formula.

17.1.1 Definition

Before giving a precise definition, let us consider the example of the KZ associator $\Phi_{KZ} \in U(\mathfrak{g})^{\otimes 3}[[h]]$ associated to a finite dimensional Lie algebra $\mathfrak{g}$ and to an invariant element $\Omega \in (S^2\mathfrak{g})^{\mathfrak{g}}$. Recall that Φ_{KZ} is defined by the equality $F_0 = F_1\Phi_{KZ}$ where F_1 and F_0 are the unique solutions of the differential equation

$$\frac{dF}{dz} = \frac{1}{2i\pi}\left(\frac{h\Omega_{12}}{z} + \frac{h\Omega_{23}}{z-1}\right)F \qquad (0 < z < 1) \tag{17.1}$$

with expansions as in (15.1) and (15.3) respectively (see section 15.2.1). In particular, F_0 and F_1 are series in the noncommuting variables $h\Omega_{12}$ and $h\Omega_{23}$ and it is clear that $F_1 \equiv F_0 \equiv 1 \pmod h$. Hence Φ_{KZ} is also a series in $h\Omega_{12}$ and $h\Omega_{23}$. Notice that the only properties of $h\Omega_{12}$ and $h\Omega_{23}$ that we used in the whole of Lecture 16 are $\Omega_{ij} = \Omega_{ji}$ and $[\Omega_{ij}, \Omega_{ik} + \Omega_{jk}] = 0$.

This motivates the following definition. We fix a field k of characteristic zero.

Let $n \geq 0$ and let T_n be the algebra over k generated by t_{ij}, $1 \leq i, j \leq n$, $i \neq j$, with defining relations

$$t_{ij} = t_{ji}, \qquad [t_{ij}, t_{ik} + t_{jk}] = 0, \qquad (i, j, k \text{ distinct})$$

$$[t_{ij}, t_{kl}] = 0, \qquad (i, j, k, l \text{ distinct}).$$

If $P_1, \ldots P_n$ are disjoint subsets of $\{1, \ldots m\}$, we define a homomorphism:

$$\rho_{P_1,\ldots P_n} : T_n \to T_m,$$
$$t_{ij} \to \sum_{p \in P_i, q \in P_j} t_{pq}$$

and we set $\rho_{P_1,\ldots P_n}(a) = a_{P_1,\ldots P_n}$ for any $a \in T_n$. Let $B = e^{ht_{12}/2} \in T_2[[h]]$.

Definition: An invertible element $\Phi \in T_3[[h]]$ is called an associator if the following relations are satisfied

1. Pentagon relation

$$\Phi_{1,2,34}\Phi_{12,3,4} = \Phi_{2,3,4}\Phi_{1,23,4}\Phi_{1,2,3} \qquad (\text{in } T_4[[h]]), \tag{17.2}$$

2. Hexagon relations

$$B_{12,3} = \Phi_{3,1,2}B_{13}\Phi_{1,3,2}^{-1}B_{23}\Phi_{1,2,3}, \tag{17.3}$$
$$B_{1,23} = \Phi_{2,3,1}^{-1}B_{13}\Phi_{2,1,3}B_{12}\Phi_{1,2,3}^{-1}, \tag{17.4}$$

in $T_3[[h]]$.

Let $U(a, b)$ be the free algebra over k generated by two elements a and b, completed with respect to the grading $\deg a = \deg b = 1$. Consider the homomorphism

$$\Pi : U(a, b) \to T_3[[h]],$$
$$a \to ht_{12},$$
$$b \to ht_{23}.$$

Now let $F(a, b)$ be the free Lie algebra over k generated by two elements a and b, completed with respect to the grading $\deg a = \deg b = 1$ and let I be the ideal generated by a and b. It is clear that $U(a, b)$ is the (topological) universal enveloping algebra of $F(a, b)$, and is thus a (topological) Hopf algebra in a natural way.

Definition: An element $\Phi \in U(a, b)$ is called a *Lie associator* if the following conditions are satisfied:

1. $\Pi(\Phi) \in T_3[[h]]$ is an associator,
2. There exists $x \in F(a, b)$ such that $\Phi = e^x$, i.e $\Delta(\Phi) = \Phi \otimes \Phi$,
3. We have

$$\Phi_{123}^{-1} = \Phi_{321}. \tag{17.5}$$

Example (the KZ associator): Define $\Phi_{KZ} \in U(a,b)$ as the monodromy from 0 to 1 of the differential equation

$$\frac{dF}{dz} = \frac{1}{2i\pi}(\frac{a}{z} + \frac{b}{z-1})F$$

as in Lecture 16. We have already seen that $\Pi(\Phi_{KZ})$ is an associator. Moreover $\Delta(F_0) = F_0 \otimes F_0$ and $\Delta(F_1) = F_1 \otimes F_1$. Indeed, ΔF_0 and $F_0 \otimes F_0$ are both solutions of

$$\frac{dH}{dz} = \frac{1}{2i\pi}(\frac{\Delta a}{z} + \frac{\Delta b}{z-1})H$$

having expansions of the form

$$F(z) = (1 \otimes 1 + \sum_{n \geq 1} z^n F^{(n)}) z^{\hbar \Delta a} \tag{17.6}$$

and such a solution is unique. A similar reasoning holds for F_1. Thus, we have $\Delta(\Phi_{KZ}) = \Phi_{KZ} \otimes \Phi_{KZ}$, i.e Φ_{KZ} is a grouplike element of $U(a,b)$ and $\Phi_{KZ} = e^x$ for some $x \in F(a,b)$. Notice that by the same argument as in Problem 16.1, we have

$$(\Phi_{KZ}^{123})^{-1} = \Phi_{KZ}^{321}.$$

17.1.2 The space of Lie associators

Let $M_1 \subset exp(F(a,b)) \subset U(a,b)$ be the space of all Lie associators, and let $M_1^{(n)}$ be the space of all elements in $exp(F(a,b))/I^{n+1}$ satisfying (17.2-17.5). It is clear that $M_1^{(n)}$ is an algebraic variety defined over $\mathbb{Q}$, and that $M_1 = \varprojlim M_1^{(n)}$ is a pro-algebraic variety defined over $\mathbb{Q}$.

From the construction of Φ_{KZ} we deduce that M_1 has a point over $\mathbb{C}$, and therefore over any algebraically closed field $k \supset \mathbb{Q}$. We will see in the next section that in fact it has a point defined over $\mathbb{Q}$.

Problem 17.1. Show that any Lie associator Φ satisfies the congruence $\Phi \equiv 1 \pmod{I^2}$. In other terms, $M_1^{(2)}$ consists of one point.

17.2 The Grothendieck-Teichmuller group

17.2.1 Definition

Let $\mathcal{C}$ be a braided monoidal category with commutativity isomorphism β and associativity isomorphism Φ. We would like to construct from these some new commutativity and associativity isomorphisms. Recall that we have an action of B_n on $V^{\otimes n}$ for any $V \in Ob(\mathcal{C})$ (see 14.1.4) and more generally an action of B_n on $\bigoplus_{\sigma \in S_n} V_{\sigma(1)} \otimes \ldots \otimes V_{\sigma(n)}$ for any objects $V_1, \ldots V_n$ in $\mathcal{C}$. Notice that the restriction of this action to the Pure Braid group PB_n is defined on $V_1 \otimes \ldots \otimes V_n$.

In particular, for $n = 2,3$ this defines an action of $PB_2 = \mathbb{Z}\sigma^2$ on $V_1 \otimes V_2$ (given by $\Pi_{V_1,V_2}(\sigma) = \beta_{V_1,V_2}\beta_{V_2,V_1}$), and an action of $PB_3 =< b_1^2, b_2^2, (b_1b_2)^3 >$ on $V^{\otimes 3}$ (given by

$$\Pi_{V_1,V_2,V_3}(b_1^2) = \beta_{V_2,V_1}\beta_{V_1,V_2},$$
$$\Pi_{V_1,V_2,V_3}(b_2^2) = \Phi^{-1}_{V_1,V_2,V_3}\beta_{V_3,V_2}\beta_{V_2,V_3}\Phi_{V_1,V_2,V_3},$$

and a more complicated formula for $(b_1b_2)^3$).

It is clear that the most general construction of new commutativity and associativity isomorphisms which works in an arbitrary monoidal category has the form:

$$\beta'_{V_1,V_2} = \beta_{V_1,V_2}(\Pi_{V_1,V_2}(\sigma))^{\frac{\lambda-1}{2}} \qquad \lambda \in 2\mathbb{Z}+1, \tag{17.7}$$
$$\Phi'_{V_1,V_2,V_3} = \Phi_{V_1,V_2,V_3}.f(\Pi_{V_1,V_2,V_3}(b_1^2), \Pi_{V_1,V_2,V_3}(b_2^2), \Pi_{V_1,V_2,V_3}((b_1b_2)^3)), \tag{17.8}$$

where $f(X,Y,Z)$ is in the free group generated by three variables. Unfortunately, this is not at all efficient:

Proposition 17.1 (Drinfeld). *The maps 17.7 and 17.8 define new commutativity and associativity isomorphisms (i.e satisfying pentagon and hexagon relations) if and only if* $\lambda = \pm 1$ *and* $f = 1$.

Things improve however if our category $\mathcal{C}$ is K-linear, defined over $K = k[[h]]$ and *quasi-symmetric*: $\beta_{V_2,V_1}\beta_{V_1,V_2} \equiv 1 \pmod h$ for any $V_1, V_2 \in Ob(\mathcal{C})$. In this case, it makes sense to consider transformations (17.7-17.8) for arbitrary $\lambda \in k$ and f of the form $exp(P(\ln x, \ln y))$ where $P \in F(\ln x, \ln y)$ is a Lie formal series (here, $x \to b_1^2$, $y \mapsto b_2^2$; we leave it to the reader to check that $(b_1b_2)^3 = b_1^2b_2b_1^2b_2$ can then be expressed via b_1^2 and b_2^2). Let us denote the transformations (17.7-17.8) by $\beta' = (\lambda, f).\beta$ and $\Phi' = (\lambda, f).\Phi$.

Definition: Let $\overline{GT}(k)$ be the set of all pairs $(\lambda, f) \in k \times F(a,b)$ such that for any $(\lambda, f) \in k \times F(a,b)$ such that for any $k[[h]]$-linear, braided quasisymmetric category $(\mathcal{C}, \beta, \Phi)$, the elements $(\lambda, f).\beta$ and $(\lambda, f)\Phi$ define a new structure of a quasisymmetric braided category.

Notice that the action (17.7-17.8) endows $\overline{GT}(k)$ with a semigroup structure. It is called the *Grothendieck-Teichmuller semigroup*. More precisely, the semigroup structure on $\overline{GT}(k)$ is given by $(\lambda_1, f_1).(\lambda_2, f_2) = (\lambda, f)$ where

$$\lambda = \lambda_1\lambda_2,$$
$$f(a,b) = f_1(f_2(a,b)a^{\lambda_2}f_2(a,b)^{-1}, b^{\lambda_2})f_2(a,b).$$

The group of invertible elements of $\overline{GT}(k)$ is called the *Grothendieck-Teichmuller group* and is denoted $GT(k)$.

Claim:

$$GT(k) = \{(\lambda, f) \in \overline{GT}(k) \mid \lambda \neq 0\}.$$

Let us also set $GT_{\lambda_0}(k) = \{(\lambda, f) \in \overline{GT}(k) \mid \lambda = \lambda_0\}$. Notice that $GT_1(k)$ is a subgroup of $GT(k)$.

The group $GT_1(K)$ admits the following presentation:

Proposition 17.2. *The group $GT_1(k)$ is the group of elements $g \in exp(F(a,b))$ satisfying*

$$g(a,b) = g(b,a)^{-1}, \tag{17.9}$$

$$g(a,b)g(b,c)g(c,a) = 1 \qquad \text{for } abc = 1, \tag{17.10}$$

$$g(x_{12}, x_{23}x_{24})g(x_{13}x_{23}, x_{34}) = g(x_{23}, x_{34})g(x_{12}x_{13}, x_{23}x_{34})g(x_{12}, x_{23}), \tag{17.11}$$

where $x_{ij} \in PB_4$ are defined by the picture

$X_{ij} =$ 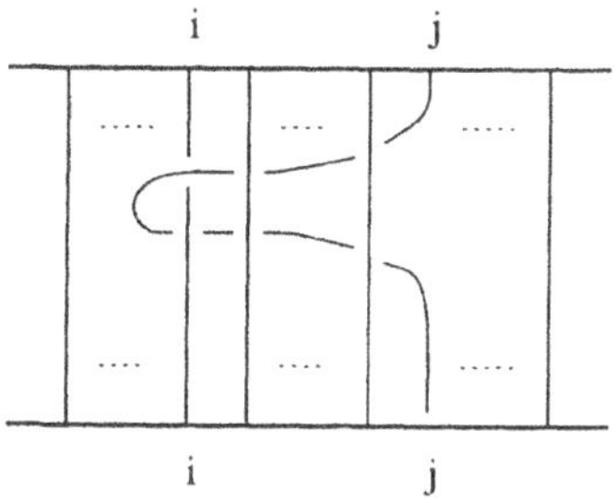

or by the formula $x_{ij} = (b_{j-1} \dots b_{i+1})b_i^2(b_{j-1} \dots b_{i+1})^{-1}$.

Proof: see [Dr4]. Relations (17.9-17.10) follow from the hexagon relations, while (17.11) follows from the pentagon relation.

Notice that $GT_1(k)$ only acts on the associator Φ without changing β. In particular, it acts on $M_1(k)$ by

$$(1, f).\Phi(a,b) = \Phi(a,b).f(a, \Phi(a,b)b\Phi(a,b)^{-1}).$$

Theorem 17.1. *The action of $GT_1(k)$ on $M_1(k)$ is free and transitive.*

Proof: see [Dr4].

Corollary 17.1. *M_1 has a point defined over $\mathbb{Q}$.*

Proof: By Theorem 17.1, the pro-unipotent group $GT_1(k)$ acts freely and transitively on $M_1(k) = \varprojlim M_1^{(n)}(k)$, therefore the maps $M_1^{(n+1)} \twoheadrightarrow M_1^{(n)}$ are affine space bundles. Since $M_1(k)$ has a point defined over $\mathbb{C}$, it has a point defined over $\overline{\mathbb{Q}}$ in any fiber of $\pi_{n+1} : M_1^{(n+1)} \twoheadrightarrow M_1^{(n)}$. We can now construct a point $x = (x_n)$ in $M_1(k)$ defined over $\mathbb{Q}$ in the following inductive way: let x_1 be the only point in $M_1^{(1)}(k)$ (it is defined over $\mathbb{Q}$) and suppose we have already constructed $x_1, \dots x_n$ all defined over $\mathbb{Q}$. Let y_{n+1} be a point in $\pi_{n+1}^{-1}(x_n)$ defined over $\overline{\mathbb{Q}}$. The element

$$x_{n+1} = \frac{1}{|Y|} \sum_{y \in Y} y$$

where Y is the orbit of y_{n+1} under $Gal(\overline{\mathbb{Q}}/\mathbb{Q})$ is defined over $\mathbb{Q}$ and $\pi_{n+1}(x_{n+1}) = x_n$. Thus $M_1(\mathbb{Q}) \neq \emptyset$.

□

Consider the semigroup map

$$\theta : \overline{GT}(k) \to k, \tag{17.12}$$
$$(\lambda, f) \mapsto \lambda. \tag{17.13}$$

Let $\tilde{\mathfrak{gt}}(k) = \{x \in \mathrm{Lie}(GT(k)) \mid d\theta(x) = 1\}$. It is an affine space parallel to $\mathfrak{gt}_1(k) = \mathrm{Lie}(GT_1(k))$. The following results are due to Drinfeld:

Theorem 17.2. *Let $x \in \tilde{\mathfrak{gt}}(k)$. The one-parameter subgroup e^{tx} defined in a formal neighborhood of 1 extends to $t \in k$, and the group homomorphism*

$$k^* \to GT(k), \tag{17.14}$$
$$e^t \mapsto e^{tx} \tag{17.15}$$

extends to a polynomial splitting $\Psi_x : k \to \overline{GT}(k)$ of θ.

Here the term "polynomial" should be understood as projective limit of polynomial maps.

Corollary 17.2. *The map $\theta : \overline{GT}(k) \to k$ is a surjection.*

In particular, we obtain in this fashion a map $\eta : \tilde{\mathfrak{gt}}(k) \to GT_0(k)$, by the assignment $x \mapsto \lim_{e^t \to 0} e^{tx}$.

Theorem 17.3. *The map $\eta : \tilde{\mathfrak{gt}}(k) \to GT_0(k)$ is a bijection.*

As a corollary, we see that $GT_0(k) \neq \emptyset$. One important consequence is the following:

Exercise 17.1. Show that the action of $GT_0(k)$ transforms any braided monoidal category into a symmetric one.

Thus an important consequence of Theorem 17.3 is that there exists a canonical structure of a symmetric category on any quasisymmetric category associated to any element of $GT_0(k)$. By canonical we mean here that this assignment respects tensor functors.

17.2.2 The action of $GT_1(k)$ on completed braid groups

Recall that a braided monoidal structure on a category induces a representation of the Braid group $\rho_{\mathcal{C}} : B_n \to \mathrm{Aut}(V^{\otimes n})$ for any $V \in Ob(\mathcal{C})$. In particular, these representations change when we change the braided monoidal structure on a quasisymmetric category $\mathcal{C}$ by acting by an element of $GT(k)$. More precisely, we obtain in this way the following action by outer automorphisms of $GT(k)$ on the $\widehat{B_n(k)}$, the pro-unipotent completion of $B_n(k)$ with respect to PB_n (recall that a pro-unipotent completion over k of a group G is the inverse limit of images of homomorphisms $G \to U$ where U is a unipotent algebraic group over k):

$$\begin{aligned}(\lambda, f) : b_1 &\mapsto b_1 (b_1^2)^{\frac{\lambda-1}{2}}, \\ b_i &\mapsto f(y_i, b_i^2)^{-1} b_i (b_i^2)^{\frac{\lambda-1}{2}} f(y_i, b_i^2),\end{aligned}$$

where y_i is given by the picture

y_i =

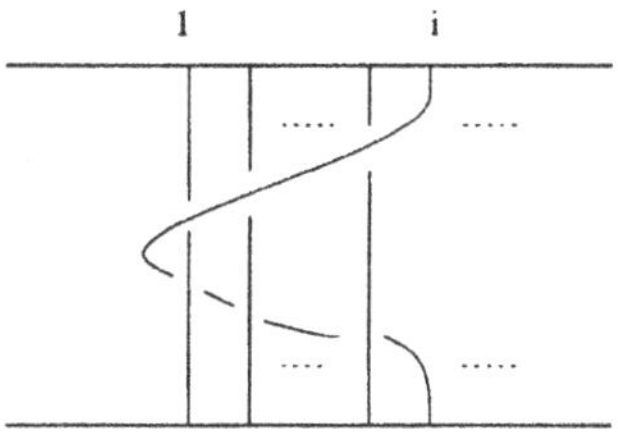

or by the formula $y_i = x_{1,i} \ldots x_{i-1,i} \in PB_n(k)$. We have $\rho_{(\lambda,f).\mathcal{D}} = (\lambda, f)\rho_{\mathcal{D}}$.

Theorem 17.4. *The action of $GT(k)$ on $\widehat{B_n(k)}$ is faithful, compatible with the inclusions $\widehat{B_n(k)} \to \widehat{B_{n+1}(k)}$, and is trivial on the symmetric group $S_n = \widehat{B_n(k)}/\widehat{PB_n(k)}$.*

Proof (sketch): It is clear that this action is compatible with the inclusions $\widehat{B_n(k)} \to \widehat{B_{n+1}(k)}$, and that it is trivial on the symmetric group. To see that the action is faithful, it is enough to check it on B_3.

□

Conversely, it is not known whether any automorphism of $\widehat{B_n(k)}$ with the above properties is of this form (up to inner automorphisms).

17.2.3 Drinfeld's conjecture

Notice that $\mathfrak{gt}_1(k)$ has a decreasing filtration $\mathfrak{gt}_1(k) = \cup_{n\geq 0}\mathfrak{gt}_1^{(n)}(k)$ where $\mathfrak{gt}_1^{(n)}(k)$ is the set of elements vanishing in degree $\leq n-1$. In other words, we have $\mathfrak{gt}_1^{(n)}(k) \supset \mathfrak{gt}_1^{(n+1)}(k)$, $[\mathfrak{gt}_1^{(n)}(k), \mathfrak{gt}_1^{(m)}(k)] \subset \mathfrak{gt}_1^{(n+m)}(k)$ and the Lie subalgebras $\mathfrak{gt}_1^{(n)}(k)$ are of finite codimension. Let $\mathfrak{grt}_1(k) = gr(\mathfrak{gt}_1(k))$. This is a $\mathbb{Z}_+$-graded Lie algebra with finite dimensional homogeneous components.

Remark: Any particular choice of $x \in \tilde{\mathfrak{gt}}_1(k)$ gives us an isomorphism $\mathfrak{grt}_1(k) \simeq \mathfrak{gt}_1(k)$, induced by the subgroup $H_x = \{e^{tx},\ t \in k\} \subset GT(k)$. Namely $\mathfrak{gt}_1^{(n)}(k) = \bigoplus_{k\geq n} E_k$ where $E_k = \{u \in \mathfrak{gt}_1(k) \mid \mathrm{Ad}(e^{tx}).u = e^{kt}u\}$. Notice however that there are no canonical elements of $\tilde{\mathfrak{gt}}_1(k)$ and hence no canonical isomorphism $\mathfrak{grt}_1(k) \simeq \mathfrak{gt}_1(k)$.

The dimensions of the homogeneous components have been computed up to order 9:

Degree	0	1	2	3	4	5	6	7	8	9
dim $\mathfrak{grt}_1^{(d)}$	0	0	0	1	0	1	0	1	1	1

Conjecture 1 (Drinfeld). $\mathfrak{grt}_1(k)$ *is the free Lie algebra in generators* (x_n) *of degree* $2n+1$ *for* $n \geq 1$.

Remark: The Grothendieck-Teichmuller group appeared in Grothendieck's "Esquisse d'un programme" ([Gro]), in relation to the group $\mathrm{Gal}(\overline{\mathbb{Q}}/\mathbb{Q})$. See [Dr4] for more details.

Lecture 18

Fiber functors and Tannaka-Krein duality

18.1 Tensor categories

So far, we have considered the notion of a (braided) monoidal structure on any category $\mathcal{C}$. We now focus on additive categories. As usual k is a field of characteristic zero.

Definition: A *tensor category* is an additive category $\mathcal{C}$ equipped with a monoidal structure for which $\otimes$ is additive, i.e there are functorial isomorphisms

$$(X \oplus Y) \otimes Z = (X \otimes Z) \oplus (Y \otimes Z),$$
$$Z \otimes (X \oplus Y) = (Z \otimes X) \oplus (Z \otimes Y)$$

for any $X, Y, Z \in \mathcal{O}b(\mathcal{C})$. Notice that the composition

$$\begin{aligned}((X_1 \oplus X_2) \otimes Y) \otimes Z &\longrightarrow ((X_1 \otimes Y) \otimes Z) \oplus ((X_2 \otimes Y) \otimes Z) \\ &\overset{\Phi_{X_1,Y,Z} \oplus \Phi_{X_2,Y,Z}}{\longrightarrow} (X_1 \otimes (Y \otimes Z)) \oplus (X_2 \otimes (Y \otimes Z)) \\ &\longrightarrow (X_1 \oplus X_2) \otimes (Y \otimes Z)\end{aligned}$$

coincides with $\Phi_{(X_1 \oplus X_2),Y,Z}$ by functoriality.

In a similar way, we define the notion of a braided tensor category (again, by functoriality, the braiding β is automatically additive).

Examples: All the (braided) monoidal categories we have considered so far are (braided) tensor categories.

18.2 Fiber functor

18.2.1 First example

Let A be an associative k-algebra with unit, and let $\operatorname{Rep} A$ be the category of (left) A-modules. There is a natural forgetful functor $F : \operatorname{Rep} A \to \operatorname{Vec}_k$ which assigns to any A-module its underlying vector space and to any morphism of A-modules the underlying linear map.

Lemma 18.1. *The functor F is representable:*

$$F(X) = \mathrm{Hom}_{\mathrm{Rep}\,A}(A, X)$$

where A acts on itself by left multiplication.

Proof: Clear.

Now consider the algebra $\mathrm{End}(F)$ of endomorphisms of F. Recall that an element $\varphi \in \mathrm{End}(F)$ is a collection of maps

$$\varphi_X : F(X) \to F(X), \qquad X \in \mathcal{O}b(\mathcal{C})$$

making the following diagram commutative

$$\begin{array}{ccc} F(X) & \xrightarrow{\varphi_X} & F(X) \\ {\scriptstyle F(a)}\downarrow & & {\scriptstyle F(a)}\downarrow \\ F(Y) & \xrightarrow{\varphi_Y} & F(Y) \end{array}$$

for any $X, Y \in \mathcal{O}b(\mathcal{C})$ and any map $a : X \to Y$.

Theorem 18.1. *There is a canonical isomorphism $\theta : A \tilde{\to} \mathrm{End}(F)$ given by*

$$\theta(a)_X = a_{|X}. \tag{18.1}$$

Proof: It is clear that 18.1 defines an element of $\mathrm{End}(F)$ and that the map θ is injective (since $a_{|A} \neq a'_{|A}$ if $a \neq a'$). Now let $\varphi \in \mathrm{End}(F)$ and consider φ_A. Set $a = \varphi_A(1)$. Any $b \in A$ defines a map $r_b \in \mathrm{Hom}_{\mathrm{Rep}\,A}(A, A)$ given by the *right* multiplication $x \mapsto xb$. Thus $\varphi_A(b) = \varphi_A(r_b(1)) = r_b(\varphi_A(1)) = ab$ and $\varphi_A = l_a$, the left multiplication (in other words, the only endomorphisms of A commuting with the left multiplication by any element are given by right multiplication by some element). A similar reasoning shows that $\varphi_{A\otimes V} = a_{|A\otimes V}$ for any free A-module $A \otimes V$ ($V \in \mathrm{Vec}_k$). But every A-module is a quotient of a free module. Thus θ is surjective.

□

We now generalize the above construction to the case of bialgebras and tensor categories. Let us start with a well-known example:

18.2.2 Tannaka-Krein duality

Let G be a finite group and set $A = k[G]$. Consider the category $\mathcal{C} = \operatorname{Rep} A$ ($= \operatorname{Rep} G$) with its usual Hopf algebra structure. The forgetful functor $F : \mathcal{C} \to \mathrm{Vec}_k$ has a natural tensor structure

$$J : F(X) \otimes F(Y) \xrightarrow{\sim} F(X \otimes Y)$$

given by the identity map.

Now consider the group $\mathrm{Aut}_{\otimes}(F)$ of tensor automorphisms of F, i.e. of elements $\gamma \in \mathrm{Aut}(F)$ making the following diagram commutative

$$\begin{array}{ccc} F(X) \otimes F(Y) & \xrightarrow{\gamma_X \otimes \gamma_Y} & F(X) \otimes F(Y) \\ J_{XY} \downarrow & & J_{XY} \downarrow \\ F(X \otimes Y) & \xrightarrow{\gamma_{X \otimes Y}} & F(X \otimes Y) \end{array}$$

for any $X, Y \in \mathcal{O}b(\mathcal{C})$.

By Theorem 18.1, $\mathrm{End}(F) \simeq A$. If $a \in A$ then $a_{|X} \otimes a_{|Y} = a \otimes a \in \mathrm{End}(F(X) \otimes F(Y))$ while $a_{|X \otimes Y} = \Delta(a) \in \mathrm{End}(F(X \otimes Y))$. Thus $a \in \mathrm{Aut}_{\otimes}(F)$ if and only if a is a grouplike element of $k[G]$, i.e if and only if $a \in G$. Thus we have reconstructed the group G from its tensor category of representations together with a forgetful functor.

This is an example of the following general theorem:

Theorem 18.2. *Let $\mathcal{C}$ be a symmetric tensor category defined over an algebraically closed field k of characteristic zero. Suppose that $\mathcal{C}$ is abelian, has finitely many indecomposable objects, and let $F : \mathcal{C} \to \mathrm{Vec}_k$ be an exact and faithful braided tensor functor. Let $G = \mathrm{Aut}_{\otimes}(F)$. Then G is a finite group and $\mathcal{C} \simeq \operatorname{Rep} G$.*

Proof: See [DM].

An exact and faithful braided tensor functor $F : \mathcal{C} \to \mathrm{Vec}_k$ from a symmetric tensor category to the symmetric category Vec_k is called a *fiber functor.*

Remark: The above theorem remains true if we drop the finiteness condition on the number of indecomposable objects in $\mathcal{C}$ and replace "finite group" by "pro-algebraic group".

Remark: The term "fiber functor" comes from the following construction of algebraic fundamental groups:

Let X be a smooth, connected, locally closed algebraic subvariety of a projective space defined over a field k of characteristic zero and suppose that $\overline{X}$ is smooth . Consider the category of algebraic vector bundles on X equipped with a flat connection with regular singularities (i.e the connection is smooth on X and has only regular singularities at $\overline{X}\backslash X$). It is a tensor category in a natural way. Fix a point $x \in X$ and consider the functor

$$\begin{aligned} F_x : \mathcal{C} &\to \mathrm{Vec}_k, \\ V &\to V_x \text{ (fiber of } V \text{ at } x). \end{aligned}$$

The group $\mathrm{Aut}_{\otimes}(F_x)$ is denoted $\pi_1^{alg}(X, x)$ (and is called the algebraic fundamental group of X based at x). If $k = \mathbb{C}$, the monodromy representation defines an action of $\pi_1(X, x)$ (usual fundamental group) on V_x for any flat vector bundle V over X, and thus a map

$$\pi_1(X, x) \to \pi_1^{alg}(X, x).$$

The *Riemann-Hilbert* correspondence states that $\pi_1^{alg}(X, x)$ is the pro-algebraic completion of $\pi_1(X, x)$.

18.2.3 Tannaka-Krein duality for bialgebras

The above principle generalizes to the case of nonsymmetric categories. Let A be an associative algebra with unit, $\mathcal{C} = \mathrm{Rep}\, A$ and denote by $F : \mathcal{C} \to \mathrm{Vec}_k$ the forgetful functor.

Theorem 18.3. *Any tensor structure on $\mathcal{C}$ together with a tensor structure on F equips A with a bialgebra structure such that $\mathcal{C} \simeq \mathrm{Rep}\, A$ as tensor categories.*

Proof: By Theorem 18.1, we have $A \simeq \mathrm{End}(F)$. To define a coproduct $A \to A \otimes A$, introduce the bifunctor:

$$\begin{aligned} F^2 : \mathcal{C} \times \mathcal{C} &\to \mathcal{C}, \\ (V, W) &\to F(V) \otimes F(W). \end{aligned}$$

Lemma 18.2. *We have* $\mathrm{End}(F^2) = \mathrm{End}(F) \otimes \mathrm{End}(F) = A \otimes A$.

Proof: It is clear that $\mathrm{End}(F) \otimes \mathrm{End}(F) \subset \mathrm{End}(F^2)$. Now let $f \in \mathrm{End}(F^2)$, and consider $f_{A,A} : A \otimes A \to A \otimes A$. As in Theorem 18.1, it is easy to see that $f_{A,A}$ is given by (left) multiplication by some $b \in A \otimes A$. Similarly, $f_{V,W} = l_b$ for any two free A-modules V, W. But any object in $\mathrm{Rep}\, A$ is a quotient of a free module. Thus $f \in A \otimes A$.

Now define a map $\Delta : \mathrm{End}(F) \to \mathrm{End}(F^2)$ in the following way: if $a \in \mathrm{End}(F)$ and $X, Y \in \mathcal{O}b(\mathcal{C})$, we set $\Delta(a)$ to be the composition:

$$F(X) \otimes F(Y) \xrightarrow{J_{X,Y}} F(X \otimes Y) \xrightarrow{a_{X \otimes Y}} F(X \otimes Y) \xrightarrow{J^{-1}_{X,Y}} F(X) \otimes F(Y)$$

or $\Delta(a)_{X,Y} = J^{-1}_{X,Y} \circ a_{|X\otimes Y} \circ J_{X,Y}$. It is an easy exercise to check that the map $\Delta : A \to A \otimes A$ is a coassociative homomorphism of algebras.

Finally, the counit $\epsilon : A \to k$ is given by the evaluation map

$$\epsilon : \operatorname{End}(F) \to \operatorname{End}(F(c)) = k,$$
$$a \to a_{F(c)}.$$

□

This gives us a new (and better) way of thinking of bialgebras: these are associative algebras together with a tensor structure on $\operatorname{Rep} A$ and an exact faithful functor $F : \operatorname{Rep} A \to \mathrm{Vec}_k$.

The above construction also admits a "braided version" where we replace "tensor structure on $\operatorname{Rep} A$" by "braided tensor structure on $\operatorname{Rep} A$" and "bialgebra" by "quasitriangular bialgebra".

Lecture 19

Quantization of finite dimensional Lie bialgebras

In this lecture, we construct a quantization of any finite dimensional Lie bialgebra defined over a field of characteristic zero, as in [EK1]. The construction depends on a choice of a Lie associator and is based on a Tannaka-Krein approach.

Let k be a field of characteristic zero, $(\mathfrak{g}_+, \delta_+)$ a finite dimensional Lie bialgebra defined over k and $\mathfrak{g}_- = \mathfrak{g}_+^*$. Let $\mathfrak{g} = D\mathfrak{g}_+$ be the Drinfeld double of $\mathfrak{g}$. Recall that $(\mathfrak{g}, \mathfrak{g}_+, \mathfrak{g}_-)$ is a Manin triple: $\mathfrak{g} = \mathfrak{g}_+ \oplus \mathfrak{g}_-$ as a vector space, the canonical form $<, >$ on $\mathfrak{g}$ is invariant, and $\mathfrak{g}_\pm$ are Lagrangian subalgebras. The Lie bialgebra structure on $\mathfrak{g}$ is given by $\delta_\mathfrak{g} = \delta_+ - \delta_-$ and it is quasitriangular with the r-matrix equal to the canonical element $r \in \mathfrak{g}_+ \otimes \mathfrak{g}_- \subset \mathfrak{g} \otimes \mathfrak{g}$. We set $\Omega = r + r^{op}$.

The first step is to construct a quantization of $\mathfrak{g}$.

19.1 Quantization of the Drinfeld double

We will use the Tannaka-Krein approach: we will construct a tensor category, a fiber functor, a tensor structure on this functor and apply the reconstruction theorems of the last lecture.

19.1.1 The Drinfeld category

Let $\mathcal{M}_\mathfrak{g}$ be the category whose objects are $\mathfrak{g}$-modules with $\mathrm{Hom}_{\mathcal{M}_\mathfrak{g}}(V, W) = \mathrm{Hom}_\mathfrak{g}(V, W)[[h]]$.

Let $\Phi(a, b) \in M_1(k)$ be a Lie associator defined over k (see Lecture 17). Then $(\mathcal{M}_\mathfrak{g}, \Phi(\Omega_{12}, \Omega_{23}), e^{h\Omega/2})$ is a braided tensor category which we will call the Drinfeld category in accordance with 15.3. As usual, we will omit the associativity morphisms when performing purely categorical constructions.

19.1.2 The forgetful functor

Definition: Let $\mathcal{A}$ denote the symmetric tensor category of topologically free K-modules $V[[h]]$, with trivial associativity and commutativity isomorphisms.

Denote by $F : \mathcal{M}_\mathfrak{g} \to \mathcal{A}$ the forgetful functor, assigning to any object in $\mathcal{M}_\mathfrak{g}$ the topologically free K-module associated to its underlying k-vector space. According to the Tannaka-Krein philosophy, we should now equip this forgetful functor with a tensor structure. For a general Lie algebra $\mathfrak{g}$ equipped with an invariant element Ω, there is no such obvious structure. However, in our case, $\mathfrak{g}$ is the double of a Lie bialgebra and we will use the polarization $\mathfrak{g} = \mathfrak{g}_+ \oplus \mathfrak{g}_-$ to produce such a structure.

19.1.3 The Verma modules

The construction of the fiber functor will be based in a vital way on the decomposition $\mathfrak{g} = \mathfrak{g}_+ \oplus \mathfrak{g}_-$. For this we introduce the following two objects of $\mathcal{M}_\mathfrak{g}$:

$$M_+ = \mathrm{Ind}_{\mathfrak{g}_+}^{\mathfrak{g}} c_+ = U(\mathfrak{g}) \otimes_{U(\mathfrak{g}_+)} c_+,$$
$$M_- = \mathrm{Ind}_{\mathfrak{g}_-}^{\mathfrak{g}} c_- = U(\mathfrak{g}) \otimes_{U(\mathfrak{g}_-)} c_-$$

where $c_\pm$ is the trivial $\mathfrak{g}_\pm$-module. By the Poincaré-Birkhoff-Witt theorem, the multiplication maps $U(\mathfrak{g}_+) \otimes U(\mathfrak{g}_-) \to U(\mathfrak{g})$ and $U(\mathfrak{g}_-) \otimes U(\mathfrak{g}_+) \to U(\mathfrak{g})$ are linear isomorphisms, and we have

$$M_+ = U(\mathfrak{g}_-)1_+,$$
$$M_- = U(\mathfrak{g}_+)1_-$$

where $1_\pm \in M_\pm$ is $\mathfrak{g}_\pm$-invariant. By analogy with the classical construction of highest weight representations of semisimple Lie algebras based on the decomposition $\mathfrak{g} = \mathfrak{n}_- \oplus \mathfrak{h} \oplus \mathfrak{n}_+$, we call $M_\pm$ the *Verma* modules and $1_\pm$ the *vacuum vectors*.

Now define the "coproduct maps" to be the $\mathfrak{g}$-homomorphisms

$$i_\pm : M_\pm \to M_\pm \otimes M_\pm$$

so that $i_\pm(1_\pm) = 1_\pm \otimes 1_\pm$ (this is possible as $\mathrm{Hom}_\mathfrak{g}(M_\pm, X) = X^{\mathfrak{g}_\pm}$, and $1_\pm \otimes 1_\pm$ is $\mathfrak{g}_\pm$-invariant). Notice that under the identification $M_\pm = U(\mathfrak{g}_\mp)$, the maps $i_\pm$ are nothing but the usual coproduct maps in universal enveloping algebras.

Lemma 19.1. *(Coassociativity of i_+ and i_-):*

$$(i_\pm \otimes 1) \circ i_\pm = (1 \otimes i_\pm) \circ i_\pm. \tag{19.1}$$

Proof: We have to show that the following diagram is commutative:

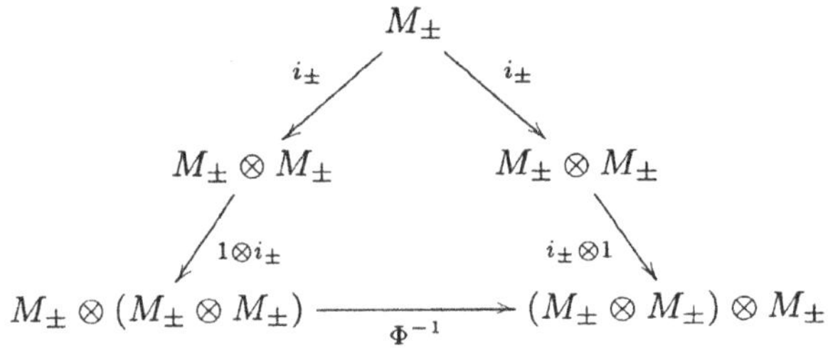

Since $i_\pm$ are identified with the usual coproduct maps $U(\mathfrak{g}_\mp) \to U(\mathfrak{g}_\mp) \otimes U(\mathfrak{g}_\mp)$, we have $(i_\pm \otimes 1) i_\pm = (1 \otimes i_\pm) i_\pm$. We claim that $\Phi \equiv 1$ on the image of $(i_\pm \otimes 1) i_\pm$. Indeed, $M_\pm$ is generated by $1_\pm$, so it is sufficient to check this equality on the vector $(i_\pm \otimes 1) i_\pm (1_\pm) = 1_\pm \otimes 1_\pm \otimes 1_\pm$. This in turn follows from the fact that Φ is a Lie associator, i.e a formal power series in Ω_{12} and Ω_{23} with constant coefficient 1, and from the fact that $\Omega(1_\pm \otimes 1_\pm) = 0$ (since $\Omega \in (\mathfrak{g}_+ \otimes \mathfrak{g}_-) \oplus (\mathfrak{g}_- \otimes \mathfrak{g}_+)$).

□

Exercise 19.1. Show that the usual counit maps $\epsilon_\pm : M_\pm \to k$ are morphisms in $\mathcal{M}_\mathfrak{g}$. In particular, the counit axioms hold: $(\epsilon_\pm \otimes 1) i_\pm = (1 \otimes \epsilon_\pm) i_\pm = 1$.

Lemma 19.2. *The assignment* $1 \to 1_+ \otimes 1_-$ *extends to an isomorphism of* $\mathfrak{g}$*-modules* $\phi : U(\mathfrak{g}) \tilde{\to} M_+ \otimes M_-$.

Proof: It is clear that ϕ preserves the standard filtration (recall that we have $M_\pm = U(\mathfrak{g}_\mp)$ as vector spaces). Hence it defines a map on the associated graded objects $S\mathfrak{g} \to S\mathfrak{g}_+ \otimes S\mathfrak{g}_-$, which is clearly the isomorphism induced by the identification $\mathfrak{g} = \mathfrak{g}_+ \oplus \mathfrak{g}_-$. This implies that ϕ itself is an isomorphism.

□

19.1.4 Tensor structure on the forgetful functor

Notice that by lemma 19.2, the forgetful functor $F : X \to X[[h]]$ is represented by the object $M_+ \otimes M_-$:

$$F(X) = \mathrm{Hom}_{\mathcal{M}_\mathfrak{g}}(M_+ \otimes M_-, X).$$

This allows us to make the following definition:

Definition: For $X, Y \in \mathcal{M}_\mathfrak{g}$, let

$$J_{X,Y} : F(X) \otimes F(Y) \to F(X \otimes Y),$$
$$x \otimes y \to J_{X,Y}(x \otimes y)$$

be the composition

$$M_+ \otimes M_- \overset{i_+\otimes i_-}{\to} M_+ \otimes M_+ \otimes M_- \otimes M_- \overset{\beta_{23}}{\to} M_+ \otimes M_- \otimes M_+ \otimes M_- \overset{x\otimes y}{\to} X \otimes Y.$$

The following proposition is crucial for the construction of quantization:

Proposition 19.1. *The collection of maps $(J_{X,Y})_{X,Y\in Ob(\mathcal{M}_{\mathfrak{g}})}$ is a tensor structure on F.*

Proof: First notice that, after the identification $F(X) \otimes F(Y) = F(X \otimes Y)$ as vector spaces, we have $J_{X,Y} \equiv 1 \pmod h$ (since $\Phi \equiv 1 \pmod h$ and $\beta \equiv \sigma \pmod h$). Hence $J_{X,Y}$ is an isomorphism. Next, it is easy to check that $J_{c,X} = J_{X,c} = 1$. We now check the relation

$$J_{X\otimes Y,Z} \circ (J_{X,Y} \otimes 1) = J_{X,Y\otimes Z} \circ (1 \otimes J_{Y,Z}). \tag{19.2}$$

As usual, we will give a pictorial proof. Represent the morphisms $i_\pm$ by

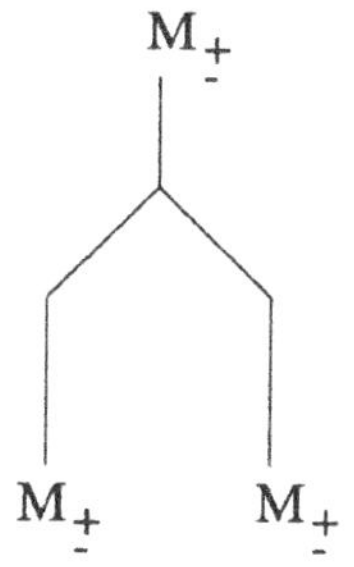

and the counit maps $M_\pm \to k$ by

By lemma 19.1, we have the coassociativity relation:

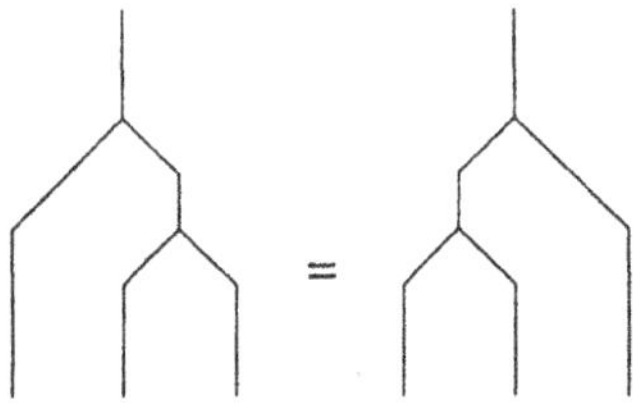

The map $J_{X,Y}$ is then represented as:

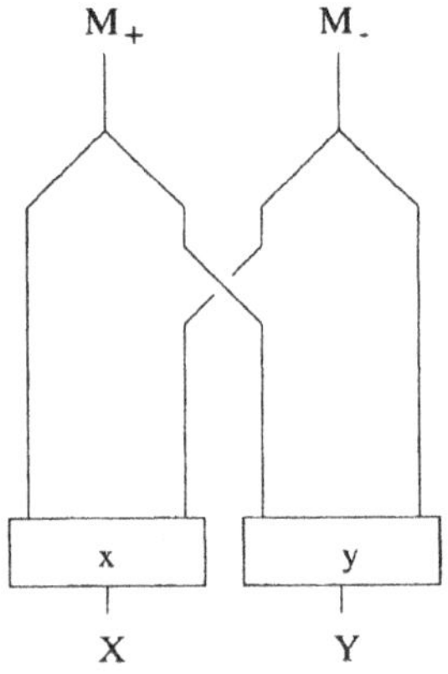

Identity (19.2) is now equivalent to the equality

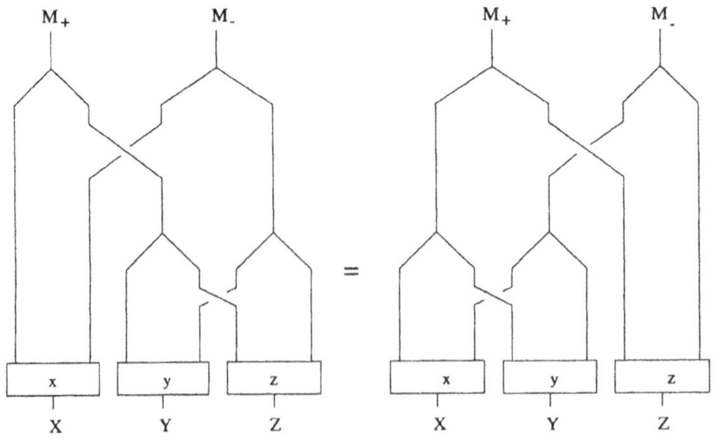

which is clear, as both sides are equal to

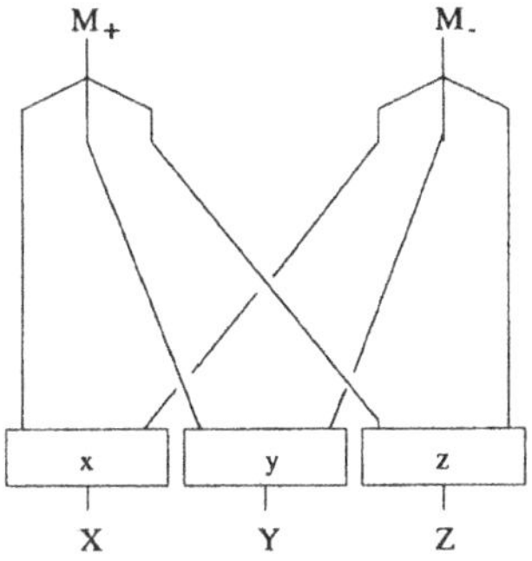

This concludes the proof of the proposition.

□

We now give an explicit formula for J (taking into account the associativity morphism). Let $x \in F(X)$, $y \in F(Y)$ and let us apply $J_{X,Y}(x \otimes y)$ to the vector $1_+ \otimes 1_-$. This means we should apply to $1_+ \otimes 1_-$ the following composition of maps:

$$M_+ \otimes M_- \overset{i_+ \otimes i_-}{\to} (M_+ \otimes M_+) \otimes (M_- \otimes M_-) \overset{\Phi_{1,2,34}}{\to} M_+ \otimes (M_+ \otimes (M_- \otimes M_-))$$
$$\overset{1 \otimes \Phi_{2,3,4}^{-1}}{\longrightarrow} M_+ \otimes ((M_+ \otimes M_-) \otimes M_-) \overset{\sigma_{23} e^{h\Omega_{23}/2}}{\longrightarrow} M_+ \otimes ((M_- \otimes M_+) \otimes M_-)$$
$$\overset{1 \otimes \Phi_{2,3,4}}{\longrightarrow} M_+ \otimes (M_- \otimes (M_+ \otimes M_-)) \overset{\Phi_{1,2,34}^{-1}}{\longrightarrow} (M_+ \otimes M_-) \otimes (M_+ \otimes M_-)$$
$$\overset{x \otimes y}{\to} X \otimes Y,$$

i.e we have

$$J_{X,Y}(x \otimes y)(1_+ \otimes 1_-)$$
$$= (x \otimes y)\big(\Phi_{1,2,34}^{-1}(1 \otimes \Phi_{2,3,4})\sigma_{23} e^{h\Omega_{23}/2}(1 \otimes \Phi_{2,3,4}^{-1})\Phi_{1,2,34}(1_+ \otimes 1_+ \otimes 1_- \otimes 1_-)\big)$$
$$= (x \otimes y)\big(\Phi_{1,2,34}^{-1}(1 \otimes \Phi_{2,3,4}) e^{h\Omega_{23}/2}(1 \otimes \Phi_{3,2,4}^{-1})\Phi_{1,3,24}(1_+ \otimes 1_- \otimes 1_+ \otimes 1_-)\big).$$

Set

$$J = (\phi^{-1} \otimes \phi^{-1})\big(\Phi_{1,2,34}^{-1}(1 \otimes \Phi_{2,3,4}) e^{h\Omega_{23}/2}(1 \otimes \Phi_{3,2,4}^{-1})\Phi_{1,3,24}(1_+ \otimes 1_- \otimes 1_+ \otimes 1_-)\big). \tag{19.3}$$

This is an element of $U(\mathfrak{g})^{\otimes 2}[[h]]$. The meaning of J is the following. If we identify $F(V)$ and V by $f \to f(1_+ \otimes 1_-)$, then the map

$$J_{V,W} : V \otimes W \simeq F(V) \otimes F(W) \to F(V \otimes W) \simeq V \otimes W$$

is given by the action of J. Thus (19.3) is an explicit formula for the tensor structure $J_{V,W}$.

19.1.5 Quantization of $\mathfrak{g}$

We now have all the ingredients to apply the Tannaka-Krein formalism. According to the reconstruction Theorem 18.3, the tensor structure J on the forgetful functor F equips $\operatorname{End} F \simeq U(\mathfrak{g})[[h]]$ with a bialgebra structure, with coproduct defined by

$$\Delta = J^{-1}\Delta_0 J, \qquad \epsilon = \epsilon_0, \tag{19.4}$$

where Δ_0 and ϵ_0 are the usual (undeformed) coproduct and counit respectively. Since $J \equiv 1 \pmod h$, this bialgebra is a deformation of the Hopf algebra $U(\mathfrak{g})$, and thus has a unique compatible Hopf algebra structure (see Proposition 8.5). Let us denote by $U_h(\mathfrak{g})$ this Hopf algebra. Notice that $U_h(\mathfrak{g})$ is nothing but the twist of $U(\mathfrak{g})[[h]]$ by the element $J \in U(\mathfrak{g})^{\otimes 2}[[h]]$. In particular, the antipode S is given by the formula

$$S(x) = QS_0(x)Q^{-1} \tag{19.5}$$

where S_0 is the usual (undeformed) antipode of $U(\mathfrak{g})[[h]]$ and $Q = m(S_0 \otimes 1)(J)$ (see Problem 9.2).

Theorem 19.1. *The Hopf algebra $U_h(\mathfrak{g})$ is a quantization of $\mathfrak{g}$.*

Proof: We will need the following lemma:

Lemma 19.3. *We have $J \equiv 1 + \frac{hr}{2} \pmod{h^2}$.*

Proof: By Problem 17.1, we have $\Phi \equiv 1 \pmod{h^2}$. Thus

$$\begin{aligned} J &\equiv (\phi^{-1} \otimes \phi^{-1})(e^{h\Omega_{23}/2}(1_+ \otimes 1_- \otimes 1_+ \otimes 1_-)) && \pmod{h^2} \\ &\equiv (\phi^{-1} \otimes \phi^{-1})((1 + \frac{h\Omega_{23}}{2})(1_+ \otimes 1_- \otimes 1_+ \otimes 1_-)) && \pmod{h^2} \\ &\equiv 1 + \frac{h}{2}(\phi^{-1} \otimes \phi^{-1})((r_{23} + r_{32})(1_+ \otimes 1_- \otimes 1_+ \otimes 1_-)) && \pmod{h^2} \end{aligned}$$

Since $r \in \mathfrak{g}_+ \otimes \mathfrak{g}_-$, $r_{21}(1_- \otimes 1_+) = 0$ and $r(1_- \otimes 1_+) = \sum_i g_i^+ 1_- \otimes g_i^- 1_+$, where g_i^+ and g_i^- are dual basis of $\mathfrak{g}_+$ and $\mathfrak{g}_-$ respectively. Thus,

$$J \equiv 1 + \frac{hr}{2} \pmod{h^2}.$$

□

Let us compute the quasiclassical limit of $U_h(\mathfrak{g})$. Let $x \in \mathfrak{g} \subset U_h(\mathfrak{g})$. We have:

$$\begin{aligned} \frac{\Delta(x) - \Delta^{op}(x)}{h} &\equiv \frac{1}{h}(J^{-1}\Delta_0(x)J - (J^{op})^{-1}\Delta_0^{op}(x)J^{op}) && \pmod h \\ &\equiv \frac{1}{2}([\Delta_0(x), r] - [\Delta_0(x), r^{op}]) && \pmod h \\ &\equiv [\Delta_0(x), r] && \pmod h \end{aligned}$$

since $r + r^{op} = \Omega$ is $\mathfrak{g}$-invariant. This proves that $U_h(\mathfrak{g})$ is a quantization of the Lie bialgebra $(\mathfrak{g}, \delta)$.

□

Corollary 19.1. *Set*

$$R = (J^{op})^{-1} e^{h\Omega/2} J. \tag{19.6}$$

Then $(U_h(\mathfrak{g}), R)$ *is a quasitriangular quantization of* $(\mathfrak{g}, r)$.

Proof: We have

$$\begin{aligned} R\Delta(x) &= (J^{op})^{-1} e^{h\Omega/2} J J^{-1} \Delta_0(x) J \\ &= (J^{op})^{-1} e^{h\Omega/2} \Delta_0(x) J \\ &= (J^{op})^{-1} \Delta_0(x) e^{h\Omega/2} J \\ &= (J^{-1}\Delta_0(x) J)^{op} (J^{op})^{-1} e^{h\Omega/2} J \\ &= \Delta^{op}(x) R. \end{aligned}$$

Exercise 19.2. Prove the two hexagon axioms:

$$(\Delta \otimes 1)R = R_{13}R_{23}, \qquad (1 \otimes \Delta)R = R_{13}R_{12}.$$

Remark: Notice that $U_h(\mathfrak{g}) = U(\mathfrak{g})[[h]]$ as an algebra (in a canonical way): only the coproduct and the antipode are deformed. Such quantizations are called *preferred quantizations.*

19.2 Quantization of finite-dimensional Lie bialgebras

In the last section, we constructed a quantization $U_h(\mathfrak{g})$ of the Drinfeld double $\mathfrak{g}$ of $\mathfrak{g}_+$. We shall now isolate inside of $U_h(\mathfrak{g})$ a Hopf subalgebra $U_h(\mathfrak{g}_+)$ which is a quantization of $\mathfrak{g}_+$, and show that $U_h(\mathfrak{g})$ is the quantum double of $U_h(\mathfrak{g}_+)$.

The first (naive) idea to set $U_h(\mathfrak{g}_+) = U(\mathfrak{g}_+)[[h]] \subset U_h(\mathfrak{g})$ fails, as this subalgebra is in general not closed under the coproduct. In order to construct the correct subalgebra, notice that since F is represented by $M_+ \otimes M_-$, there is a canonical isomorphism

$$\mathrm{End}_{\mathcal{M}_\mathfrak{g}}(M_+ \otimes M_-) \xrightarrow{\sim} \mathrm{End}(F) = U_h(\mathfrak{g})$$

The correct idea is to use this polarization.

Definition: Let $U_h(\mathfrak{g}_+) = F(M_-) \overset{i}{\hookrightarrow} \mathrm{End}_{\mathcal{M}_\mathfrak{g}}(M_+ \otimes M_-)$ where the embedding is defined in the following way: for $x \in F(M_-) = \mathrm{Hom}_{\mathcal{M}_\mathfrak{g}}(M_+ \otimes M_-, M_-)$, let $i(x)$ be the composition

$$M_+ \otimes M_- \overset{i_+\otimes 1}{\to} M_+ \otimes M_+ \otimes M_- \overset{1\otimes x}{\to} M_+ \otimes M_-.$$

Exercise 19.3. Check that $U_h(\mathfrak{g}_+)$ is a subalgebra of $U_h(\mathfrak{g})$.

Exercise 19.4. Show that the above embedding can be drawn as:

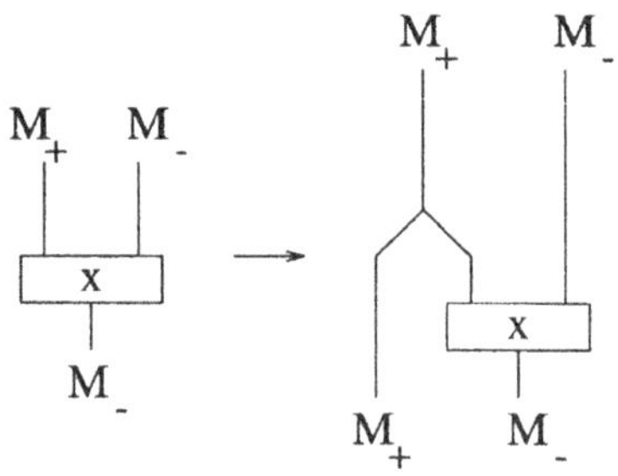

As a vector space, $U_h(\mathfrak{g}_+) = F(M_-) = U(\mathfrak{g}_+)[[h]]$, but the product is deformed. Indeed, it is explicitly given by the following rule: if $x, y \in F(M_-)$ then $m(x, y)$ is the composition:

$$M_+ \otimes M_- \overset{i_+\otimes 1}{\to} (M_+ \otimes M_+) \otimes M_- \overset{\Phi}{\to} M_+ \otimes (M_+ \otimes M_-) \overset{1\otimes y}{\to} M_+ \otimes M_- \overset{x}{\to} M_-.$$

Since $\Phi \equiv 1 \pmod{h^2}$, we have $m(x, y) \equiv m_0(x, y) \pmod{h^2}$. Therefore $U_h(\mathfrak{g}_+)$ is a deformation of $U(\mathfrak{g}_+)$. In particular, the image of the embedding of $U_h(\mathfrak{g}_+)$ in $U_h(\mathfrak{g}) = U(\mathfrak{g})[[h]]$ coincides with $U(\mathfrak{g}_+)[[h]]$ only modulo h^2.

In a similar way, we define the subalgebra $U_h(\mathfrak{g}_-) = F(M_+) \hookrightarrow U_h(\mathfrak{g})$ which is a deformation of $U(\mathfrak{g}_-)[[h]]$.

Theorem 19.2. *The subalgebras $U_h(\mathfrak{g}_+)$ and $U_h(\mathfrak{g}_-)$ are Hopf subalgebras.*

Proof: We start with the following result.

Lemma 19.4 (Polarization of the R-matrix). *We have*

$$R \in U_h(\mathfrak{g}_+) \otimes U_h(\mathfrak{g}_-) \subset U_h(\mathfrak{g}) \otimes U_h(\mathfrak{g}).$$

Proof: Recall the formula $R = J^{op-1} e^{h\Omega/2} J$, which we rewrite here in the more convenient way as $JR^{op} = \sigma e^{h\Omega/2} J\sigma$. In pictures, this means that R is uniquely defined by the relation:

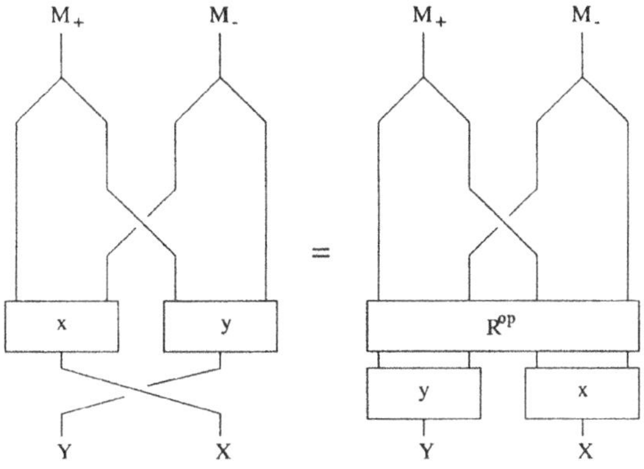

(19.7)

Notice that, by functoriality of the braiding, we have

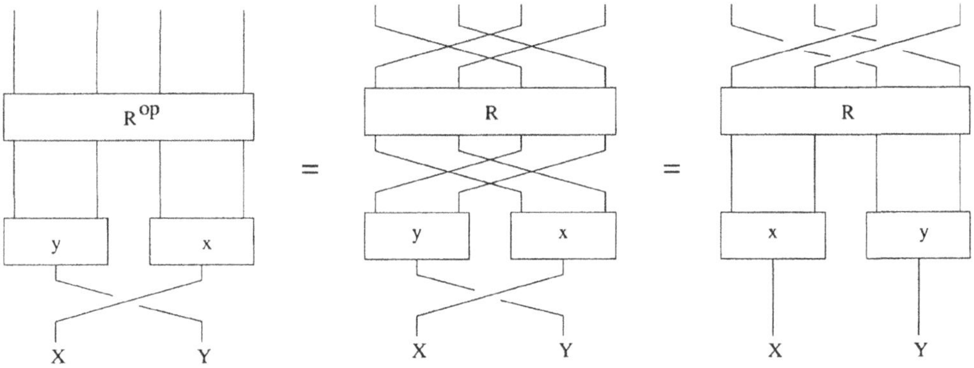

where the crossing lines denote the usual permutation of components. Therefore, by functoriality (19.7) is equivalent to:

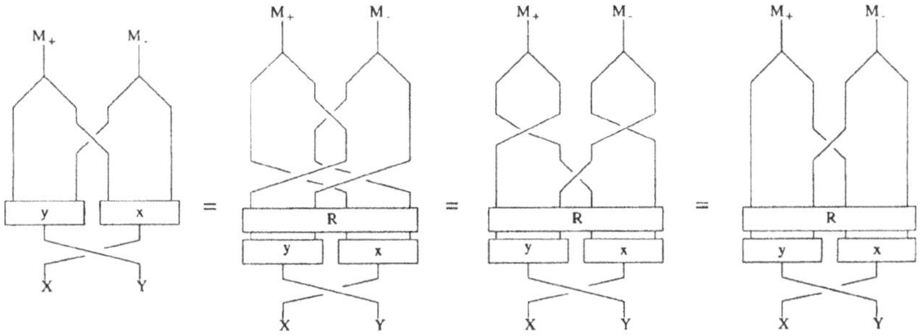

(19.8)

where we use the easily checked relation:

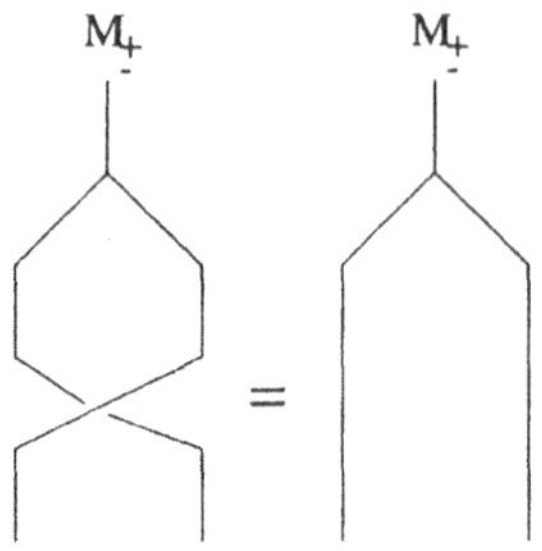

Thus we can rewrite the defining relation (19.8) for R as:

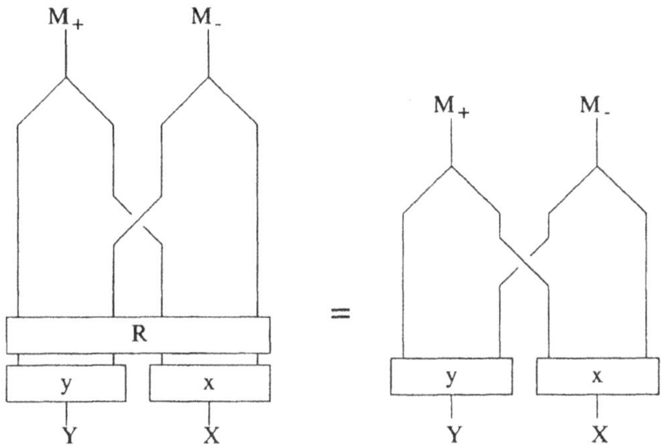

(19.9)

To prove the polarization of R, we use the following trick: notice that by the unit axioms, we have:

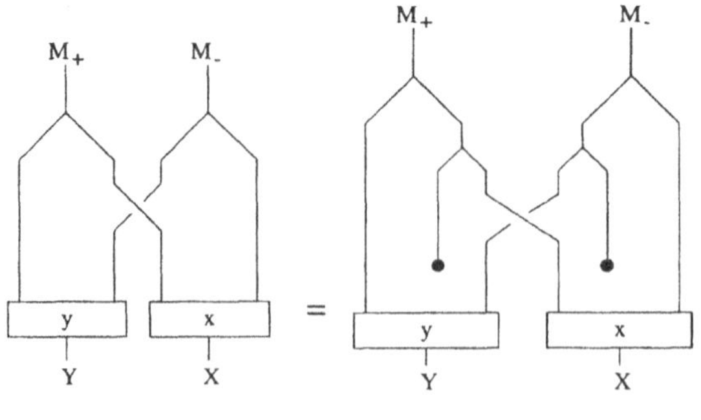

(19.10)

Now, applying (19.7) to $V = M_+$, $W = M_-$, $v = \epsilon_+ \otimes 1$, $w = 1 \otimes \epsilon_-$, we see that the r.h.s of (19.10) is equal to

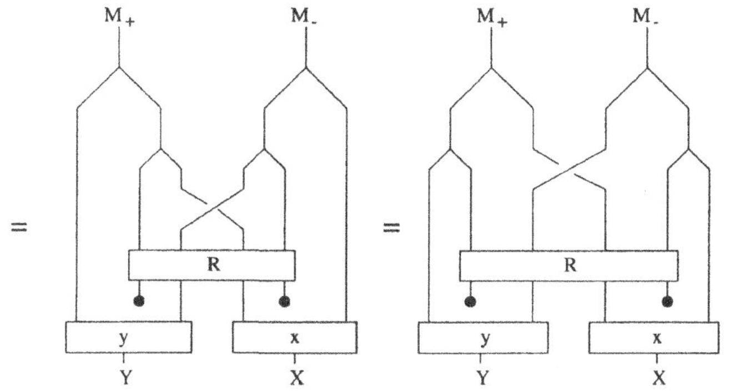

(19.11)

where we used coassociativity of $i_\pm$ in the last equality. Comparing the l.h.s of (19.9) and the r.h.s of (19.11), we see that we have proved

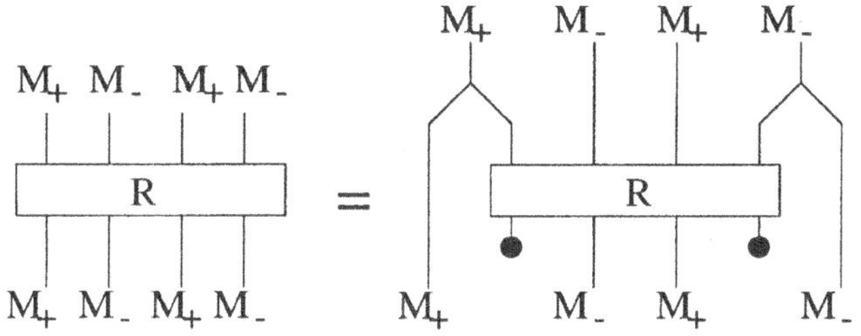

□

In particular, $R \in U_h(\mathfrak{g}_+) \otimes U_h(\mathfrak{g}_-)$.

Consider the two linear maps:

$$\mathcal{P}_\pm : (U_h(\mathfrak{g}_\mp))^* \to U_h(\mathfrak{g}_\pm),$$
$$\mathcal{P}_+(f) = (1 \otimes f)(R),$$
$$\mathcal{P}_-(f) = (f \otimes 1)(R)$$

and let $\mathcal{U}_\pm$ be the algebra generated by $\operatorname{Im} \mathcal{P}_\pm$. The hexagon relations for R

$$(\Delta \otimes 1)\Delta(R) = R_{13}R_{23},$$
$$(1 \otimes \Delta)\Delta(R) = R_{13}R_{12}$$

imply that $\mathcal{U}_\pm$ is closed under coproduct. Similarly, from $(S \otimes 1)R = R^{-1}$, we see that $\mathcal{U}_\pm$ is closed under antipode. Thus, Theorem 19.2 will follow from the following lemma:

Lemma 19.5. *We have*

$$U_h(\mathfrak{g}_\pm) \otimes_{k[[h]]} k((h)) = \widehat{\mathcal{U}_\pm} \otimes_{k[[h]]} k((h))$$

where $\widehat{\mathcal{U}_\pm}$ *is the* h*-adic completion of* $\mathcal{U}_\pm$ *and* $\otimes$ *is understood in the* h*-adically complete sense.*

Proof: By corollary 19.1, we have $R \equiv 1 + hr \pmod{h^2}$. Thus, for each $x \in \mathfrak{g}_\pm$, there exists $\overline{x} \in \widehat{\mathcal{U}_\pm} \otimes k((h)) \cap U_h(\mathfrak{g}_\pm)$ such that $\overline{x} \equiv x \pmod h$, and $h\overline{x} \in \widehat{\mathcal{U}_\pm}$. More generally, for any $x \in U(\mathfrak{g}_\pm)$ of degree $\leq m$, there exists $\overline{x} \in \widehat{\mathcal{U}_\pm} \otimes k((h)) \cap U_h(\mathfrak{g}_\pm)$ such that $\overline{x} \equiv x \pmod h$, and $h^m\overline{x} \in \widehat{\mathcal{U}_\pm}$. Now let $T \in U_h(\mathfrak{g}_\pm)$ and let x_0 be its reduction modulo h. Then $\frac{1}{h}(T - \overline{x_0}) \in U_h(\mathfrak{g}_\pm)$ and we can repeat the procedure. This yields a series $T = \sum_{m \geq 0} \overline{x}_m h^m$, $\overline{x_m} \in \widehat{\mathcal{U}_\pm} \otimes_{k[[h]]} k((h))$. This proves the lemma, and the theorem.

□

Remark: In fact, it can be proved that $\mathcal{U}_\pm = \widehat{\mathcal{U}_\pm}$, but we will not use this statement.

Corollary 19.2. *The Hopf algebras* $U_h(\mathfrak{g}_+)$ *and* $U_h(\mathfrak{g}_-)$ *are quantizations of* $\mathfrak{g}_+$ *and* $\mathfrak{g}_-$ *respectively.*

Proof: It is clear that $U_h(\mathfrak{g}_+)$ and $U_h(\mathfrak{g}_-)$ are deformations of $U(\mathfrak{g}_+)$ and $U(\mathfrak{g}_-)$. The fact that they are quantizations follows from the fact that $U(\mathfrak{g})$ is a quantization of $\mathfrak{g}$.

□

Notice that, as opposed to the case of $U_h(\mathfrak{g})$, the Hopf algebras $U_h(\mathfrak{g}_\pm)$ are not manifestly preferred quantizations of $\mathfrak{g}_\pm$. The question of existence of a preferred quantization for Lie bialgebras is unsettled. We do not expect that such quantizations exist for all Lie bialgebras.

Finally, let us show that the described quantization procedure commutes with the double construction:

Proposition 19.2. *The Hopf algebra* $U_h(\mathfrak{g})$ *is the quantum double of* $U_h(\mathfrak{g}_+)$.

Proof (sketch): It follows from the proof of Lemma 19.5 that $\widehat{\mathcal{U}_\pm}$ is the h-formal group dual to the QUE algebra $U_h(\mathfrak{g}_\mp)$, and that $U_h(\mathfrak{g}_\pm) = U_h(\mathfrak{g}_\mp)^*$ (as QUE algebras). Furthermore, the multiplication map $U_h(\mathfrak{g}_+) \otimes U_h(\mathfrak{g}_-) \to U_h(\mathfrak{g})$ is an isomorphism since it is a deformation of the isomorphism $U(\mathfrak{g}_+) \otimes U(\mathfrak{g}_-) \stackrel{\sim}{\to} U(\mathfrak{g})$. Finally, $U_h(\mathfrak{g})$ is quasitriangular, with R-matrix equal to the canonical element $R \in U_h(\mathfrak{g}_+) \otimes U_h(\mathfrak{g}_-) \subset U_h(\mathfrak{g}) \otimes U_h(\mathfrak{g})$. By Proposition 12.1, this identifies $U_h(\mathfrak{g})$ with the quantum double of $U_h(\mathfrak{g}_+)$.

□

19.3 Quasitriangular quantization

In this section, we adapt the quantization procedure of the double of a Lie bialgebra to an arbitrary quasitriangular Lie bialgebra. More precisely, we prove:

Theorem 19.3. *Let $(\mathfrak{g}, r)$ be a quasitriangular Lie bialgebra. There exists a preferred quantization $(U_h^{qt}(\mathfrak{g}), R)$ of $(\mathfrak{g}, r)$.*

Proof: Recall the construction of Proposition 4.1: set

$$\mathfrak{g}_+ = \{(1 \otimes f)r \,|f \in \mathfrak{g}^*\},$$
$$\mathfrak{g}_- = \{(f \otimes 1)r \,|f \in \mathfrak{g}^*\}.$$

Then $\mathfrak{g}_- = \mathfrak{g}_+^*$ and there is a canonical map of quasitriangular Lie bialgebras $\pi : D\mathfrak{g}_+ \to \mathfrak{g}$. Let $\mathcal{M}_\mathfrak{g}$ be the category of $\mathfrak{g}$-modules, with $\mathrm{Hom}_{\mathcal{M}_\mathfrak{g}}(V, W) = \mathrm{Hom}_\mathfrak{g}(V, W)[[h]]$, and let $\mathcal{M}_{D\mathfrak{g}_+}$ be the Drinfeld category as in section 1. The map π gives rise to a functor

$$\pi^* : \mathcal{M}_\mathfrak{g} \to \mathcal{M}_{D\mathfrak{g}_+}$$

which enables us to equip $\mathcal{M}_\mathfrak{g}$ with the pull-back braided monoidal structure (notice that $\pi(\Omega) = r + r^{op}$ is $\mathfrak{g}$-invariant by definition.). Let M_+ and M_- in $\mathcal{M}_{D\mathfrak{g}_+}$ be the Verma modules, and consider the forgetful functor $F : \mathcal{M}_\mathfrak{g} \to \mathcal{A}$, given by $F(V) = \mathrm{Hom}_{\mathcal{M}_{D\mathfrak{g}_+}}(M_+ \otimes M_-, \pi^*(V))$. This functor inherits a tensor structure from $\mathcal{M}_{D\mathfrak{g}_+}$. We set $U_h^{qt}(\mathfrak{g}) = \mathrm{End}\, F \simeq U(\mathfrak{g})[[h]]$ (as an algebra), and define R and J by formulas (19.3) and (19.6). It is easy to see from the previous results that this is a quasitriangular Hopf algebra, which is a quantization of $(\mathfrak{g}, r)$.

□

Remark: Clearly, if $(\mathfrak{g}, r)$ is triangular, then $(U_h^{qt}(\mathfrak{g}), R)$ is also triangular.

19.4 Quantization of r-matrices

We give the last application of the methods developed in this lecture. Recall that an r-matrix in an associative algebra is an element $r \in A \otimes A$ satisfying the CYBE. An R-matrix is an element $R \in A \otimes A[[h]]$ satisfying the QYBE. We prove:

Theorem 19.4. *Let A be an associative algebra, and let $r \in A \otimes A$ be an r-matrix. There exists an R-matrix $R \in A \otimes A[[h]]$ such that $R \equiv 1 + hr \pmod{h^2}$.*

Proof: Let us consider A as a Lie algebra. Notice that the space spanned by the components of r is a finite-dimensional Lie subalgebra. By Proposition 4.2, there exists a quasitriangular Lie bialgebra $(\mathfrak{g}, \bar{r})$, and a Lie algebra map $\pi : \mathfrak{g} \to A$ such that $(\pi \otimes \pi)(\bar{r}) = r$. By Theorem 19.3, there exists a preferred quantization $(U_h^{qt}(\mathfrak{g}), \overline{R})$ of $(\mathfrak{g}, \bar{r})$. The map π can thus be extended to a map $\pi : U_h^{qt}(\mathfrak{g}) \to A[[h]]$, and we can set $R = (\pi \otimes \pi)\overline{R}$.

□

Exercise 19.5. Show that we can take R to be unitary (i.e $RR_{21} = 1$) if r is unitary (i.e $r + r^{op} = 0$).

Here is a refinement of the preceding theorem. Let $\mathcal{R}$ be the ring of algebraic functions of the variable h which are regular at zero.

Theorem 19.5. *Let A be an associative, finite dimensional algebra, and let $r \in A \otimes A$ be an r-matrix. There exists an R-matrix $R \in A \otimes A \otimes \mathcal{R}$ such that $R = 1 + hr + O(h^2)$.*

Proof: Notice that the QYBE for $R = 1 + hr + h^2X$ is a system of polynomial equations on X defined over $k[h]$. The theorem is thus a consequence of the following result of M. Artin (see [Ar]):
Theorem: Any system of polynomial equations in $x_1, \dots x_n$ defined over $k[h]$ which admits solutions over $k[[h]]$ also admits solutions over $\mathcal{R}$.

Lecture 20

Universal constructions

In this lecture, we describe a rigorous mathematical background suitable to deal with questions of functoriality and universality of constructions in the categories of Lie bialgebras, Hopf algebras, etc.

20.1 Cyclic categories

20.1.1 Definition

Definition: A *cyclic category* is a symmetric $\mathbb{Q}$-linear monoidal category $(\mathcal{C}, \otimes)$ whose objects $[0], [1], \ldots$ are indexed by nonnegative integers such that $[1]^{\otimes n} = [n]$ (we then have $[0] = c$).

In particular, there is a natural map $\mathbb{Q}[S_n] \to \mathrm{Hom}_{\mathcal{C}}([n],[n])$ (permutation morphisms). We will always suppose that $\mathbb{Q}[S_n] \hookrightarrow \mathrm{Hom}_{\mathcal{C}}([n],[n])$ is an embedding.

Definition: Let $S = \cup S_{m,n}$ be a set of morphisms of a cyclic category $(\mathcal{C}, \otimes)$, $S_{m,n} \subset \mathrm{Hom}_{\mathcal{C}}([m],[n])$. We say that $\mathcal{C}$ is generated by S if any morphism in $\mathcal{C}$ can be obtained from S and permutation morphisms by a sequence of the following operations: composition, tensor products and $\mathbb{Q}$-linear combination.

Remark. Our terminology "cyclic category" is motivated by the fact that the set of objects of such category is a free cyclic monoid. We note that cyclic categories are also called PROPs, after MacLane.

20.1.2 Basic notions related to cyclic categories

We will now briefly develop the basic theory of cyclic categories:

Definition: Let $Z = \bigcup_{m,n \geq 0} Z_{m,n}$ be a set. The *free cyclic category generated by* Z is the cyclic category $\mathcal{F}_Z$ generated by Z which is the solution to the following universal problem: for any cyclic category $\mathcal{C}$ generated by Z, there exists a symmetric tensor functor $F_{\mathcal{C}} : \mathcal{F}_Z \to \mathcal{C}$ such that $F_{\mathcal{C}}([n]) = [n]$ and $F_{\mathcal{C}}$ acts trivially on morphisms in Z.

This is an analogue of the notion of a free algebra generated by a set Z.

Now let $\mathcal{C}$ be any cyclic category.

Definition: A collection $I = (I_{m,n})$ of $\mathbb{Q}$-subspaces $I_{m,n} \subset \mathrm{Hom}_{\mathcal{C}}([m],[n])$ is called a tensor ideal if it is stable under composition and tensor products with morphisms in $\mathcal{C}$.

In particular, if $J = (J_{m,n})$ is any collection of morphisms in $\mathcal{C}$, we denote by $< J >$ the tensor ideal generated by J (the intersection of all tensor ideals containing J).

Let $\mathcal{C}$ be a cyclic category and $I = (I_{m,n})$ a tensor ideal.

Definition: The quotient cyclic category $\mathcal{C}/I$ is the cyclic category defined by

$$\mathrm{Hom}_{\mathcal{C}/I}([m],[n]) = \mathrm{Hom}_{\mathcal{C}}([m],[n])/I_{m,n}.$$

We leave it to the reader to check that this indeed defines a cyclic category.

Proposition 20.1. *Any cyclic category generated by a set Z is of the form $\mathcal{F}_Z/I$ for some tensor ideal I in $\mathcal{F}_Z$.*

Proof: Left to the reader.

If I is a tensor ideal in $\mathcal{C}$ and $n \geq 0$, let I^n be the tensor ideal generated by morphisms $z_1 \circ z_2 \ldots \circ z_n$, where $z_i \in I$.

Definition: The completion of $\mathcal{C}$ with respect to I is the cyclic category $\mathcal{C}_I$ with the same objects as in $\mathcal{C}$ and morphisms defined by

$$\mathrm{Hom}_{\mathcal{C}_I}([m],[n]) = \varprojlim_k \mathrm{Hom}_{\mathcal{C}}([m],[n])/I^k_{m,n}.$$

20.1.3 Linear algebraic structures

The following notion first appeared in [La].

Let $\mathcal{C}$ be a cyclic category, and let $\mathcal{N}$ be a $\mathbb{Q}$-linear symmetric monoidal category.

Definition: A *linear algebraic structure of type $\mathcal{C}$* on an object $X \in \mathcal{O}b(\mathcal{N})$ is a symmetric tensor functor

$$G_X : \mathcal{C} \to \mathcal{N}$$

such that $G_X([1]) = X$.

In other terms, a linear algebraic structure of type $\mathcal{C}$ on X is a collection of morphisms between tensor powers of X satisfying certain relations "of type $\mathcal{C}$". We will denote by $\mathcal{C}(\mathcal{N})$ the following category: objects are pairs (X, G_X) as above, and morphisms consist of all morphisms in $\mathcal{N}$ which are functorial, i.e which commute with the morphisms coming from $\mathcal{C}$. In particular, if $\mathcal{N} = \mathrm{Vec}_k$, we will denote $\mathcal{C}(\mathcal{N})$ by $\mathcal{C}(k)$.

Let us now give examples of some cyclic categories which will be important for us.

Associative algebras: Let $AA = \mathcal{F}_{(i,m)}/I$, where i is of degree $(0,1)$ (the unit) and m is of degree $(2,1)$ (the multiplication), denoted by the graphs:

and I is the ideal generated by the following relations:

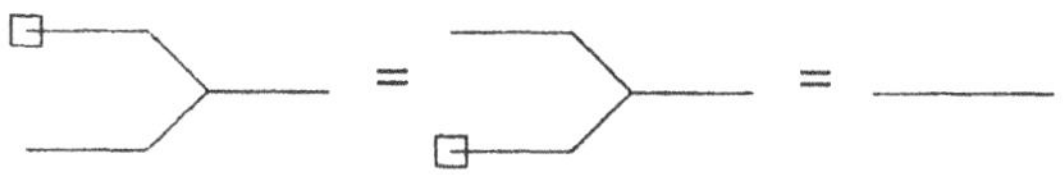

It is clear that this cyclic category describes associative algebras. For instance, if $\mathcal{N} = \mathrm{Vec}(k)$, a linear algebraic structure of type AA on $V \in \mathcal{O}b(\mathcal{N})$ is an associative algebra structure on V.

Lie algebras: Let $LA = \mathcal{F}_{([,])}/I$ where

and I is generated by

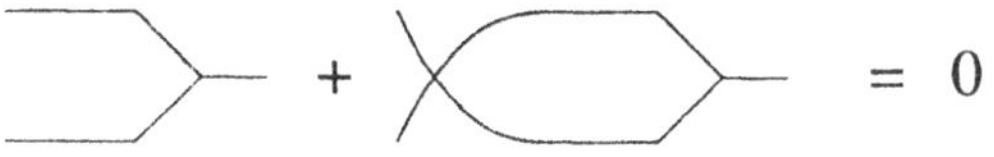

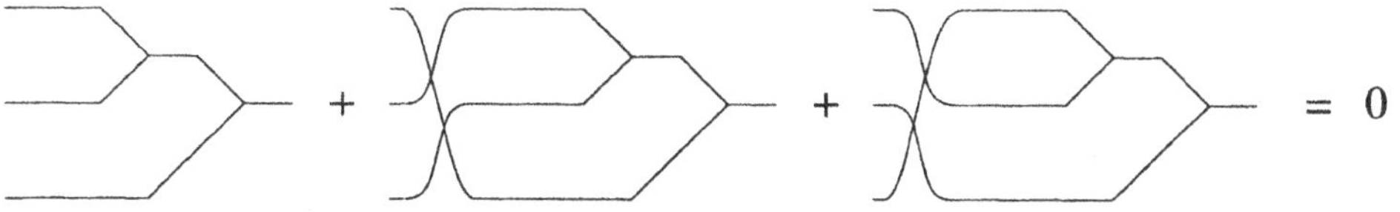

Lie bialgebras: Let $LBA = \mathcal{F}_{([\,,\,],\delta)}/I$ where

$[\,,\,]=$ 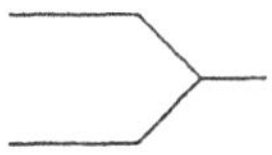 $\delta =$ 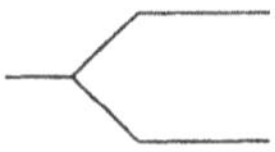

and I is generated by skew-symmetry and Jacobi identity as above, co-skew symmetry and coJacobi identity, and the cocycle condition (see Proposition 2.2).

In a similar way, it should be clear now how to define the following cyclic categories

1. $QTLBA$ (quasitriangular Lie bialgebra): $\mathcal{C}$ is generated by two elements of degrees $(2,1)$ and $(0,2)$ (commutator and r-matrix) satisfying the usual axioms of a Lie algebra, the CYBE for r and the invariance of $r + r^{op}$.

2. HA (Hopf algebras): $\mathcal{C}$ is generated by six elements of degrees $(2,1)$, $(1,2)$, $(0,1)$, $(1,0)$, $(1,1)$ and $(1,1)$ (the product, coproduct, unit, counit, antipode and inverse antipode) satisfying the relations of Lecture 8).

3. $QTHA$ (quasitriangular Hopf algebras): $\mathcal{C}$ is generated by eight elements of degrees $(2,1)$, $(1,2)$, $(0,1)$, $(1,0)$, $(1,1)$, $(1,1)$, $(0,2)$ and $(0,2)$ (the product, coproduct, unit, counit, antipode inverse antipode, R-matrix and inverse R-matrix) satisfying the usual axioms.

4. $CYBA$ (classical Yang-Baxter algebra): $\mathcal{C}$ is generated by three elements of degrees $(2,1)$, $(0,1)$, $(0,2)$ (product, unit, and r-matrix), satisfying the axioms of an associative algebra and the CYBE for r.

5. $QYBA$ (quantum Yang-Baxter algebra): $\mathcal{C}$ is generated by three elements of degrees $(2,1)$, $(0,1)$, $(0,2)$ (the product, unit and R-matrix), satisfying the usual axioms of an associative algebra and the QYBE for R.

If $\mathcal{N}$ is a $\mathbb{Q}$-linear symmetric category, by a Lie algebra in $\mathcal{N}$ (resp. Hopf algebra, etc.) we will mean an object X in $\mathcal{N}$ with a linear algebraic structure of type LA (resp. HA, etc.).

20.2 Universal constructions

Let us first start with a naive definition: if $\mathcal{C}_1$ and $\mathcal{C}_2$ are two cyclic categories, a universal construction from $\mathcal{C}_1$ to $\mathcal{C}_2$ is a symmetric tensor functor $G : \mathcal{C}_2 \to \mathcal{C}_1$. For any $\mathbb{Q}$-linear symmetric monoidal category $\mathcal{N}$, the functor G induces a functor

$$G_{\mathcal{N}} : \mathcal{C}_1(\mathcal{N}) \to \mathcal{C}_2(\mathcal{N}).$$

Example: Any associative algebra is naturally a Lie algebra. In terms of cyclic categories, this corresponds to the functor

$$G : LA \to AA$$

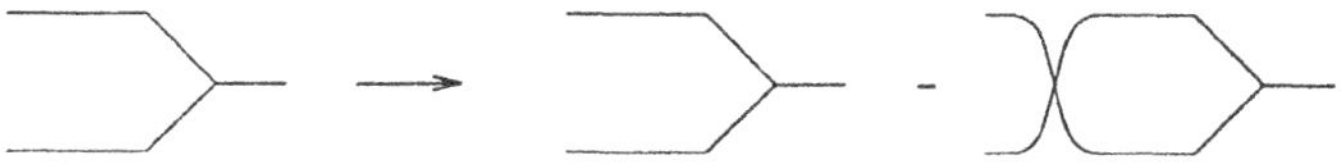

However, the above definition is not well suited for example to the following basic construction: if $\mathfrak{g}$ is a Lie algebra, then the universal enveloping algebra $U(\mathfrak{g})$ of $\mathfrak{g}$ is $U(\mathfrak{g}) = S\mathfrak{g}$ as a vector space (by the symmetrization map), with multiplication maps $m^p_{m,n} : S^n\mathfrak{g} \otimes S^m\mathfrak{g} \to S^p\mathfrak{g}$ given by the Campbell-Hausdorff series: for instance

$$\begin{aligned} m^1_{1,1} : S^1\mathfrak{g} \otimes S^1\mathfrak{g} &\to S^1\mathfrak{g}, \\ x \otimes y &\to \frac{1}{2}[x,y], \\ m^2_{1,1} : S^1\mathfrak{g} \otimes S^1\mathfrak{g} &\to S^2\mathfrak{g}, \\ x \otimes y &\to \frac{1}{2}(xy + yx). \end{aligned}$$

In other terms, we need to add the objects $S^n\mathfrak{g} \subset \mathfrak{g}^{\otimes n}$. This motivates the following definition:

Definition: Let $\mathcal{C}$ be a cyclic category. Denote by $\hat{\mathcal{C}}$ the following extended category:

Objects are pairs $([n], P)$, where $n \geq 0$ and $P \in \mathbb{Q}[S_n] \subset \mathrm{Hom}_{\mathcal{C}}([n],[n])$ is an idempotent.

Morphisms are defined by

$$\mathrm{Hom}_{\hat{\mathcal{C}}}(([n], P), ([m], Q)) = \{f \in \mathrm{Hom}_{\mathcal{C}}([n],[m]) \mid fP = Qf = f\}.$$

In other words, we add all the kernels of the idempotents of $\mathbb{Q}[S_n]$ acting in $[n]$. For instance, $S^n[1]$ is represented by $([n], \mathrm{Sym}_n)$ where

$$\mathrm{Sym}_n = \frac{1}{n!} \sum_{\sigma \in S_n} \sigma$$

is the symmetrizer in $\mathbb{Q}[S_n]$.

Definition: Let $\overline{\mathcal{C}}$ be the closure of $\hat{\mathcal{C}}$ under (possibly infinite) direct sums.

For instance, $S[1] = \bigoplus_{n \geq 0} S^n[1] \in \mathcal{O}b(\overline{\mathcal{C}})$.

Remark: It is clear that indecomposable objects X_D of $\overline{\mathcal{C}}$ are in one-to-one correspondence with irreducible representations of GL_∞ (i.e with Young diagrams) and all objects are of the form

$$V = \bigoplus V_D \otimes X_D, \qquad V_D \in \mathrm{Vec}_k.$$

We can now formulate the correct definition of a universal construction. Let $\mathcal{C}_1$ and $\mathcal{C}_2$ be two cyclic categories.

Definition: A *universal construction* from $\mathcal{C}_1$ to $\mathcal{C}_2$ is a symmetric tensor functor

$$G : \mathcal{C}_2 \to \overline{\mathcal{C}_1}.$$

This gives rise to a functor $G_{\mathcal{N}} : \mathcal{C}_1(\mathcal{N}) \to \mathcal{C}_2(\overline{\mathcal{N}})$ where $\overline{\mathcal{N}}$ is the closure of $\mathcal{N}$ under (possibly infinite) direct sums.

20.2.1 Acyclic tensor calculus

Here is a more down-to-earth explanation of the above definition. Let us represent a given algebraic structure (such as associative algebras, Lie bialgebras, etc.) by a set of oriented graphs and a certain set of relations among them, corresponding to the basic operations and the axioms defining the algebraic structure (in the same way as above). A universal construction from an algebraic structure $\mathcal{C}_1$ to an algebraic structure $\mathcal{C}_2$ should assign to any basic operation in $\mathcal{C}_2$ an oriented graph composed of basic operations from $\mathcal{C}_1$. For instance, in the universal construction of the Lie algebra structure on an associative algebra, we have:

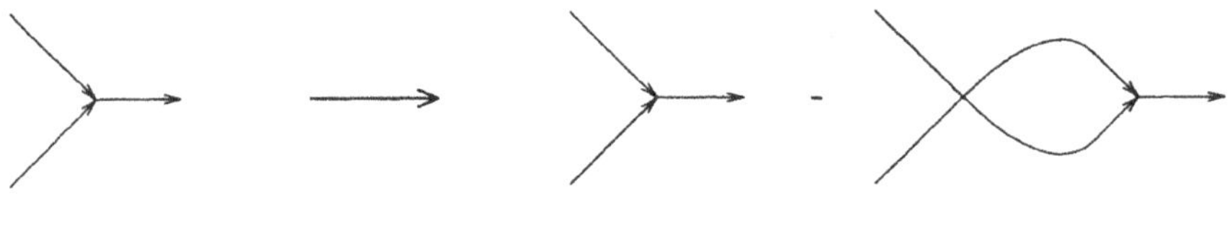

We want to do this in such a way that makes sense when we replace vertices of the graph by vector spaces and edges by maps between these vector spaces. This may not be the case if some of the above graphs contain *oriented cycles*, and the vector spaces are infinite dimensional: for instance, the graph

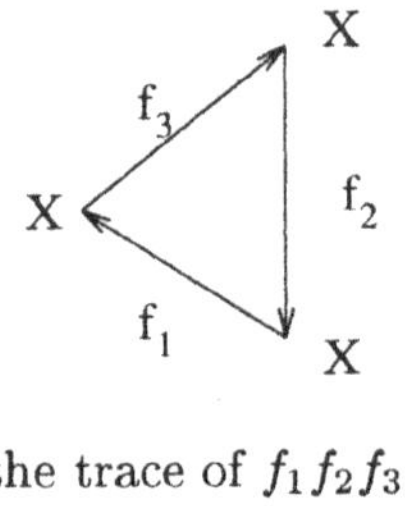

assigns to the maps f_1, f_2, f_3 the trace of $f_1 f_2 f_3$, and this may not exist in the case dim $X = \infty$. Thus the condition for a construction to be universal is that it can be expressed in terms of ***acyclic graphs*** ("acyclic tensor calculus").

Exercise 20.1. Show that a universal construction in the sense of section 20.2 is expressed in terms of acyclic graphs.

Lecture 21

Universal quantization

21.1 Statement of the theorem

The aim of this lecture is to describe a universal construction of quantization of Lie bialgebras, quasitriangular Lie bialgebras, and r-matrices.

Recall the notation $\mathcal{C}_I$ for the completion of the cyclic category $\mathcal{C}$ with respect to the tensor ideal I. For instance, morphisms in the category $\overline{LBA}_{<\delta>}$ are power series in $< \delta >$.

Theorem 21.1. *There exists universal construction functors*

1.
$$Q : HA \to \overline{LBA}_{<\delta>}$$
such that, for any Lie bialgebra $(\mathfrak{g}, [\, , \,], \delta)$ *defined over* k, $Q_{\mathrm{Vec}_K}(\mathfrak{g}[[h]], [\, , \,], h\delta)$ *is a quantization of* $(\mathfrak{g}, [\, , \,], \delta)$.

2.
$$Q^{qt} : QTHA \to \overline{QTLBA}_{<r>}$$
such that, for any quasitriangular Lie bialgebra $(\mathfrak{g}, r)$ *defined over* k, $Q^{qt}_{\mathrm{Vec}_K}(\mathfrak{g}[[h]], hr)$ *is a quantization of* $(\mathfrak{g}, r)$.

3.
$$Q^{YB} : QYBA \to CYBA_{<r>}$$
such that, for any classical Yang-Baxter algebra (A, r) *defined over* k, $Q^{YB}_{\mathrm{Vec}_K}(A[[h]], hr)$ *is a quantization of* (A, r).

In particular, Theorem 21.1 defines quantization functors:

$$\tilde{Q} : \quad LBA(k) \to HA(K),$$
$$\tilde{Q}^{qt} : QTLBA(k) \to QTHA(K),$$

(for not necessarily finite dimensional Lie bialgebras), and

$$\tilde{Q}^{YB} : CYBA(k) \to QYBA(K)$$

which proves Theorems 9.1, 9.2, 9.3.

The basic idea of the construction of the functor is the same as in Lecture 20, where we dealt with finite-dimensional Lie bialgebras: construct a tensor structure on the forgetful functor from the Drinfeld category of representations of the double, and extract a Hopf subalgebra. However, this construction cannot be applied as such for the following reasons:

1. The construction involves the Verma modules M_+ and M_- which are very different in nature. For instance, consider the Casimir element $\Omega = \sum_i g_i^+ \otimes g_i^- + g_i^- \otimes g_i^+$, where g_i^+ and g_i^- are dual basis of $\mathfrak{g}_+$ and $\mathfrak{g}_-$ respectively. This expression may not be well defined as an operator on the module $M_- \otimes M_+$ if $\mathfrak{g}$ is infinite dimensional:
$$\Omega(1_- \otimes 1_+) = \sum_i g_i^+ 1_- \otimes g_i^- 1_+$$
and this does not converge in any reasonable sense. The same situation arises when considering categories of representation of Kac-Moody algebras (the category $\mathcal{O}$): M_- is "highest weight", i.e it is generated by an $\mathfrak{g}_-$-invariant element, and the subalgebra $\mathfrak{g}_-$ acts locally nilpotently, while the dual module M_+ is "lowest weight", i.e generated by an $\mathfrak{g}_+$-invariant element and $\mathfrak{g}_+$ acts locally nilpotently. For instance, when $\mathfrak{g}$ is infinite-dimensional, the Casimir element $\Omega = \sum_i g_i^+ \otimes g_i^- + g_i^- \otimes g_i^+$ only acts in tensor products of two highest weight (or lowest weight) representations and not in $M_- \otimes M_+$.

2. Considering the double of a Lie bialgebra is not a functorial operation; $\mathfrak{g} \to \mathfrak{g} \oplus \mathfrak{g}^*$ mixes the covariant functor $\mathfrak{g} \to \mathfrak{g}$ and the contravariant functor $\mathfrak{g} \to \mathfrak{g}^*$.

This explains why one has to modify the construction of lecture 19 in order to obtain a functorial quantization, which can be applied to infinite-dimensional Lie bialgebras.

21.2 Quantization of finite-dimensional Lie bialgebras revisited

In this section, we indicate how to deal with the first obstruction. Let $\mathfrak{g}_+$, $\mathfrak{g} = D\mathfrak{g}_+$, $\mathcal{M}_\mathfrak{g}$, etc. have the same meaning as in Lecture 19. Our aim is to construct a forgetful functor $F : \mathcal{M}_\mathfrak{g} \to \mathcal{A}$ and a tensor structure on it without using the module M_+. For this we set
$$F(V) = \mathrm{Hom}_{\mathcal{M}_\mathfrak{g}}(M_-, M_+^* \otimes V)$$
where M_+^* is the dual of M_+. Notice that M_+^* is now morally of "highest weight". For instance, if $\mathfrak{g}$ is a Kac-Moody algebra with a "doubled" Cartan subalgebra as in section 4.4.2, i.e $\mathfrak{g} = (\mathfrak{n}_- \oplus \mathfrak{h}) \oplus (\mathfrak{n}_+ \oplus \mathfrak{h})$, and $\mathfrak{g}_\pm = \mathfrak{n}_\pm \oplus \mathfrak{h}$ then M_+ is the lowest weight Verma module $U(\mathfrak{n}_+ \oplus \mathfrak{h})$ while M_+^* is the highest weight Verma module isomorphic to $U(\mathfrak{n}_- \oplus \mathfrak{h})$.

The tensor structure on F is defined in the following way: if V, W are two objects in $\mathcal{M}_{\mathfrak{g}}$, $v \in F(V)$, $w \in F(W)$, we set

$$J_{V,W}(v \otimes w) = (i_+^* \otimes 1 \otimes 1) \circ (1 \otimes \beta_{23}^{-1} \otimes 1) \circ (v \otimes w) \circ i_-$$

i.e,

$$M_- \xrightarrow{i_-} M_- \otimes M_- \xrightarrow{v \otimes w} M_+^* \otimes V \otimes M_+^* \otimes W \xrightarrow{\beta_{23}^{-1}} M_+^* \otimes M_+^* \otimes V \otimes W \xrightarrow{i_+^*} M_+^* \otimes V \otimes W.$$

In terms of pictures, we have made the following transformation on the tensor structure defined in Lecture 19:

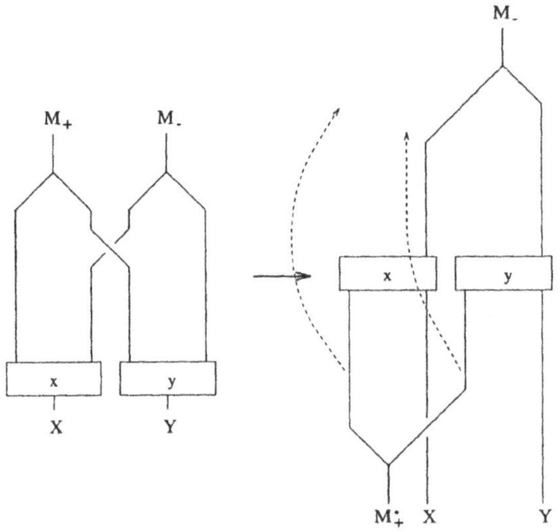

Notice that these pictures are isotopic: the M_+-part is "pulled around" into the M_+^*-part. This means that this construction is formally equivalent to the one made in lecture 19.

The fact that it is indeed a tensor structure is proved in the exact same way as in Lecture 19. We define the element $J \in U(\mathfrak{g})^{\otimes 2}[[h]]$ in a manner similar to (19.3). It is easy to see that

$$J \equiv 1 + \frac{hr}{2} \pmod{h^2}. \tag{21.1}$$

Set $U_h(\mathfrak{g}) = \text{End}\,(F)$. It is a Hopf algebra in a natural way, and from (21.1), we conclude as in Lecture 19 that $U_h(\mathfrak{g})$ is a quantization of $\mathfrak{g}$.

We now quantize the Lie bialgebra $\mathfrak{g}_+$. Let $i:\ F(M_-) \hookrightarrow U_h(\mathfrak{g})$ be the embedding defined as follows; for $V \in Ob(\mathcal{M}_{\mathfrak{g}})$ and $x \in F(M_-)$, set

$$i(x):\ F(V) \to F(V),$$
$$v \to (i_+^* \otimes 1) \circ (1 \otimes v) \circ x$$

in terms of pictures, the action of $i(x)$ is given by:

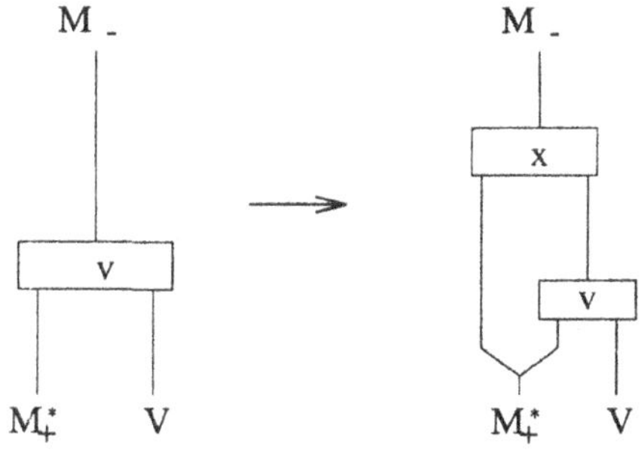

Exercise 21.1. Give a pictorial proof of the following formula:

$$i(x) \circ i(y) = i((i_+^* \otimes 1) \circ (1 \otimes y) \circ x). \tag{21.2}$$

for $x, y \in F(M_-)$.

We set $U_h(\mathfrak{g}_+) = i(F(M_-))$. The preceding exercise gives the formula for the product in $U_h(\mathfrak{g}_+)$. Let us now show that $U_h(\mathfrak{g}_+)$ is closed under coproduct and compute it explicitly.

By definition, for any $y \in U_h(\mathfrak{g})$, the element $\Delta(y) \in U_h(\mathfrak{g}) \otimes U_h(\mathfrak{g})$ is determined by the commutativity of the diagram

$$\begin{array}{ccc}
F(V) \otimes F(W) & \xrightarrow{J_{V,W}} & F(V \otimes W) \\
\Big\downarrow{\scriptstyle \Delta(y)} & & \Big\downarrow{\scriptstyle y} \\
F(V) \otimes F(W) & \xrightarrow[J_{V,W}]{} & F(V \otimes W)
\end{array}$$

If $y = i(x)$, this relation has the following graphical representation:

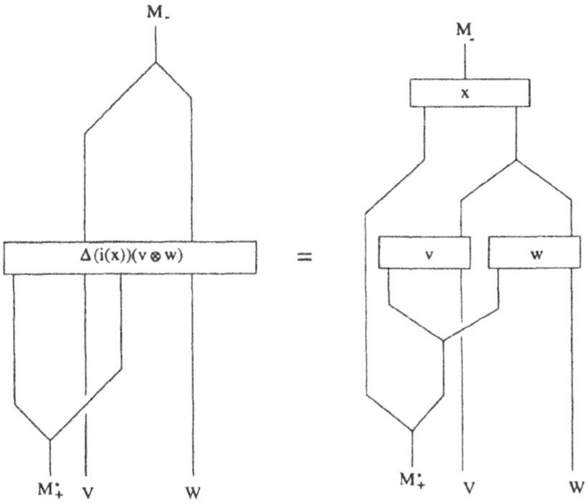

(21.3)

Claim:

$$\Delta(i(x)) = (i \otimes i)(J^{-1}_{M_-,M_-}(1 \otimes i_-) \circ x). \tag{21.4}$$

Proof: First notice that the element $D = (i \otimes i)(J^{-1}_{M_-,M_-}(1 \otimes i_-) \circ x)$ is characterized by the property $(1 \otimes i_-)x = J_{M_-,M_-}D$, or, in pictures:

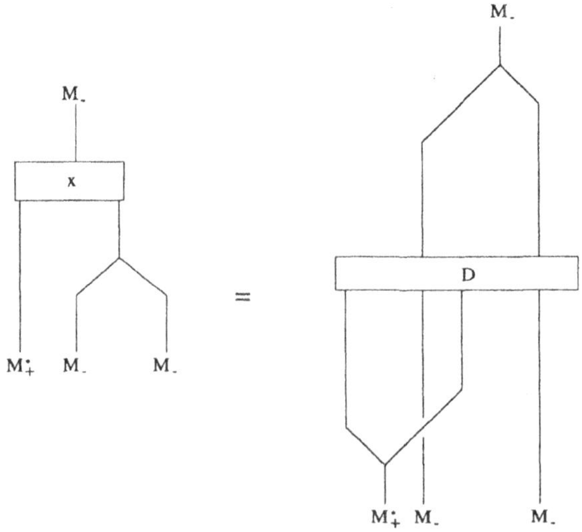

then substitute (21.4) into the l.h.s of (21.3):

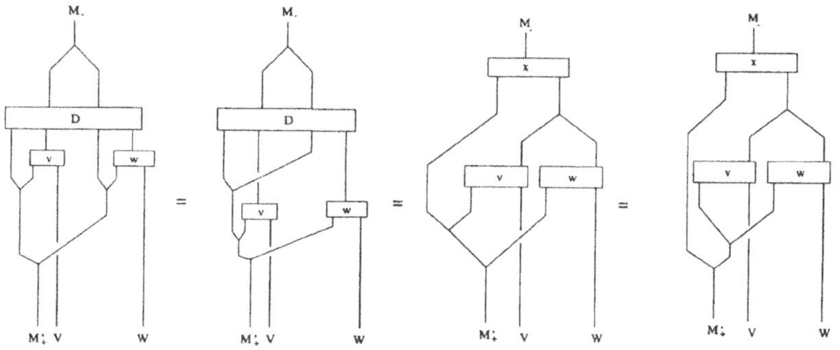

This proves the claim.

In particular, this shows that $U_h(\mathfrak{g}_+)$ is a Hopf subalgebra of $U_h(\mathfrak{g})$. It is clear that it is a quantization of $\mathfrak{g}_+$.

Let us write down explicitly formulas (21.2) and (21.4) for the product and coproduct in $U_h(\mathfrak{g}_+)$. For this, let us first identify $U_h(\mathfrak{g}_+) = F(M_-)$ with $M_-[[h]] = U(\mathfrak{g}_+)[[h]]$ by

$$\begin{aligned} F(M_-) &\stackrel{\sim}{\to} M_-[[h]], \\ f &\mapsto (1_+ \otimes 1)(f(1_-)). \end{aligned}$$

We introduce the following two maps. For $X \in \mathcal{O}b(\mathcal{M}_{\mathfrak{g}})$, let X^0 denote the vector space X equipped with the trivial $\mathfrak{g}$-action. By Frobenius reciprocity, the map

$$\begin{aligned} \theta_X : \mathrm{Hom}_{\mathfrak{g}}(M_- \otimes X^0, M_+^* \otimes X) &\to \mathrm{Hom}_k(X^0, X) \\ f &\mapsto (1_+ \otimes 1) \circ f \circ (1_- \otimes 1) \end{aligned}$$

is an isomorphism. We set

$$\psi = \theta_{M_-}^{-1}(1) : \ M_- \otimes M_- \to M_+^* \otimes M_-$$

and

$$\eta = \psi(1_- \otimes \cdot) : \ M_- \to M_+^* \otimes M_-.$$

Lemma 21.1. *The formulas for the product and coproduct in $U_h(\mathfrak{g}_+)$ can be written as follows:*

$$\begin{aligned} m : M_- \otimes M_- &\stackrel{\sigma}{\to} M_- \otimes M_- \stackrel{\eta\otimes 1}{\to} (M_+^* \otimes M_-) \otimes M_- \\ &\stackrel{\Phi}{\to} M_+^* \otimes (M_- \otimes M_-) \stackrel{1\otimes\psi}{\to} M_+^* \otimes (M_+^* \otimes M_-) \qquad (21.5) \\ &\stackrel{1_+\otimes 1\otimes 1}{\longrightarrow} M_+^* \otimes M_- \stackrel{1_+\otimes 1}{\longrightarrow} M_-, \end{aligned}$$

$$\Delta : M_- \overset{i_-}{\to} M_- \otimes M_- \overset{J^{-1}}{\to} M_- \otimes M_-, \tag{21.6}$$

where J is the composition

$$\begin{aligned} M_- \otimes M_- &\overset{\eta\otimes\eta}{\to} (M_+^* \otimes M_-) \otimes (M_+^* \otimes M_-) \overset{\Phi^{-1}_{2,3,4}\Phi_{1,2,34}}{\longrightarrow} M_+^* \otimes ((M_+^* \otimes M_-) \otimes M_-) \\ &\overset{\beta_{23}^{-1}}{\to} M_+^* \otimes ((M_- \otimes M_+^*) \otimes M_-) \overset{\Phi^{-1}_{1,2,34}\Phi_{2,3,4}}{\longrightarrow} (M_+^* \otimes M_+^*) \otimes (M_- \otimes M_-) \\ &\overset{(1_+\otimes 1_+)\otimes 1}{\longrightarrow} M_- \otimes M_-. \end{aligned} \tag{21.7}$$

Proof: We leave it to the reader to check that these indeed coincide with formulas (21.2) and (21.4).

21.3 Drinfeld double in any symmetric monoidal category

We now deal with the second obstruction, and describe the category of representations of $\mathfrak{g} = D\mathfrak{g}_+$ purely in terms of $\mathfrak{g}_+$, in the general framework of a symmetric tensor category.

Let $\mathcal{N}$ be any $\mathbb{Q}$-linear, symmetric, monoidal category.
Definition: Let $(\mathfrak{g}_+, [\,,\,])$ be a Lie algebra in $\mathcal{N}$. A *left module* over $\mathfrak{g}_+$ is an object X in $\mathcal{N}$ endowed with a morphism

$$\pi : \mathfrak{g}_+ \otimes X \to X$$

such that

$$\pi(1 \otimes \pi_{23}) - \pi(\pi_{13} \otimes 1) = \pi([\,,\,] \otimes 1) \tag{21.8}$$

on $\mathfrak{g}_+^{\otimes 2} \otimes X$. In pictures, we will represent the action of $\mathfrak{g}_+$ on X by

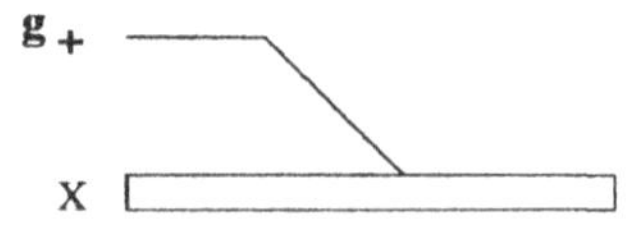

Thus (21.8) can be rewritten as

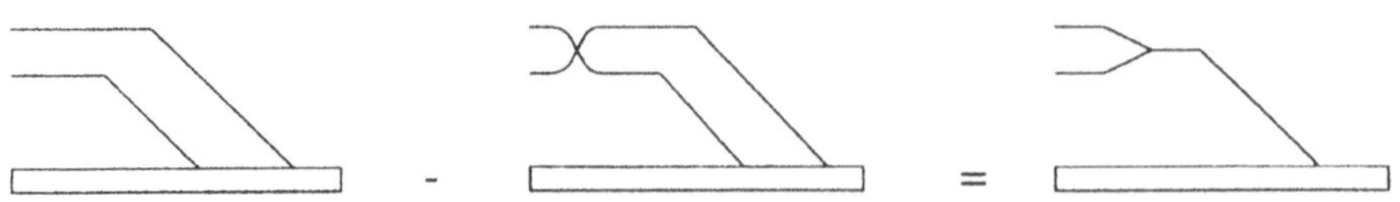

Definition: Let $(\mathfrak{g}_+, \delta)$ be a Lie coalgebra. A *right comodule* over $\mathfrak{g}_+$ is an object X in $\mathcal{N}$ endowed with a morphism $\pi^* : X \to \mathfrak{g}_+ \otimes X$ such that

$$\sigma_{23}(1 \otimes \pi^*)\pi^* - (1 \otimes \pi^*)\pi^* = (\delta \otimes 1)\pi^* \tag{21.9}$$

as maps $X \to \mathfrak{g}_+ \otimes \mathfrak{g}_+ \otimes X$. In pictures, we represent the coaction by

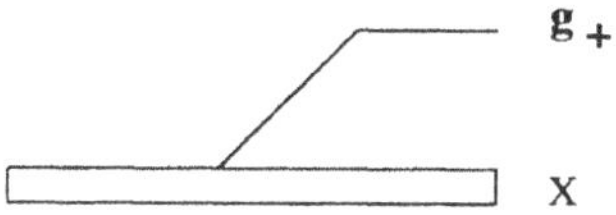

and (21.9) is then

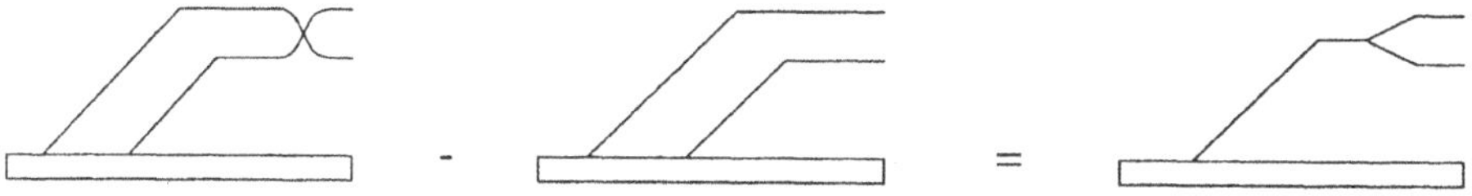

Of course, when $\mathcal{N} = \mathrm{Vec}(k)$, the above definitions correspond to the usual notions of (co)modules over a Lie (co)algebra defined over k.

Definition: Let $(\mathfrak{g}_+, [\,,\,], \delta)$ be a Lie bialgebra in $\mathcal{N}$. A *dimodule* over $\mathfrak{g}_+$ is an object X in $\mathcal{N}$ endowed with both the structure of a module and a comodule, with the following consistency condition:

$$\pi^*\pi - (\pi \otimes 1)\sigma_{12}(1 \otimes \pi^*) = ([\,,\,] \otimes 1)(1 \otimes \pi^*) - (1 \otimes \pi)(\delta \otimes 1) \tag{21.10}$$

as maps $\mathfrak{g}_+ \otimes X \to \mathfrak{g}_+ \otimes X$. In pictures this is expressed as,

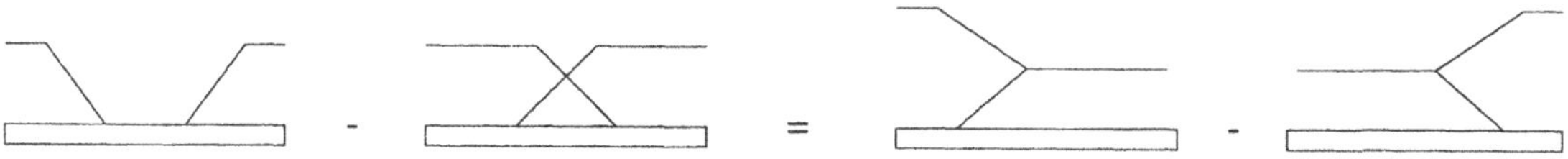

When $\mathcal{N} = \mathrm{Vec}_k$ and $\mathfrak{g}_+$ is *finite dimensional*, this corresponds to the usual notion of a module over the Drinfeld double $\mathfrak{g} = D\mathfrak{g}_+$ of $\mathfrak{g}_+$.

Exercise 21.2. Check the last statement.

Remark: When $\mathcal{N} = \mathrm{Vec}_k$ and $\mathfrak{g}_+$ is not necessarily finite dimensional, a dimodule over a Lie bialgebra $\mathfrak{g}_+$ corresponds to a module X over $D\mathfrak{g}_+$ for which the action $D\mathfrak{g}_+ \otimes X \to X$ is "equicontinuous" in some sense. See [EK1] for a more precise statement.

Remark. Dimodules are also called Drinfeld-Yetter modules.

The collection of modules (comodules, dimodules) over a Lie algebra (coalgebra, bialgebra) $\mathfrak{g}_+$ clearly forms a tensor category denoted by Mod$(\mathfrak{g}_+)$, Comod$(\mathfrak{g}_+)$, Dimod$(\mathfrak{g}_+)$ respectively: morphisms are required to preserve all structures, and the tensor product is defined in the usual way. For instance, if X and Y are two dimodules, the action and coaction on $X \otimes Y$ are given by the pictures

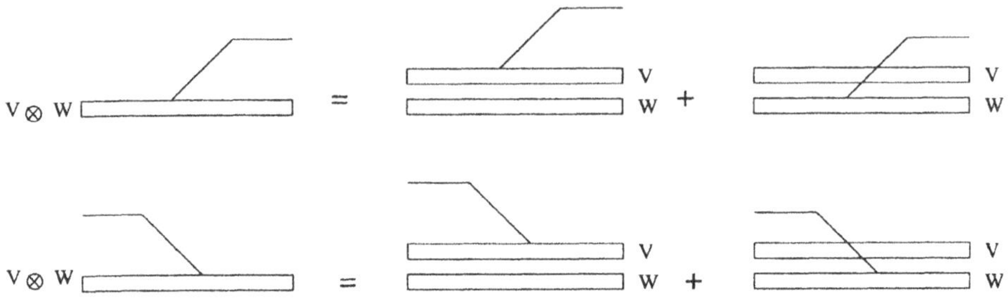

The unit is the trivial $\mathfrak{g}_+$-module (comodule, dimodule) (with zero $\mathfrak{g}_+$-action and coaction) realized on the unit object $c = [0]$.

Exercise 21.3. Check the consistency relation (21.10).

We can now define the analogue of the Casimir operator. Recall that in the finite dimensional double construction, the r-matrix is given by $r = \sum x_i^+ \otimes x_i^-$, where (x_i^+) and (x_i^-) are dual basis of $\mathfrak{g}_+$ and $\mathfrak{g}_- = \mathfrak{g}_+^*$ respectively. If V and W are dimodules over $\mathfrak{g}_+$, it is thus natural to set

$$r = (\pi \otimes 1)\sigma_{12}(1 \otimes \pi^*)$$

or, in pictures

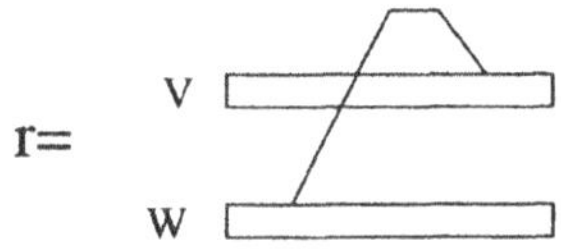

and $\Omega = r + r^{op}$. It is a well defined element of End$(V \otimes W)$.

21.4 The Drinfeld category

Let us now fix $\mathcal{N} = \overline{LBA}_{<\delta>}$. Consider $\mathfrak{g}_+ = [1]$, with (universal) commutator $[\,,\,]$ and cocommutator δ. This is a Lie bialgebra and it is natural to call it the "universal Lie bialgebra". We will put a braided monoidal structure on the category Dimod $\mathfrak{g}_+$ as follows:

Consider the category Dimod$_h(\mathfrak{g}_+)$ whose objects are those of Dimod$(\mathfrak{g}_+)$ with

$$\mathrm{Hom}_{\mathrm{Dimod}_h(\mathfrak{g}_+)}(V, W) = \mathrm{Hom}_{\mathrm{Dimod}(\mathfrak{g}_+)}(V, W)[[h]].$$

For each choice of a Lie associator defined over k, we can equip $\text{Dimod}_h(\mathfrak{g}_+)$ with a structure of a braided monoidal category by setting $\beta = e^{h\Omega/2}$ and $\Phi = \Phi(h\Omega_{12}, h\Omega_{23})$. We will denote this category by $\mathcal{M}_{\mathfrak{g}_+}$ (the Drinfeld category).

21.5 Quantization of $\mathfrak{g}_+$

We now proceed to construct a "quantization" of $\mathfrak{g}_+$, by equipping the object $U(\mathfrak{g}_+) = S\mathfrak{g}_+$ with a Hopf algebra structure which agrees with the standard Hopf algebra structure modulo h and which yields the Lie bialgebra structure of $\mathfrak{g}_+$ modulo h^2. The idea here is to use the "universal formulas" (21.5) and (21.6) in this general setting.

Let us first define Verma dimodules M_- and M_+^* in $\mathcal{M}_{\mathfrak{g}_+}$:

As an object, $M_- = S\mathfrak{g}_+ = c \oplus \mathfrak{g}_+ \oplus S^2\mathfrak{g}_+ \oplus \ldots$. The module structure is given by the (left) multiplication $\pi_- : \mathfrak{g}_+ \otimes M_- \to M_-$, expressed by the usual Campbel-Hausdorff series. The coaction $\pi_-^* : M_- \to \mathfrak{g}_+ \otimes M_-$ is determined by $\pi^*_{-|c} = 0$ and the consistency condition between action and coaction for dimodules.

As an object, $M_+^* = \prod_{n \geq 0} S^n\mathfrak{g}_+$ (since $\mathcal{N}$ is not closed under projective limits, we will formally adjoin this object). The action and coaction of $\mathfrak{g}_+$ on M_+^* are deduced from the coaction and action in M_- respectively by interchanging m and δ, reversing the order of composition and changing the signs.

The maps $\psi : M_- \otimes M_- \to M_+^* \otimes M_-$ and $\eta : M_- \to M_+^* \otimes M_-$ are defined in exactly the same way as in section 21.2 (notice that Frobenius reciprocity holds in this general setting). We can thus define (via formulas (21.5) and (21.6)) the maps

$$\tilde{m} : M_- \otimes M_- \to M_-,$$
$$\tilde{\Delta} : M_- \to M_- \otimes M_-.$$

Lemma 21.2. *The maps $\tilde{m}, \tilde{\Delta}$ endow M_- with a bialgebra structure.*

Proof: This follows from the construction of $U_h(\mathfrak{g}_+)$ of section 21.2.

Moreover, the standard argument of existence of an antipode in a bialgebra deformation of a Hopf algebra shows:

Lemma 21.3. *There exists a unique map $\tilde{S} : M_- \to M_-$ endowing the bialgebra $(M_-, \tilde{m}, \tilde{\Delta})$ with a Hopf algebra structure.*

We will need one more result, which is an easy consequence of formulas (21.5) and (21.6)). Define $\tilde{\mathcal{M}}_{\mathfrak{g}_+} \subset \mathcal{M}_{\mathfrak{g}_+}$ to be the image of the functor

$$P: \text{Dimod}(\mathfrak{g}_+) \to \mathcal{M}_{\mathfrak{g}_+}$$

defined by $[1] \mapsto [1]$, $[\,,\,] \mapsto [\,,\,]$ and $\delta \mapsto h\delta$.

Lemma 21.4. *The morphisms m, Δ and S belong to $\tilde{\mathcal{M}}_{\mathfrak{g}_+}$.*

We are now in a position to prove part 1) of Theorem 21.1. The functor Q is uniquely determined by $Q([1])$ and $Q(m)$, $Q(\Delta)$, $Q(S)$. We set $Q([1]) = U(\mathfrak{g}_+) = S[1]$. By lemma 21.4, there exists (unique) morphisms

$$m' : M_- \otimes M_- \to M_-,$$
$$\Delta' : M_- \to M_- \otimes M_-,$$
$$S' : M_- \to M_-$$

such that $P(m') = \tilde{m}$, $P(\Delta') = \tilde{\Delta}$, $P(S') = \tilde{S}$. We set $Q(m) = m'$, $Q(\Delta) = \Delta'$ and $Q(S) = S'$. By lemma 21.2, this extends to a functor $Q : HA \to \overline{LBA}_{<\delta>}$, which satisfies the requirements of Theorem 21.1.

The remaining statements of the theorem are proved along the same lines; for quasitriangular Lie bialgebras, we use formulas (19.4), (19.5) and (19.6); for classical r-matrices, we use formula (19.6).

We have defined several quantizations $U_h(\mathfrak{g}_+)$ of a Lie bialgebra $\mathfrak{g}_+$: one in Lecture 19 if $\mathfrak{g}_+$ is finite dimensional, one in Theorem 21.1, and one more if $\mathfrak{g}_+$ is quasitriangular.

Proposition 21.1. *Let $(\mathfrak{g}_+, r)$ be a quasitriangular Lie bialgebra. Then $U_h(\mathfrak{g}_+) \simeq U_h^{qt}(\mathfrak{g}_+)$.*

Proof: Recall that by Proposition 4.2, there exists a morphism of quasitriangular Lie bialgebras $\pi : D\mathfrak{g}_+ = \mathfrak{g} \to \mathfrak{g}_+$, which is the identity on $\mathfrak{g}_+$. By functoriality of quantization, this induces a morphism of QUE algebras: $Q^{qt}(U_h^{qt}(\mathfrak{g})) \to U_h^{qt}(\mathfrak{g}_+)$. But by construction, $U_h(\mathfrak{g}_+)$ is a Hopf subalgebra of $U_h^{qt}(\mathfrak{g})$. Restricting $Q^{qt}(\pi)$ to $U_h(\mathfrak{g}_+)$, we obtain a map of QUE algebras $U_h(\mathfrak{g}_+) \to U_h^{qt}(\mathfrak{g}_+)$. This is an isomorphism since it is the identity modulo h.

□

Proposition 21.2. *Let $(\mathfrak{g}_+, \delta)$ be a finite dimensional Lie bialgebra. The quantizations of $(\mathfrak{g}_+, \delta)$ defined in Lecture 19 and in Lecture 21 are isomorphic.*

Proof: This follows from the fact that the pictures describing the constructions in both cases are "isotopic" (see also [EK1] for an algebraic proof).

Remark: Notice that all these constructions depend a priori on a choice of a Lie associator Φ defined over k. Our statements relate to situations where the same associator is used.

21.6 Quantization of Poisson-Lie groups

Let $PLG(k)$ (resp $PAG(k)$, $PFG(k)$) be the category of Poisson-Lie groups (resp Poisson-algebraic groups, Poisson-formal groups). For PLG, we assume $k = \mathbb{R}$ or $\mathbb{C}$. Let $\mathcal{D}_G$ and $\mathcal{D}_{G\times G}$ denote the algebra of differential operators on G and $G \times G$. Recall that a quantization H of a Poisson manifold M is said to be local if the coefficients $c_i(f,g)$ of the h-expansion of the product are given by bidifferential operators in $\mathcal{C}^\infty(M)$. Similarly, a local coproduct on $\mathcal{C}^\infty(M)$ is a coproduct whose coefficients $d_i(f)$ are of the form $f \to D.f(xy)$, where D is a differential operator on M^2. In this section, we apply the previous results to prove the following theorem:

Theorem 21.2. *There exists quantization functors*

$$\begin{aligned} Q_{PLG} &: PLG(k) \to HA(K), \\ Q_{PAG} &: PAG(k) \to HA(K), \\ Q_{PFG} &: PFG(k) \to HA(K) \end{aligned}$$

such that $Q_{PFG}(G)$ (resp $Q_{PAG}(G)$, $Q_{PFG}(G)$) is a local quantization of G.

Remark: Strictly speaking, we need to consider topological Hopf algebras in the cases of Poisson-Lie groups and Poisson-Formal groups, since the coproduct maps $\Delta : \mathcal{C}^\infty(G) \to \mathcal{C}^\infty(G \times G)$ take value in a completion of $\mathcal{C}^\infty(G) \otimes \mathcal{C}^\infty(G)$. We ignore this minor point in the notation.

Proof(sketch): We will prove the result for PAG and PFG, the case of PLG being similar. Let us first consider the case of Poisson formal groups. Let $(\mathfrak{g}_+, [\,,\,], \delta)$ be a Lie bialgebra and $(U(\mathfrak{g}_+)^*, \{\,,\,\})$ the corresponding Poisson formal group. The functor Q of Theorem 21.1 provides us with a QUE algebra $U_h(\mathfrak{g}_+)$, a quantization of $\mathfrak{g}_+$. It is clear from Theorem 21.1 that the h-formal group $U_h(\mathfrak{g}_+)^*$ is a quantization of $U(\mathfrak{g}_+)^*$ and that this quantization is functorial. We identify $U_h(\mathfrak{g}_+)$ and $U_h(\mathfrak{g}_+)^*$ with $U(\mathfrak{g}_+)[[h]]$ and $U(\mathfrak{g}_+)^*[[h]]$ respectively (as vector spaces). Let us now show that this quantization is in fact local, i.e the coefficients of the h-expansion of the $*$ product

$$\begin{aligned} &U_h(\mathfrak{g}_+)^* \otimes U_h(\mathfrak{g}_+)^* \to U_h(\mathfrak{g}_+)^*, \\ &\qquad f \otimes g \to f * g = fg + hc_1(f,g) + h^2 c_2(f,g) + \ldots \end{aligned}$$

and of the coproduct

$$\begin{aligned} &U_h(\mathfrak{g}_+) \to U_h(\mathfrak{g}_+) \otimes U_h(\mathfrak{g}_+), \\ &\qquad f \to \Delta(f) = \Delta_0(f) + hd_1(f) + \ldots \end{aligned}$$

are of the form $c_i(f,g) = \sum_i D_i f.D'_i g$ and $d_i(f) = D''_i.\Delta_0(f)$ for some differential operators $D_i, D'_i \in \mathcal{D}_{U(\mathfrak{g}_+)^*}$ and $D''_i \in \mathcal{D}_{U(\mathfrak{g}_+)^*\otimes U(\mathfrak{g}_+)^*}$.

We will use the notations of section 21.2. Recall that the coproduct in $U_h(\mathfrak{g}_+)$ is defined by the formula

$$\Delta(x) = J^{-1} i_-(x) \tag{21.11}$$

where $i_- : U(\mathfrak{g}_+) \to U(\mathfrak{g}_+) \otimes U(\mathfrak{g}_+)$ is the usual coproduct, and where the element $J \in U(\mathfrak{g}) \otimes U(\mathfrak{g})[[h]]$ is given by formula (21.7). The product is given by

$$yx = X(1 \otimes Y)\Phi^{-1}(1_+ \otimes 1_+ \otimes 1_-) \tag{21.12}$$

where $X, Y \in \mathrm{Hom}_{\mathfrak{g}}(M_+ \otimes M_-, M_-)$ are determined by $X(1_+ \otimes 1_-) = x$ and $Y(1_+ \otimes 1_-) = y$. Formulas (21.11) and (21.12) imply that the product and coproduct in $U_h(\mathfrak{g}_+)^*$ are given by the action of some elements in $U(\mathfrak{g}) \otimes U(\mathfrak{g})$ or $U(\mathfrak{g})$. Thus Theorem 21.2 is a consequence of the following lemma (for more detail, see [EK2]) :

Lemma 21.5. *The algebra $U(\mathfrak{g})$ acts in $U(\mathfrak{g}_+)^* = M_-^*$ by differential operators. Moreover, if $\mathfrak{g}_+$ is the Lie algebra of a Poisson algebraic group G_+ then these differential operators belong to $\mathcal{D}_{G_+}$.*

Proof: It is sufficient to show that $\mathfrak{g}$ acts in $U(\mathfrak{g}_+)^*$ by differential operators from $\mathcal{D}_{G_+}$. This is clear for the subalgebra $\mathfrak{g}_+$. Now, we have

$$\{f, g\} = -m(r(f \otimes g)) = -\sum_i x_i^+ f.x_i^- g$$

Hence, for $g \in \mathcal{O}_{G_+}$, the Hamiltonian vector field $\sum_i (x_i^- g)x_i^+$ belongs to $\mathcal{D}_{G_+}$. But since (x_i^+) is a basis of left-invariant vector fields on G_+, the functions $x_i^- g$ belong to $\mathcal{O}_{G_+}$. Therefore $x_j^- \in \mathcal{D}_{G_+}$. This proves the lemma and the theorem.

Lecture 22

Dequantization and the equivalence theorem

Let $LBA_0(K)$ be the category of Lie bialgebras defined over $K = k[[h]]$ for which $\delta \equiv 0 \pmod h$. Consider the functor

$$Q_{0,K} : LBA_0(K) \to QUE(K)$$

induced by Q. The aim of this lecture is to sketch a proof of the following theorem:

Theorem 22.1. *The functor $Q_{0,K}$ is an equivalence of categories.*

Remark: Notice that the functor $Q_{0,K}$ assigns a Hopf algebra defined over K to a Lie bialgebra defined over the *same ring* K, and it shouldn't be confused with the functor $\tilde{Q} : LBA(k) \to QUE(K)$. In particular, to define the inverse functor to $Q_{0,K}$ (dequantization functor), we need more than the usual operation of taking the quasiclassical limit, because we want to obtain a Lie bialgebra defined over K rather than over k, i.e we want to recover more information.

We will describe the construction of the inverse functor $DQ_{0,K}$. First notice a technical difficulty: the operation of assigning to a universal enveloping algebra $U(\mathfrak{g})$ the Lie algebra $\mathfrak{g}$ is not a universal construction in the sense of Lecture 20 (it consists in taking primitive elements). We will thus replace the notion of a Lie bialgebra by the equivalent notion coPoisson Hopf algebra (see exercise 9.1).

In order to reverse the construction of quantization, we need the following notion:

22.1 The quantum double in a symmetric tensor category

Let A_+ be a Hopf algebra in a $\mathbb{Q}$-linear symmetric tensor category $\mathcal{N}$.

Definiton: An object X in $\mathcal{N}$ is a (left) *module* over A_+ if it is endowed with a map $\pi : A_+ \otimes X \to X$, represented by

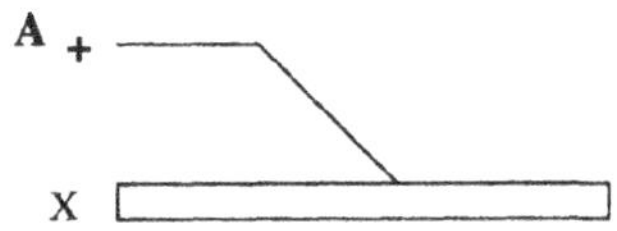

such that

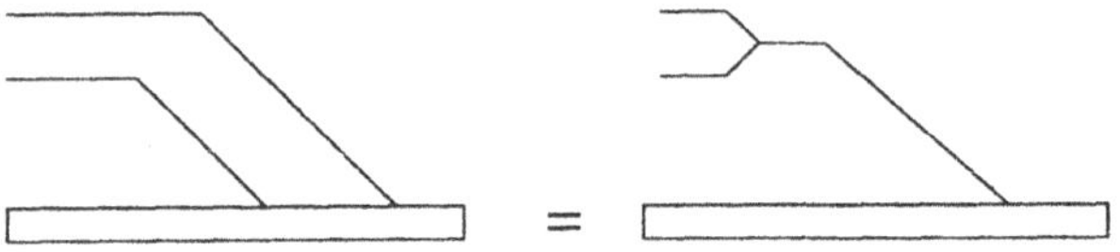

Definition: An object X in $\mathcal{N}$ is a (right) *comodule* over A_+ if it is endowed with a map $\pi^* : X \to A_+ \otimes X$ represented by

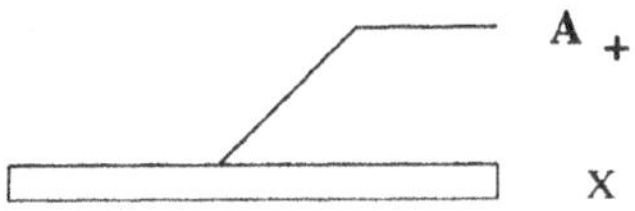

such that

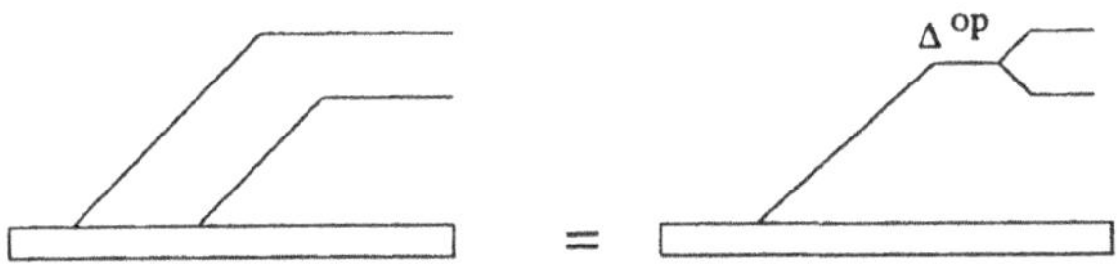

Definition: An object X in $\mathcal{N}$ is a (left) *dimodule* over A_+ if it is both a left module and a right comodule, and satisfies the following consistency condition:

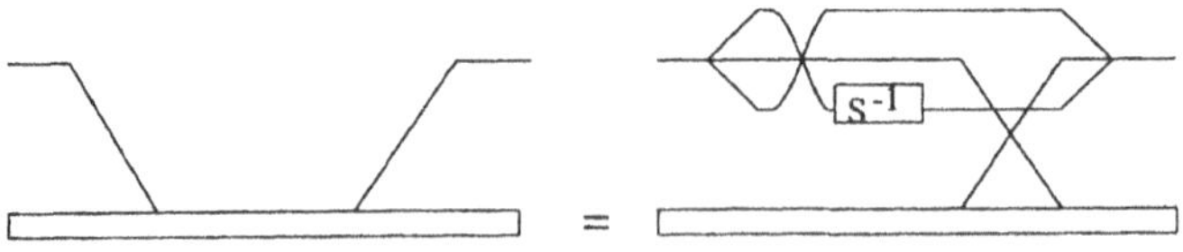

Remark. Dimodules over a Hopf algebra are also called Drinfeld-Yetter modules.

Exercise 22.1. Let $\mathcal{N} = \mathrm{Vec}_k$ and let A_+ be a finite dimensional Hopf algebra. Show that the notion of dimodule over A_+ coincides with the notion of module over the quantum double of A_+.

The collection of dimodules over A_+ forms a tensor category $\mathrm{Dimod}(A_+)$: if V, W are two dimodules, then $V \otimes W$ is a dimodule with action and coaction given by

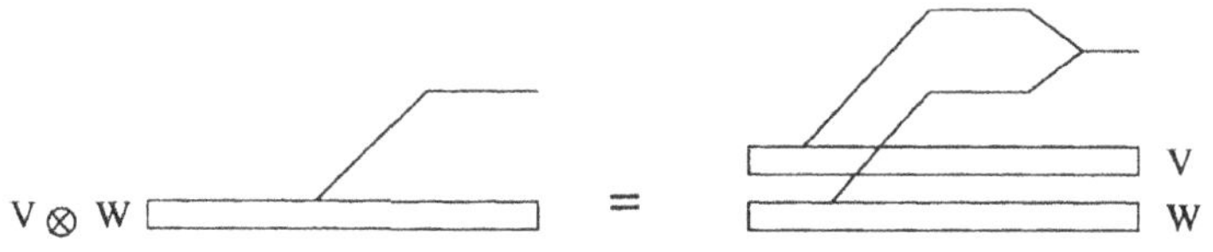

Since the quantum double of a Hopf algebra is quasitriangular, it is not surprising that this category admits the braiding defined by the R-matrix:

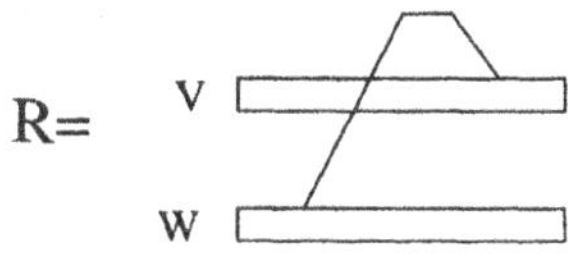

,

via $\beta = \sigma R$.

Let us now construct Verma modules M_- and M_+^* in $\mathrm{Dimod}(A_+)$. As objects, set $M_- = M_+^* = A_+$. The action on M_- is given by the (left) multiplication

$$\pi_- = m : A_+ \otimes A_+ \to A_+.$$

The coaction is given by the formula

$$\pi_-^* = (m^{op} \otimes 1)(S^{-1} \otimes \sigma_{23})\Delta_3$$

or by the picture

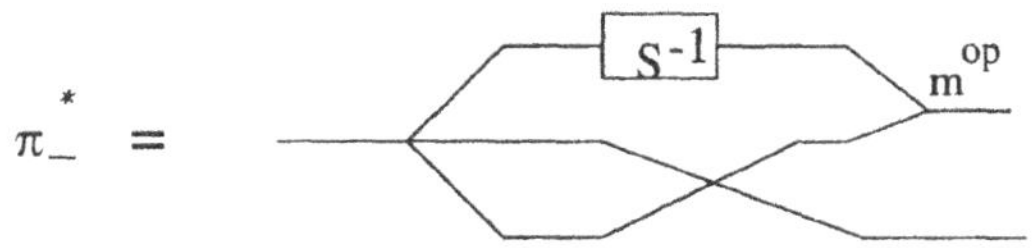

These formulas are obtained from the consistency condition. The action and coaction on M_+^* are obtained from the coaction and action on M_- by exchanging m and Δ and reversing the order of composition. More precisely, we set

$$\pi_+^* = \Delta^{op}, \qquad \pi_+ = m \circ \sigma_{23} \circ (1 \otimes S^{-1} \otimes 1)(\Delta^{op} \otimes 1).$$

Finally, the construction of the maps ψ and η is the same as in section 21.2. Notice that the Hopf algebra structure on A_+ is given by the formulas (21.5) and (21.6), where J can now be expressed as

$$J = (1_+ \otimes 1 \otimes 1_+ \otimes 1)R^{op}(\eta \otimes \eta)(1_- \otimes 1_-).$$

22.2 Dequantization

We want to define on $A_+ = M_-$ the structure of a coPoisson Hopf algebra. For this purpose, we will use the Grothendieck-Teichmuller semigroup $\overline{GT}(k)$. Recall that $\overline{GT}(k)$ acts on the set of braided tensor structures on any quasisymmetric tensor category. In our case, A_+ is a QUE algebra, i.e a deformation of a cocommutative Hopf algebra, and the category Dimod(A_+) is quasisymmetric. Moreover, GT_0 transforms a braided tensor category into a symmetric one. Furthermore, for every $a_0 \in GT_0$, there exists a (unique) one-parameter sub(semi)group $a(t)$, $t \in k$ such that $a(t) \in GT(t)$ (see Theorem 17.2).

Now choose $a_0 \in GT_0(k)$ and $a(t)$, $t \in k$ as above. Denote by Dimod(A_+, t) the braided tensor category obtained by twisting Dimod(A_+) by $a(t)$. Formulas (21.5) and (21.6) thus define a family of Hopf algebras $A_+^{(t)}$, where we now have

$$J = (1_+ \otimes 1 \otimes 1_+ \otimes 1)\Phi_{1,3,24}\Phi_{3,2,4}^{-1}R^{op}\Phi_{2,3,4}^{-1}\Phi_{1,2,34}(\eta \otimes \eta)(1_- \otimes 1_-).$$

Notice that Dimod$(A_+, 0)$ is symmetric. From this it can be deduced that $A_+^{(0)}$ is cocommutative. The family $A_+^{(t)}$ therefore induces a coPoisson structure on $A_+^{(0)}$, by

$$\delta = \lim_{t \to 0} \frac{1}{t}(\Delta_t - \Delta_t^{op}).$$

This is the desired coPoisson Hopf algebra. The reader is referred to [EK3] for more details.

Appendix 1

KZ associator and multiple zeta functions.

1.1 The multiple zeta function

In this appendix we discuss the relationship between multiple zeta values and associators. A similar discussion can be found in [Kass].

One of the most important functions in mathematics is the Riemann zeta function:

$$\zeta(s) = \sum_{n>0} n^{-s}, \mathrm{Re}\, s > 1.$$

This function has a multivariable generalization, which is called the multiple zeta function.

Definition: The multiple zeta function of depth k is the function

$$\zeta_k(s_1, ..., s_k) = \sum_{n_1>...>n_k>0} n_1^{-s_1}...n_k^{-s_k}, \mathrm{Re}(s_1 + ... + s_m) > m, 1 \leq m \leq k. \tag{1.1}$$

Thus, the Riemann zeta function is of depth 1: $\zeta(s) = \zeta_1(s)$.

In number theory, people are interested in values of the multiple zeta function at positive integer sequences $s_1, ..., s_k$ $(s_1 > 1)$.

Definition: A multiple zeta value of weight N is a value of ζ_k at a positive integer sequence $s_1, ..., s_k$ such that $\sum_j s_j = N$, and $s_1 > 1$. It is clear that the number of such possibilities (for all k) is 2^{N-2}.

Remark. If $s_1 = 1$, the series (1.1) is divergent.

Multiple zeta values are linked by many polynomial relations with rational coefficients, which are homogeneous with respect to the weight. For example, $\zeta(2) = \pi^2/6$, $\zeta(4) = \pi^4/90$, so $\zeta(4) = 2\zeta(2)^2/5$. Also, $2\zeta(2,2) + \zeta(4) = \zeta(2)^2$, which implies that $\zeta(2,2) = \pi^4/120 = 3\zeta(4)/4$. The purpose of this appendix is to give a conjectured description of all such relations.

Let us put this question in a more precise framework. Let $V_N \subset \mathbb{C}$ be the $\mathbb{Q}$-vector space generated by multiple zeta values of weight N, and W_N be the subspace of V_N consisting of all elements representable as polynomials with rational coefficients of multiple zeta values of lower weights, in which all terms have total weight exactly N. Let $D_N = \dim_{\mathbb{Q}}(V_N/W_N)$.
The following conjecture is due to D. Zagier:

Conjecture 2. $D_N = D_N^*$, *where* D_N^* *are given by the formula*

$$\prod_{N=1}^{\infty}(1-x^N)^{D_N^*} = 1 - x^2 - x^3. \tag{1.2}$$

D. Kreimer and D. Broadhurst [BrKr],[Br] gathered a lot of computational evidence for this conjecture.

Conjecture 2 consists of 2 parts: the upper bound $D_N \geq D_N^*$, and the lower bound $D_N \leq D_N^*$. The upper bound involves proving that $\zeta(2n+1)/(2\pi i)^{2n+1}$ is transcendental, and is well beyond the range of accessibility for today's number theory. On the other hand, the lower bound reduces to a purely algebraic problem, which can be solved modulo Drinfeld's conjecture 1, which is equivalent to another conjecture of Deligne [Del].

1.2 Multiple zeta values and the KZ equation

Let $s = (s_1, ..., s_k)$ be as above, and $I_s = \{s_1, s_1 + s_2, ..., s_1 + ... + s_k\} \subset \{1, ..., N\}$. It is well known (and easy to check) that

$$\zeta_k(s_1, ..., s_k) = \int_{\Delta_N} \prod_{i \notin I_s} z_i^{-1} \prod_{i \in I_s} (1-z_i)^{-1} dz_1...dz_N, \tag{1.3}$$

where Δ_N is the simplex $1 > z_1 > ... > z_N > 0$.

Following Drinfeld, consider the Knizhnik-Zamolodchikov equation

$$\frac{dF}{dz} = \left(\frac{a}{z} - \frac{b}{z-1}\right)F, \tag{1.4}$$

where a, b are noncommuting variables (see Lecture 15; in this appendix it is convenient to us to use $-b$ instead of b). We look for solutions with values in the algebra $U = U(a,b)$ of noncommutative formal series in a and b.

Let F_0, F_1 be two solutions of this equation on $(0,1)$ of the form

$$F_0(z) = (1 + \sum_{n\geq 1} F_0^n z^n) z^a, \qquad F_1(z) = (1 + \sum_{n \geq 1} F_1^n (1-z)^n)(1-z)^{-b}. \tag{1.5}$$

(Here for $z > 0$ $z^a = \sum_{m\geq 0} a^m (\ln z)^m / m!$).
Let $\Phi(a, -b) \in U$ be the monodromy operator, defined by $F_1(z)\Phi = F_0(z)$ (for any $z \in (0,1)$). As the reader remembers from Lecture 15, this operator is called the Knizhnik-Zamolodchikov associator.

Let W be the set of all words made out of letters a, b. For $w \in W$, let $|w|$ be the length of the word w, and $w(i)$, $i \leq |w|$, be the i-th letter of w.

Define the U-valued function of two variables u, v $(0 < v \leq u < 1)$ by

$$G(u,v) = \sum_{w \in W} w \int_{\Delta_{|w|}(v,u)} \prod_{i:w(i)=a} z_i^{-1} \prod_{i:w(i)=b} (1-z_i)^{-1} dz_1 ... dz_{|w|}, \tag{1.6}$$

where $\Delta_{|w|}(v,u) = \{(z_1, ..., z_{|w|}) : u > z_1 > ... > z_{|w|} > v\}$. It is easy to see that $G(v,v) = 1$, and $G(u,v)$ satisfies the KZ equation in u. Therefore,

$$G(u,v) = (P\exp \int_v^u)(\frac{a}{z} - \frac{b}{z-1}), \tag{1.7}$$

where $P\exp$ denotes the ordered exponential. Therefore,

$$\Phi = \lim_{v \to 0, u \to 1} (1-u)^b G(u,v) v^a. \tag{1.8}$$

In particular,

$$\Phi = \sum_{w \in W} C_w w, \tag{1.9}$$

where for any w such that $w(1) = a$ and $w(|w|) = b$ we have

$$C_w = \int_{\Delta_{|w|}} \prod_{i:w(i)=a} z_i^{-1} \prod_{i:w(i)=b} (1-z_i)^{-1} dz_1 ... dz_{|w|}. \tag{1.10}$$

Thus, for such w the number C_w is the multiple zeta value corresponding to some combination $s_1, ..., s_k$, and each such combination is obtained exactly once.

Remark: Other coefficients of Φ express explicitly in terms of these multiple zeta values.

Let $F = F(a,b)$ be the Lie algebra freely generated by a, b and completed with respect to the grading. Then U is its universal enveloping algebra. Let $F_{\geq 2}$ be the sum of components of degree ≥ 2 in F.
The following easy proposition is due to Drinfeld.

Proposition 1.1. *$\Phi = e^X$, where $X \in F_{\geq 2}$.*

Remark: In fact, $X = \frac{\pi^2}{6}[a,b]$+elements of degree ≥ 3.

Let $U_{\geq 2} \subset U$ be the universal enveloping algebra of $F_{\geq 2}$. By the Poincaré-Birkhoff-Witt theorem, the multiplication map identifies $\mathbb{C}[[b]] \otimes U_{\geq 2} \otimes \mathbb{C}[[a]]$ with U (here by $\otimes$ we mean the completed tensor product). Therefore, $U_{\geq 2}$ is naturally identified with $U/(bU + Ua)$. The latter space is spanned by words w such that $w(1) = a$ and $w(|w|) = b$. Denote the set of such words by W_0. Let Z_w be the element of $U_{\geq 2}$ corresponding to $w \in W_0$. Then

$$\Phi = \sum_{w \in W_0} C_w Z_w, \tag{1.11}$$

where C_w is given by formula (1.10). Thus, the Knizhnik-Zamolodchikov associator is a "noncommutative generating function" for multiple zeta values.

For us it will be more convenient to use a renormalized associator $\Phi_*(a, -b) = \Phi(\frac{a}{2\pi i}, \frac{-b}{2\pi i})$. This associator is the generating function for renormalized multiple zeta values $\zeta_k(s_1, ..., s_k)/(2\pi i)^{\sum s_j}$. However, the dimensions D_n for such values are clearly the same as for the usual ones.

1.3 The relations between multiple zeta values

The element $\Phi_*(a, b)$ satisfies some remarkable algebraic equations (equations (17.2,17.3,17.4)), coming from the fact that it is a Lie associator (see Lecture 15). Recall that for any field k of characteristic zero, we denoted by $M_1(k)$ the space of all elements $X \in F_{\geq 2}(k)$ such that $\Psi = e^X$ is a Lie associator, and by $M_1^{(N)}(k)$ be the same space modulo terms of degree $> N$ in the formal series.

Remark: The space $M_1^{(2)}$ consists of one point, as for any $\Psi \in M_1(k)$, $\Psi = 1 - \frac{[ab]}{24} + \dots$.

According to Theorem 17.1, the natural map $M_N \to M_{N-1}$ is an affine space bundle. The variety M_1 is the inverse limit of $M_1^{(N)}$ as $N \to \infty$.

Set $d_N = \dim(M_1^{(N)}) - \dim(M_1^{(N-1)})$ for $N \geq 3$. Thus, d_N is the dimension of the fiber of the affine space bundle $M_N \to M_{N-1}$.

Theorem 1.1. $D_N \leq d_N$.

Proof: For any $X \in F_{\geq 2}$, write $e^X = \sum_{w \in W_0} f_w(X) Z_w$. Let h_w be the restriction of f_w to the variety M_1. For $N \geq 3$, the system of equations (13.12) with respect to $\Psi = e^X$, $X \in F_{\geq 2}$, can be written in the form

$$\sum_{w \in W_0 : |w| = N} R_i^w h_w(X) = \sum_{j=0}^{N} P_i^j(X), i \in I. \tag{1.12}$$

where I is some finite index set, R is a matrix of rational numbers, and $P_i^j(X)$ are homogeneous polynomials of $\{h_w(X) : |w| < N\}$ of degree j with rational coefficients (here the degree of h_w is defined to be $|w|$). Since equations (17.2) are invariant under the change $A, B \to -A, -B$ (i.e. $h_w \to (-1)^{|w|}h_w$), the polynomials P_i^j are zero unless $N - j$ is even. Since $h_{ab} = -1/24$, we can always assume that P_i^j is of degree N.

It is clear that the dimension of the kernel of the matrix R (i.e. the dimension of the space of solutions of (1.12)) is equal to d_N. This is true because the space of solutions of (1.12) is the fiber of the bundle $M_1^{(N)} \to M_1^{(N-1)}$. Since Φ_* is a complex-valued point of the variety M_1, any $\mathbb{Q}$-rational polynomial equation satisfied identically on M_1 has to be satisfied for Φ_*. Therefore, $D_N \le d_N$.

□

As we explained in Lecture 17, Drinfeld showed that the variety $M_1^{(N)}$ is isomorphic to the space $\mathfrak{g}_3 \oplus ... \oplus \mathfrak{g}_N$, where $\mathfrak{g} = \oplus_{i \ge 3}\mathfrak{g}_i$ is a certain graded Lie algebra, called the Grothendieck-Teichmuller Lie algebra (denoted by $\mathfrak{grt}_1$ in lecture 18), and $\mathfrak{g}_i$ is the i-th homogeneous component of $\mathfrak{g}$. Thus, $d_N = \dim \mathfrak{g}_N$.

Finally, recall that Drinfeld's conjecture 1 says

Conjecture: *The Lie algebra $\mathfrak{g}$ has no more than one generator in odd degrees, and no generators in even degrees.*

Remark: This conjecture is in fact weaker than Conjecture 1.

This conjecture implies that the Hilbert series of $U(\mathfrak{g})$ is dominated by

$$(1 - x^3 - x^5 - x^7 - ...)^{-1} = (1 - \frac{x^3}{1 - x^2})^{-1} = \frac{1 - x^2}{1 - x^2 - x^3}.$$

Thus, the conjecture implies that $d_N \le D_N^*$.

Now it is clear that this conjecture implies the estimate $D_N \le D_N^*$.

Remark: As we remarked, computer simulations [BrKr] show that it is likely that $D_N = D_N^*$. If one believes Drinfeld's conjecture and the prediction of these simulations, then $\mathfrak{grt}_1$ has exactly one generator in odd degrees and no generators in even degrees. This would mean that Φ_* is a "Weil generic point" of M_1, i.e. that it does not belong to any proper subvariety of M_1 defined over $\mathbb{Q}$. Another way of saying this would be that all homogeneous polynomial relations with rational coefficients follow from the axioms of a Lie associator.

Unfortunately, as we said, at present such a statement is beyond reach.

Appendix 2

Solutions to Problems and Exercises

Lecture 1

The three problems follow each other closely:

Problem 1.1: Let A_0 be a Poisson algebra without zero divisors, with a non-trivial Poisson bracket $\{,\}$. We show that A_0 has transcendence degree at least 2. Let $x, p \in A_0$ such that $\{x,p\} \neq 0$. Then x is transcendental. Indeed, if not, let P be its minimal polynomial. We then have $0 = \{P(x), p\} = P'(x)\{x,p\}$, which contradicts the minimality of P. In a similar fashion, p is transcendental over $k(x)$, and so A_0 has transcendence degree at least 2.

Problem 1.2: Let A be a deformation of A_0, and write

$$f * g = fg + hc_1(f,g) + h^2 c_2(f,g) + \dots$$

By Problem 1.1, the Poisson structure $\{,\}_1$ induced on A_0 by A is trivial. Hence $c_1(f,g) = c_1(g,f)$. In this case, we can define a Poisson structure from c_2,

$$\{f,g\}_2 = c_2(f,g) - c_2(g,f)$$

and we obtain $c_2(f,g) = c_2(g,f)$. In a similar fashion, we get $c_i(f,g) = c_i(g,f)$ for all i, i.e A is commutative.

Problem 1.3: Let us write $Z(L) = \bigcup_{n \geq 0} Z(L)^n$ for the filtration on $Z(L)$ induced from the filtration of D_X, and let us consider the following algebra:

$$A = (\bigoplus h^n Z(L)^n)[[h]] = \{z_0 + hz_1 + h^2 z_2 + \dots \mid i \geq deg(z_i),\ i - deg(z_i) \underset{i \to \infty}{\longrightarrow} \infty\}.$$

Then just as in the case of the canonical quantization (of T^*X), A is a deformation of $A_0 = Gr(Z(L))$. We claim that the Poisson structure $\{,\}$ induced on A_0 is trivial. Indeed, the restriction of this Poisson structure to the subalgebra A_L of $A_0 = Gr(Z(L))$ generated by L is trivial, and since $L \notin k$ and X is a curve, A_0 is algebraic over A_L. By the same arguments as in Problem

1.1, $\{,\}$ vanishes. Similarly, $\{,\}_2$, $\{,\}_3, \ldots$ vanish, and by Problem 1.2, $Z(L)$ is commutative. The generalization is proved along the same lines. The fact that $Z(L_1, \ldots L_n)$ is algebraic over $k[L_1, \ldots L_n]$ follows from Bernstein's inequality for the dimension of modules over rings of differential operators (D-modules)-see [Bo].

Lecture 2

Problem 2.1 Let $x, y \in G_0$. Then, by (2.2), we have

$$\Pi(xy) = (d(\rho_y)_{|x} \otimes d(\rho_y)_{|x})\Pi(x) + (d(\lambda_x)_{|y} \otimes d(\lambda_x)_{|y})\Pi(y) = 0.$$

Similarly, $\Pi(x^{-1}) = -(di_x \otimes di_x)\Pi(x) = 0$ so that G_0 is a subgroup, which is closed by definition.

Problem 2.2: Let Π be the Poisson bivector of G. The map $\pi : G \to G/H$ induces a Poisson structure on G/H if and only if

$$(d\pi_{|g} \otimes d\pi_{|g})\Pi(g) = (d\pi_{|gh} \otimes d\pi_{|gh})\Pi(gh) \qquad \forall g \in G,\ h \in H. \tag{2.1}$$

Now, G is a Poisson-Lie group, so we have

$$\Pi(gh) = (d(\lambda_g)_{|h} \otimes d(\lambda_g)_{|h})\Pi(h) + (d(\rho_h)_{|g} \otimes d(\rho_h)_{|g})\Pi(g).$$

Since $(d\pi_{|gh} \otimes d\pi_{|gh})(d(\rho_h)_{|g} \otimes d(\rho_h)_{|g}) = (d\pi_{|g} \otimes d\pi_{|g})$, we see that (2.1) reduces to

$$(d\pi_{|gh} \otimes d\pi_{|gh})(d(\lambda_g)_{|h} \otimes d(\lambda_g)_{|h})\Pi(h) = 0,$$

which is equivalent to $(d\pi_{|h} \otimes d\pi_{|h})\Pi(h) = 0$. Viewing the Poisson bivector as a map $\Pi : G \to \Lambda^2\mathfrak{g}$ (using right translations), we thus have that Π descends to a Poisson structure on G/H if and only if

$$\Pi(h) \subset \mathfrak{g} \otimes \mathfrak{h} + \mathfrak{h} \otimes \mathfrak{g} \qquad h \in H.$$

In particular, $\delta(\mathfrak{h}) \subset \mathfrak{h} \otimes \mathfrak{g} + \mathfrak{g} \otimes \mathfrak{h}$.

Lecture 3

Problem 3.1: Consider the following (topological) basis of $\mathfrak{a}$: $a_n = -z^{n+1}\frac{d}{dz}$ $(n \geq -1)$, for which the relations are $[a_n, a_m] = (n-m)a_{n+m}$. To show that any coboundary structure on $\mathfrak{a}$ is triangular, it suffices to show that $(\Lambda^3\mathfrak{a})^{\mathfrak{a}}$ is trivial. But since $[a_0, a_n] = -na_n$, any $u \in (\Lambda^3\mathfrak{a})^{\mathfrak{a}}$ is a sum of terms of total degree 0, and hence is proportional to $u_0 = a_{-1} \wedge a_0 \wedge a_1$, and it is easy to see that u_0 is not $\mathfrak{a}$-invariant.

Now, by Theorem 3.2, to classify all triangular structures on $\mathfrak{a}$, we must first look for nontrivial finite dimensional Lie subalgebras. Let $\mathfrak{h} \subset \mathfrak{a}$ be a Lie subalgebra, and suppose that there exists two linearly independent elements $c_1, c_2 \in \mathfrak{h}$ such that

$$\begin{aligned} c_1 =& x_i a_i + x_{i+1} a_{i+1} + \dots \qquad (i > 0,\ x_i \neq 0), \\ c_2 =& y_j a_j + y_{j+1} a_{j+1} + \dots \qquad (j > 0,\ y_j \neq 0). \end{aligned}$$

Without loss of generality, we can suppose $i \neq j$. Then it is easy to see that $[c_1, c_2] = x_i y_j (i - j) a_{i+j} + \dots$, and similarly $ad(c_1)^n c_2 = \lambda_n a_{ni+j} + \dots$ with $\lambda_n \neq 0$. In particular, $\mathfrak{h}$ is not finite dimensional. Therefore if $\mathfrak{h}$ is finite dimensional, $\mathfrak{h} \cap \bigoplus_{i>0} \mathbb{C} a_i$ is at most one dimensional and $\mathfrak{h}$ is at most three dimensional. There are no nondegenerate elements in $\Lambda^2 \mathfrak{h}$ if dim $\mathfrak{h} = 3$. Thus all triangular structures on $\mathfrak{a}$ come from a Lie subalgebra generated by

$$\begin{aligned} c_1 &= x_i a_i + \dots \qquad (i \geq 0), \\ c_0 &= y_{-1} a_{-1} + y_0 a_0 + \dots \end{aligned}$$

It is clear that if $i > 0$ then $y_{-1} = 0$ and $y_0 \neq 0$ (or $[c_1, c_0] \notin \mathbb{C} c_1 \oplus \mathbb{C} c_0$). Notice that the group Aut $\mathbb{C}[[z]]$ acts on $\mathfrak{a}$, and conjugating by a formal diffeomorphism $x \mapsto \alpha_1 x + \alpha_2 x^2 + \dots$, we can suppose $c_1 = a_0$ or $c_0 = a_0$. In this case, it is easy to see that $\mathfrak{h} = \mathfrak{h}_n = \mathbb{C} a_0 \oplus \mathbb{C} a_n$ for some $n = 1$ or $n > 0$. Hence, all triangular structures on $\mathfrak{a}$ come from the image under a formal diffeomorphism of a nondegenerate one on $\mathfrak{h}_n$ for some $n \neq 0$.

Remark: in fact, Kuperschmidt and Stoyanov prove more: they show that all coboundary structures, even of infinite rank, come from a two-dimensional Lie bialgebra (see [KS]).

Problem 3.2: (a) Let $e_0 = 1, e_1, \dots e_n$ be a basis of A and set $e_i e_j = \sum_k c_{ij}^k e_k$. Let $\Pi(x) = \sum_{i,j} \Pi_{ij}(x) e_i \otimes e_j$ be the Poisson bivector of a Poisson structure on A compatible with the multiplication. By definition, we have

$$(dm_{|x,y} \otimes dm_{|x,y})(\Pi(x), \Pi(y)) = \Pi(xy)$$

where $dm_{|x,y} : T_x A \times T_y A \simeq A \times A \to A$ is the differential of the multiplication map. In our coordinates system, where we set $x = \sum_i x_i e_i$ and $y = \sum_i y_i e_i$, this gives us

$$\sum_{i,j,k,l} c_{il}^r c_{jk}^s \left(\Pi_{ij}(x) y_k y_l + \Pi_{lk}(y) x_i x_j \right) = \Pi_{rs}(xy). \tag{2.2}$$

Setting $x = y = t.1$ ($t \in \mathbb{R}$) in (2.2), we see that $\Pi_{rs}(t^2) = 2t^2 \Pi_{rs}(t)$, which implies that $\Pi_{rs}(t) = \lambda_{rs} t^2 log(t)$ for some constants λ_{rs}. But since Π is smooth, we must have $\lambda_{rs} = 0 = \Pi_{rs}(t)$. Now, from (2.2) with $x = t$, we get $\Pi_{rs}(ty) = t^2 \Pi_{rs}(y)$. Smooth functions satisfying this are necessarily quadratic.
(b) It is clear that G_A is a Lie group, with Lie algebra Lie$(G_A) = A$, as G contains some neighborhood of 1. Since G is coboundary, the Poisson bivector of G is given by

$$\Pi(g) = (d\rho_g \otimes d\rho_g) r - (d\lambda_g \otimes d\lambda_g) r$$

for some $r \in A \otimes A$. Extending Π to the whole A by the same formula yields a Poisson structure on A compatible with the multiplication, hence quadratic by part (a).

Remark: The condition of smoothness (or at least of regularity of class C^2) is essential here. Consider the algebra $\mathbb{R}^2$ (with componentwise addition and multiplication). Then the Poisson bracket

$$\{f(x,y), g(x,y)\} = xy\log|x|(\frac{\partial f}{\partial x}\frac{\partial g}{\partial y} - \frac{\partial f}{\partial y}\frac{\partial g}{\partial x})$$

is compatible with the algebra structure, but is not quadratic. Note also that the restriction of this Poisson bracket to the multiplicative group is a Poisson-Lie structure, so that the word "coboundary" in (b) is essential.

Lecture 4

Exercise 4.1: The isomorphism $\tilde{\mathfrak{g}} \to \mathfrak{g} \oplus \mathfrak{h}$ is given by

$$\begin{aligned} e_\alpha &\mapsto (e_\alpha, 0), \\ f_\alpha &\mapsto (f_\alpha, 0), \\ h^{(1)} &\mapsto (h, h), \\ h^{(2)} &\mapsto (h, -h). \end{aligned}$$

Lecture 5

Problem 5.1: By construction (the double construction), $\mathfrak{b}_\pm$ are subbialgebras for the standard structure on $\mathfrak{g}$. Now consider the Lie bialgebra structure on $\mathfrak{g} = \mathfrak{n}_- \oplus \mathfrak{h} \oplus \mathfrak{n}_+$ corresponding to a non-empty Belavin-Drinfeld triple $(\Gamma_1, \Gamma_2, \tau)$. Let θ be a maximal root of $\mathfrak{g}_{\Gamma_1}$ and let β be a root of $\mathfrak{g}_{\Gamma_2}$, $\beta > \alpha$. Since $\Gamma_2 \neq \Gamma$, β is not the maximal root of $\mathfrak{g}$. Thus there exists a simple root $\gamma \in \Gamma$ such that $[x_\gamma, x_\beta] \neq 0$. But then the term $[x_\gamma \otimes 1 + 1 \otimes x_\gamma, x_{-\alpha} \otimes x_\beta]$ is the only contribution to the $\mathfrak{g}_{-\alpha} \otimes \mathfrak{g}_{\beta+\gamma}$-part of $\delta(x_\gamma) = [x_\gamma \otimes 1 + 1 \otimes x_\gamma, r]$. In particular, $\delta(x_\gamma) \notin \Lambda^2(\mathfrak{b}_+)$.

Lecture 6

Problem 6.1: This construction of a Manin triple $(\mathfrak{a}, \mathfrak{g}, \mathfrak{g}_-)$ inducing the standard Lie bialgebra structure on $\mathfrak{g} = \mathfrak{n}_- \oplus \mathfrak{h} \oplus \mathfrak{n}_+$ works in fact for any Kac-Moody algebra (but has a geometric interpretation only for affine Kac-Moody algebras): set

$$\begin{aligned} \mathfrak{a} &= \mathfrak{g} \oplus \mathfrak{g}, \qquad \text{(direct sum of Lie algebras)}, \\ \mathfrak{g}_- &= \{x \oplus y,\ x \in \mathfrak{b}_+, y \in \mathfrak{b}_- \mid x + y \in \mathfrak{n}_+ \oplus \mathfrak{n}_- \subset \mathfrak{g}\} \end{aligned}$$

and let $\mathfrak{g} \hookrightarrow \mathfrak{a}$ by the embedding $x \mapsto x \oplus x$. Consider the bilinear form

$$< x_1 \oplus y_1, x_2 \oplus y_2 > = < x_1, x_2 > - < y_1, y_2 > .$$

Then $<,>$ is invariant, nondegenerate and $\mathfrak{g}, \mathfrak{g}_-$ are Lagrangian subalgebras such that $\mathfrak{g} \oplus \mathfrak{g}_- = \mathfrak{a}$. Thus $(\mathfrak{a}, \mathfrak{g}, \mathfrak{g}_-)$ is a Manin triple.

To see that the Lie bialgebra structure $(\mathfrak{g}, [\,,\,], \delta)$ induced on $\mathfrak{g}$ is indeed the standard one $(\mathfrak{g}, [\,,\,], \delta_s)$ corresponding to the polarization $\mathfrak{g} = \mathfrak{n}_- \oplus \mathfrak{h} \oplus \mathfrak{n}_+$, it is enough to identify $(\mathfrak{a}, \mathfrak{g}, \mathfrak{g}_-)$ with the Manin triple associated to $(\mathfrak{g}, [\,,\,], \delta_s)$. We will just sketch the construction, leaving the details to the readers (in particular the "doubling" of the Cartan subalgebra): recall that $(\mathfrak{g}, [\,,\,], \delta_s)$ is itself defined as (almost) the double of $\mathfrak{b}_+$. In particular, $\mathfrak{b}_\pm$ are subbialgebras. Thus the dual of $(\mathfrak{g}, [\,,\,], \delta_s)$ is (almost) a direct sum of $\mathfrak{b}_+$ with the usual Lie algebra structure and $\mathfrak{b}_-$ with the opposite Lie coalgebra structure (since the cocommutator of a Drinfeld double $\mathfrak{g}$ is defined as $\delta_\mathfrak{g} = \delta_{\mathfrak{g}_+} - \delta_{\mathfrak{g}_-}$). This identifies $\mathfrak{g}_-$ with $\mathfrak{g}^*$ and $(\mathfrak{a}, \mathfrak{g}, \mathfrak{g}_-)$ with the Manin triple associated to $(\mathfrak{g}, [\,,\,], \delta_s)$.

Lecture 8

Exercise 8.1: Let us prove the identity:

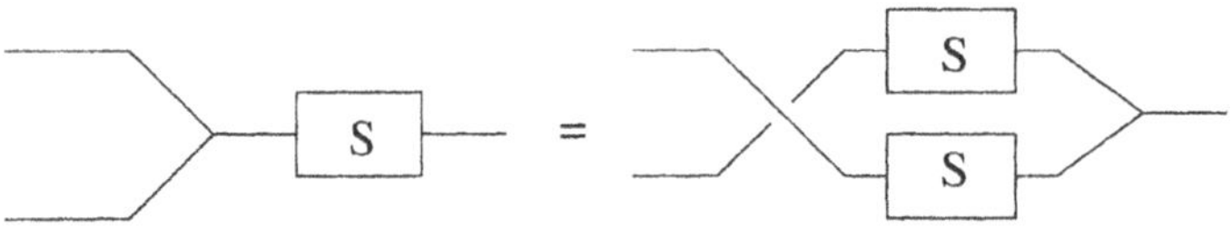

In the same fashion as in Proposition 8.1, let us introduce the linear map η : Hom $(A \otimes A, A) \to$ Hom $(A \otimes A, A)$ defined by the picture

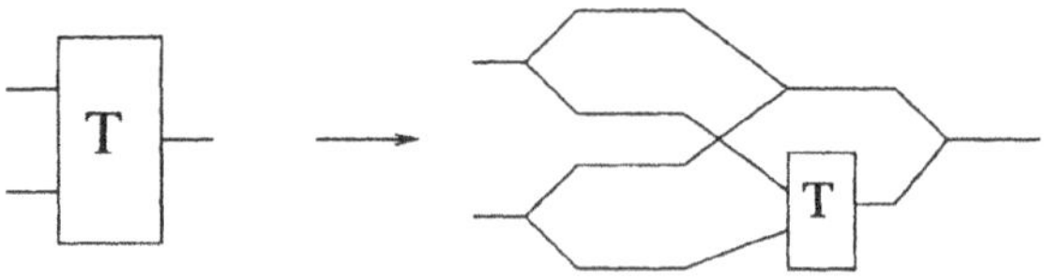

It is easy to check that η is invertible and that its inverse is given by the picture

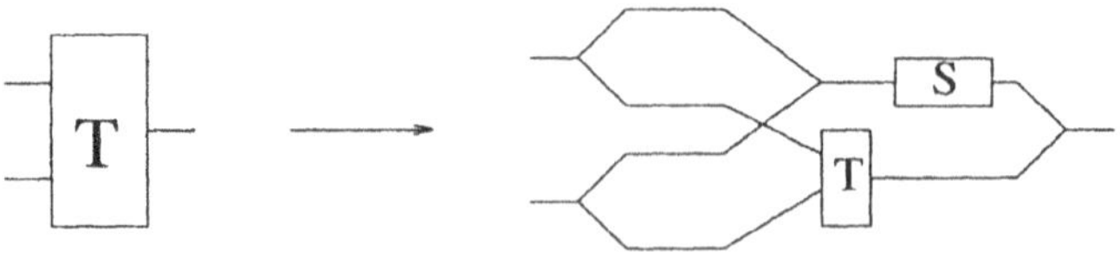

Consider the two morphisms

$$\begin{aligned} \nu_1 &: A \otimes A \to A, \\ &\quad f \otimes g \mapsto S(fg), \\ \nu_2 &: A \otimes A \to A, \\ &\quad f \otimes g \mapsto S(g)S(f). \end{aligned}$$

We have

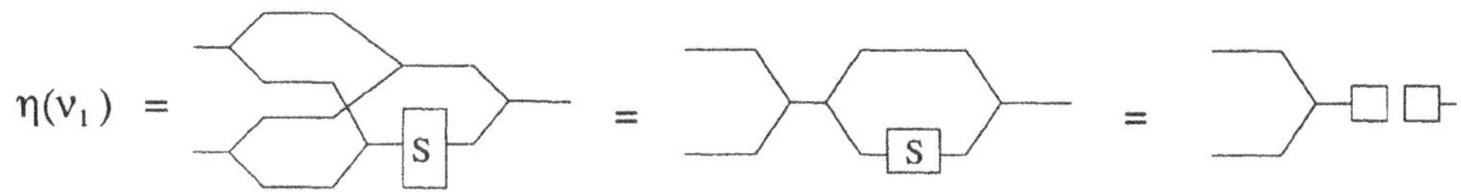

and

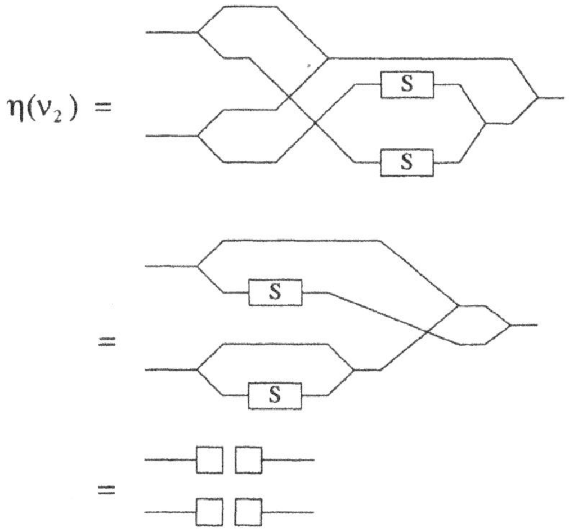

Thus $\eta(\nu_1) = \eta(\nu_2) = \epsilon \otimes \epsilon$. We conclude as in Prop 8.1 that $\nu_1 = \nu_2$. The second identity in 6. is proved in the same way, using the map

$$\eta : \operatorname{Hom}(A, A \otimes A) \to \operatorname{Hom}(A, A \otimes A),$$

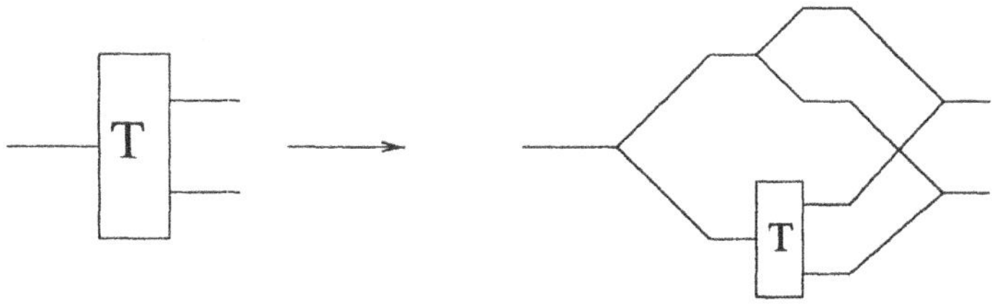

Exercise 8.2: Suppose that A is commutative. Then

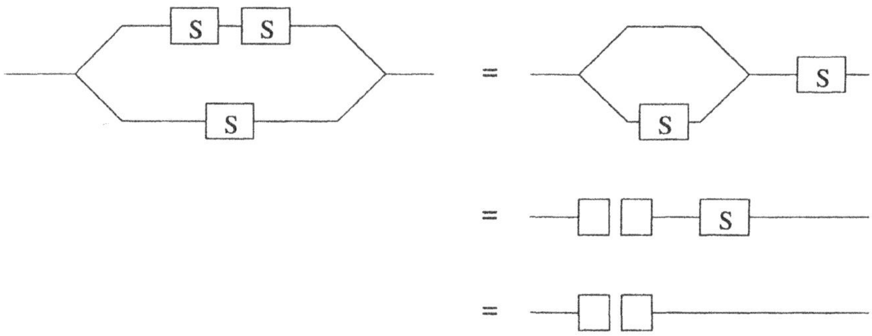

Since the map End $A \to$ End A,

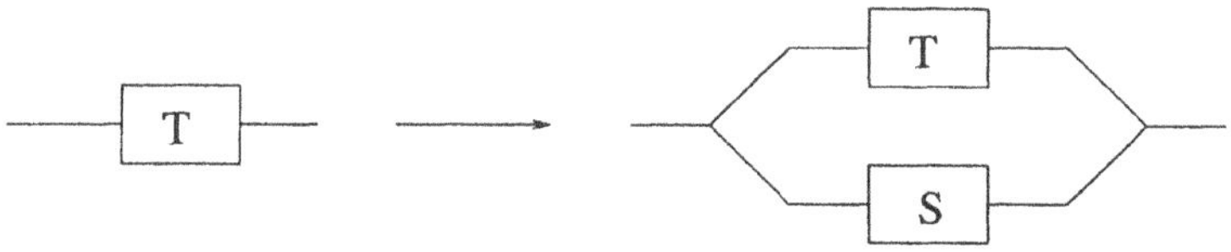

is invertible, this implies that $S^2 = Id$. The case when A is cocommutative is treated in a similar way (or by considering the dual Hopf algebra).

Exercise 8.3: By definition, $\mathcal{F}_k(G) = k[G]^*$ as a vector space. The multiplication

$$k[G] \otimes k[G] \to k[G],$$
$$g_1 \otimes g_2 \mapsto g_1 g_2$$

is dual to the comultiplication

$$\mathcal{F}_k(G) \to \mathcal{F}_k(G) \otimes \mathcal{F}_k(G) = \mathcal{F}_k(G \times G),$$
$$f(\cdot) \mapsto f(\cdot \times \cdot).$$

Conversely, the comultiplication

$$k[G] \to k[G] \otimes k[G],$$
$$g \mapsto g \otimes g$$

is dual to the multiplication

$$\mathcal{F}_k(G) \otimes \mathcal{F}_k(G) \to \mathcal{F}_k(G),$$
$$f(\cdot) \otimes h(\cdot) \mapsto f(\cdot)h(\cdot).$$

Lecture 9

Exercise 9.1: Let $\mathfrak{g}$ be a Lie algebra, $U(\mathfrak{g})$ its enveloping algebra, and let $\delta : U(\mathfrak{g}) \to \Lambda^2 U(\mathfrak{g})$ be a coPoisson structure on $U(\mathfrak{g})$. As in Proposition 9.1, it follows from the coLeibniz identity that δ restricts to a map $\mathfrak{g} \to \Lambda^2\mathfrak{g}$. The coJacobi and cocycle identities for a Lie bialgebra are just the restriction of the coJacobi and cocycle identities for a coPoisson Hopf algebra. Thus δ defines a Lie bialgebra structure on $\mathfrak{g}$.

Conversely, let $(\mathfrak{g}, [\,,\,], \delta)$ be a Lie bialgebra, and let $U(\mathfrak{g})$ be its enveloping algebra. From the cocycle condition,

$$\delta([a,b]) = [\delta(a), \Delta(b)] + [\Delta(a), \delta(b)]. \qquad (*)$$

we see that it is possible to extend the map $\delta : \mathfrak{g} \to \Lambda^2\mathfrak{g}$ to $\delta : U(\mathfrak{g}) \to \Lambda^2 U(\mathfrak{g})$ by the rule

$$\delta(ab) = \delta(a)\Delta(b) + \Delta(a)\delta(b). \qquad (**)$$

The coLeibniz identity follows from $(*)$ and $(**)$, while the coJacobi identity follows from the coJacobi identity for δ on $\mathfrak{g}$ and $(**)$. Thus δ defines a coPoisson Hopf algebra structure on $U\mathfrak{g}$.

Exercise 9.2: This follows from $\begin{bmatrix} n \\ n \end{bmatrix}_q = 1$ and the easily checked recursion formula

$$\begin{bmatrix} n+1 \\ k \end{bmatrix}_q = q^{-k} \begin{bmatrix} n \\ k \end{bmatrix}_q + q^{n-k+1} \begin{bmatrix} n \\ k-1 \end{bmatrix}_q .$$

Exercise 9.3: It is clear that $U_h(\mathfrak{g})/hU_h(\mathfrak{g}) = U(\mathfrak{g})$ as a Hopf algebra (the relations defining $U_h(\mathfrak{g})$ agree with the relations defining $U(\mathfrak{g})$ modulo h where $E_i \equiv e_i \pmod h$, $F_i \equiv f_i \pmod h$, $H_i \equiv h_i \pmod h$). Let us compute the induced Lie bialgebra structure on $\mathfrak{g}$:

$$\begin{aligned} \delta(e_i) &\equiv \frac{1}{h}(\Delta(E_i) - \Delta^{op}(E_i)) \pmod h \\ &\equiv d_i E_i \wedge H_i \pmod h \\ &= d_i e_i \wedge h_i. \end{aligned}$$

In the same way, we have $\delta(f_i) = d_i f_i \wedge h_i$ and $\delta(h_i) = 0$, which proves that the quasiclassical limit of $U_h(\mathfrak{g})$ is indeed (twice) the standard structure on $\mathfrak{g}$.

Problem 9.1: Let us first show that $S^2(x)u = ux$. We will use the following trick:

Claim: We have $1 \otimes S^2(x) = (1 \otimes S^2)(m \otimes 1)(S \otimes 1 \otimes 1)\Delta_2(x)$.

In pictures, this is:

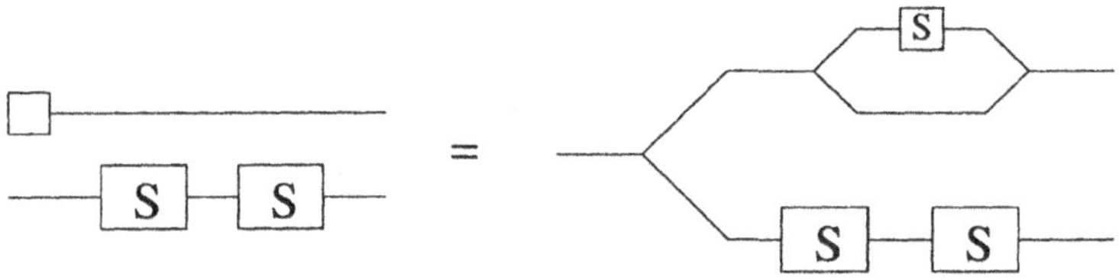

Proof: Apply $1 \otimes S^2$ to the picture

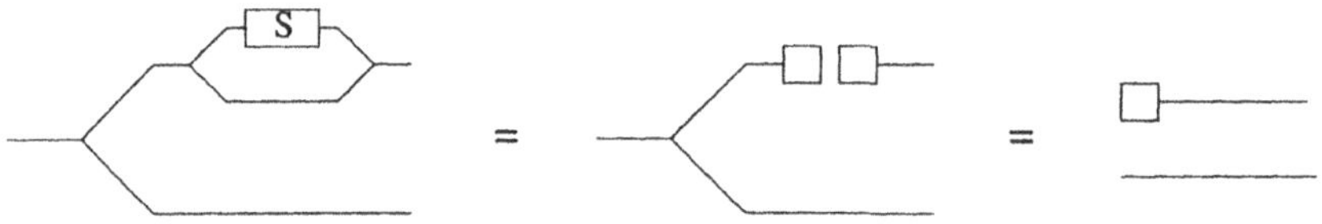

□

Thus, we have the following expression for $S^2(x)u$:

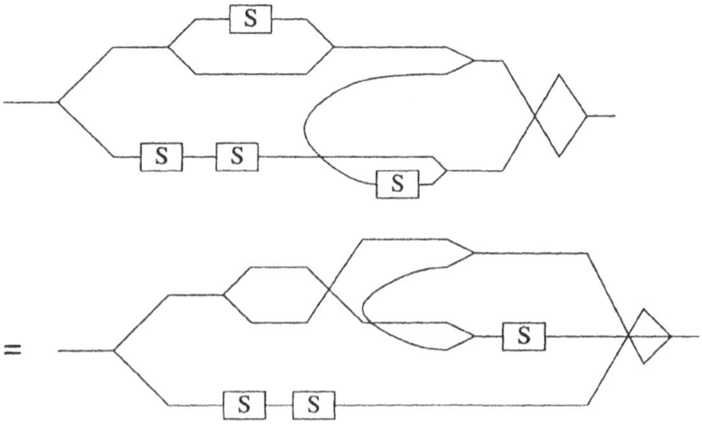

From $R\Delta = \Delta^{op}R$, we have

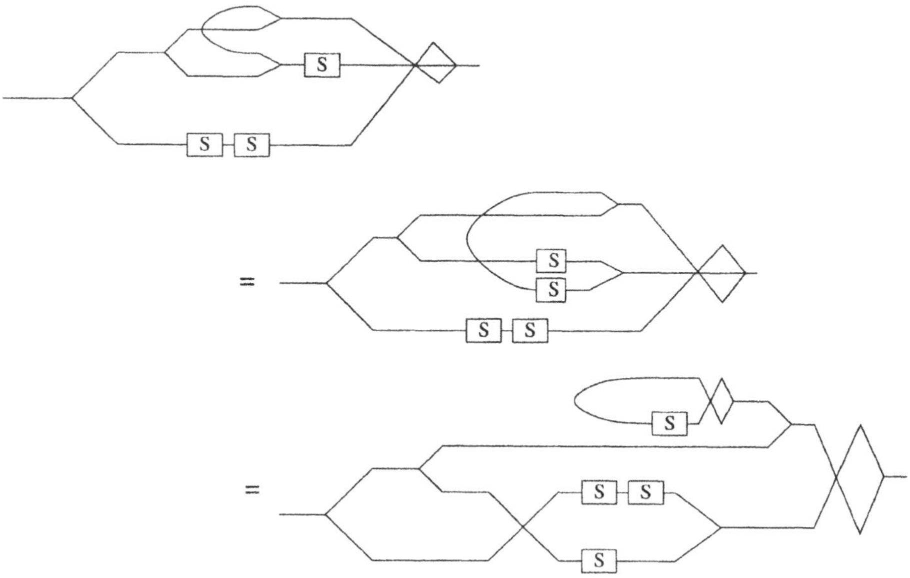

(2.3)

But from the equality

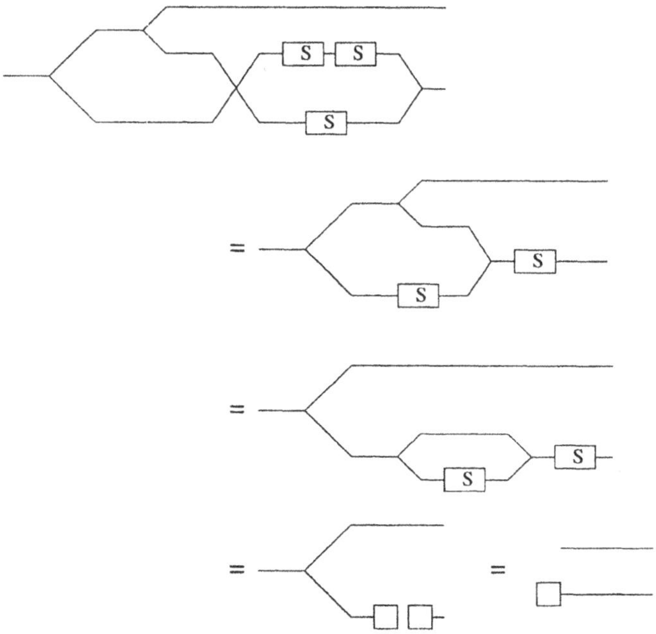

we see that (2.3) is equal to ux.

We now show that u is invertible by exhibiting its inverse: by the first part of the proof,

$$\begin{aligned}u.m((S^{-1}\otimes 1)(R^{21})^{-1}) &= m((S\otimes 1)(R^{21})^{-1}.u\otimes 1)\\ &= m((S\otimes 1)R^{21}(R^{21})^{-1})\\ &= 1.\end{aligned}$$

Let us now prove that $uS(u)$ is a central element. Notice that $S^2(u) = u$. Thus, $S(u) = S^{-1}(u)$. Applying S to (9.8) yields

$$S(u^{-1})S(x)S(u) = S^3(x) \tag{2.4}$$

Replacing x by $S^{-1}(x)$ gives

$$S(u)^{-1}xS(u) = S^2(x) \tag{2.5}$$

Combining (2.4) and (2.5) shows that $uS(u)$ is central.

Problem 9.2: Let us show that $\tilde{S}$ is the antipode of $\tilde{H}$: we have

$$\begin{aligned}m(\tilde{S}\otimes 1)\tilde{\Delta}(x) &= m((\tilde{S}\otimes 1)J^{-1}\Delta(x)J)\\ &= m((Q^{-1}SQ\otimes 1)J^{-1}\Delta(x)J)\\ &= m((Q^{-1}\otimes 1)(S\otimes 1)JJ^{-1}\Delta(x)J)\\ &= m((Q^{-1}\otimes 1)(S\otimes 1)\Delta(x)J)\\ &= m((Q^{-1}\otimes 1)(Q\otimes 1)(S\otimes 1)\Delta(x))\\ &= m(S\otimes 1)\Delta(x)\\ &= i\circ\epsilon(x).\end{aligned}$$

The equality $m(1\otimes\tilde{S})\tilde{\Delta}(x) = i\circ\epsilon(x)$ is proved in the same way. Finally, the counit axioms follows from $(\epsilon\otimes 1)J = (1\otimes\epsilon)J = 1$.

Problem 9.3: Set $\tilde{R} = (J^{21})^{-1}RJ$. From

$$((J^{21})^{-1}RJ)(J^{-1}\Delta J) = (J^{-1}\Delta J)^{op}(J^{21})^{-1}RJ$$

we deduce $\tilde{\Delta}^{op}\tilde{R} = \tilde{R}\tilde{\Delta}$. Suppose that R is a coboundary structure on H, i.e we have $R_{12}(\Delta\otimes 1)R = R_{23}(1\otimes\Delta)R$. Then,

$$\begin{aligned}\tilde{R}_{12}(\tilde{\Delta}\otimes 1)\tilde{R} &= J_{21}^{-1}R_{12}J_{12}J_{12}^{-1}(\Delta\otimes 1)(J_{21}^{-1}RJ)J_{12}\\ &= J_{21}^{-1}R_{12}J_{23,1}^{-1}(\Delta\otimes 1)RJ_{12,3}J_{12}\\ &= J_{21}^{-1}J_{32,1}^{-1}R_{12}(\Delta\otimes 1)RJ_{12,3}J_{12}\\ &= J_{21}^{-1}J_{32,1}^{-1}R_{23}(1\otimes\Delta)RJ_{12,3}J_{12}.\end{aligned}$$

A similar computation shows that

$$\tilde{R}_{23}(1\otimes\tilde{\Delta})\tilde{R} = J_{32}^{-1}J_{3,21}^{-1}R_{23}(1\otimes\Delta)J_{1,23}J_{23}.$$

The relation $J_{12,3}J_{12}J_{23}^{-1}J_{1,23}^{-1} = 1$ shows that these two expressions are equal, i.e that $\tilde{R}$ is a coboundary structure on the twist of H by J. The (quasi)triangular case is treated in a similar fashion.

Problem 9.4: Notice that $(H, 1)$ is a quasitriangular Hopf algebra. Therefore, $(\tilde{H}, (J^{21})^{-1}J)$ is a quasitriangular Hopf algebra. But

$$(J^{21})^{-1}J \equiv 1 + h\frac{r - r^{21}}{2} \ (\mod h^2),$$

so $(\tilde{H}, (J^{21})^{-1}J)$ is a quantization of $(\mathfrak{g}, \frac{r-r^{21}}{2})$. In particular, $\frac{r-r^{21}}{2} \in \mathfrak{g}\otimes\mathfrak{g}$.

Lecture 10

Exercise 10.1 The dual of A in the quantized enveloping algebra sense is generated by F, H with relation $[H, F] = -2F$ and

$$\begin{aligned} \Delta(F) &= F\otimes 1 + q^{-H}\otimes F, & \Delta(H) &= H\otimes 1 + 1\otimes H,\\ S(E) &= q^H F, & S(H) &= -H,\\ \epsilon(F) &= 0, & \epsilon(H) &= 0. \end{aligned}$$

This is a special case of the Drinfeld pairing-see Lecture 12.

Lecture 12

Exercise 12.1: We have

$$\begin{aligned} < e^{hH_i}, e^{hH_j} > &= \sum_{m,n} < \frac{h^n}{n!}H_i^n, \frac{h^m}{m!}H_j^m > \\ &= \sum_{m,n} \frac{h^{n+m}}{n!m!} < H_i^n, H_j^m > \\ &= \sum_{m,n} \frac{h^{n+m}}{n!m!} < \Delta_m(H_i)^n, H_j^{\otimes m} > . \end{aligned}$$

It is clear that $< H_j, H_i^l >= 0$ if $l \neq 1$. Thus the only terms in the above sum which are non-zero are those for which $m = n$. Moreover,

$$\Delta_n(H_i)^n = (\sum_{l=1}^{n} H_i^{(l)})^n.$$

In particular, the coefficient of $H_i^{\otimes n}$ in $(\Delta_n(H_i))^n$ is $n!$. Hence,

$$\begin{aligned}
< e^{hH_i}, e^{hH_j} > &= \sum_n \frac{h^{2n}}{n!^2} n! < H_i^{\otimes n}, H_j^{\otimes n} > \\
&= \sum_n \frac{h^{2n}}{n!} \frac{1}{h^n} < h_i, h_j >^n \\
&= e^{h<h_i,h_j>}.
\end{aligned}$$

Exercise 12.2: This follows from lemma 12.2 and from the fact that the multiplication map

$$U_h(\mathfrak{b}_+) \otimes U_h(\mathfrak{b}_-) \to D(U_h(\mathfrak{b}_+))$$

is an isomorphism of vector spaces.

Problem 12.1: Let E, F, H be the generators of $U_h(sl_2)$ as in Section 9.3. We have $< H, H >= 2$. Thus,

$$R = e^{h\frac{H \otimes H}{2}} \sum_i a_i \otimes a^i$$

where (a_i) is a basis of $U_h(\mathfrak{n}_+)$ and (a^i) is the dual basis of $U_h(\mathfrak{n}_-)$. It is natural to choose $a_i = E^i$ for a basis of $U_h(\mathfrak{n}_+)$. From the definition of the Drinfeld pairing, it is clear that $< E^p, F^q >= 0$ if $p \neq q$. Thus we only need to compute $< E^p, F^p >$ for $p \geq 1$.
For $p = 1$, by definition, we have $< E, F >= \frac{1}{q-q^{-1}}$.
For $p = 2$, we have

$$\begin{aligned}
< E^2, F^2 > &=< E \otimes E, \Delta^{op}(F^2) > \\
&=< E \otimes E, 1 \otimes F^2 + F^2 \otimes q^{-2H} + F \otimes (Fq^{-H} + q^{-H}F) > \\
&=< E, Fq^{-H} + q^{-H}F >< E, F > \\
&=< \Delta(E), F \otimes q^{-H} + q^{-H} \otimes F >< E, F > \\
&= (< E, F >< q^H, q^{-H} > + < 1, q^{-H} >< E, F >) < E, F > \\
&= (1 + q^{-2})(q - q^{-1})^{-2} \\
&= \frac{[2]_q!}{q}(q - q^{-1})^{-2}.
\end{aligned}$$

Let us now compute $< E^p, F^p >$ by induction on p:

$$\begin{aligned}< E^p, F^p > &=< E \otimes E^{p-1}, \Delta^{op}(F^p) > \\ &= < E^{p-1}, (\sum_{i=0}^{p-1} F^i q^{-H} F^{p-1-i}) >< E, F > \\ &=< E^{p-1}, (\sum_{i=0}^{p-1} q^{-2i}) q^{-H} F^{p-1} >< E, F > \\ &= \frac{[p]_q}{q^{p-1}}(q - q^{-1}) < E^{p-1}, F^{p-1} > .\end{aligned}$$

Thus $< E^p, F^p >= (q - q^{-1})^{-p} \frac{[p]_q!}{q^{p(p-1)/2}}$ and formula (9.9) is proved.

Problem 12.2: Let $A = D(H)$ be the quantum double of H. It is a quasitriangular, finite dimensional Hopf algebra, with R-matrix given by the canonical element in $H \otimes H^* \subset A \otimes A$. Define $u = m((S \otimes 1)R^{21})$ as in Problem 9.1. Recall that $S^2 = Ad\ u$ and $uS(u)$ is a central element of A. In particular, $S^4 = Ad\ g$ where $g = uS(u)^{-1}$. We will now show that g is a grouplike element, i.e $\Delta(g) = g \otimes g$.

We first show that

$$\Delta(u) = (u \otimes u)(R^{21}R)^{-1}. \tag{2.6}$$

Let us write $R = \sum_i a_i \otimes b_i$. Thus we have $u = \sum_i S(b_i)a_i$ and $\Delta(u) = \sum_i (S \otimes S)\Delta^{op}(b_i)\Delta(a_i)$. From $R\Delta = \Delta^{op}R$, we have $\Delta RR^{21} = RR^{21}\Delta$, and

$$\Delta(u)R^{21}R = \sum_i (S \otimes S)\Delta^{op}(b_i)R^{21}R\Delta(a_i).$$

Consider the following right action of $A^{\otimes 4}$ on $A^{\otimes 2}$:

$$(x_1 \otimes x_2) \cdot (y_1 \otimes y_2 \otimes y_3 \otimes y_4) = S(y_3)x_1y_1 \otimes S(y_4)x_2y_2.$$

We have

$$\begin{aligned}\Delta(u)R^{21}R &= R^{21} \cdot (R^{12}(\Delta \otimes \Delta^{op})R), \\ &= R^{21} \cdot (R^{12}R^{13}R^{23}R^{24}), \\ &= R^{21} \cdot (R^{23}R^{13}R^{12}R^{24})\end{aligned}$$

where we have used the hexagon relations for R and the Quantum Yang-Baxter equation. Now,

$$\begin{aligned}R^{21} \cdot R^{23} &= \sum_{i,j} S(b_j)b_i \otimes a_i a_j \\ &= (S \otimes 1)(R^{21})^{-1}R^{21} \\ &= 1.\end{aligned}$$

Furthermore, $1 \cdot R^{13} = u \otimes 1$ and

$$(u \otimes 1) \cdot R^{12}R^{14} = \sum_{i,j} ua_i a_j \otimes S(b_j)b_i = u \otimes 1.$$

Finally, $(u \otimes 1) \cdot R^{24} = u \otimes u$. Combining these relations, we obtain (2.6). Applying S to (2.6), we deduce

$$\Delta(S(u)) = (S(u) \otimes S(u))(R^{21}R)^{-1}. \tag{2.7}$$

Combining (2.6) and (2.7), we obtain $\Delta(uS(u)^{-1}) = uS(u)^{-1} \otimes uS(u)^{-1}$ as desired.

To conclude Problem 12.2, notice that since g is a grouplike element in a finite dimensional Hopf algebra, it is of finite order (consider the cocommutative Hopf subalgebra generated by g). In particular, $S^4 = Ad\ g$ is also of finite order.

Note: This result is due to Radford [R]. The above beautiful proof (which is much simpler than Radford's original proof) is due to Drinfeld, [Dr6].

Lecture 13

Exercise 13.1: The fact that the tensor product $V \otimes W$ of two A-modules V and W is again an A-module comes from the relation $\Delta(xy) = \Delta(x)\Delta(y)$. The fact that Φ is a morphism comes from the coassociativity of Δ.

Exercise 13.2: We have to check that $(ab).f = a.(b.f)$ for all $a, b \in A$ and $f \in \mathrm{Hom}\,(X, Y)$. This follows from the fact that Δ is an algebra homomorphism and S an algebra anti-homomorphism. If A is a group algebra $k[G]$ then the action is given by

$$(a.f)(x) = af(a^{-1}x), \qquad a \in G.$$

If A is an enveloping algebra $U(\mathfrak{g})$, then the action is given by

$$(u.f)(x) = u.f(x) - f(u.x), \qquad u \in \mathfrak{g}.$$

Lecture 14

Exercise 14.1: Let $x = (z_1, \ldots z_n) \in X$. A loop with base point x can be represented in $\mathbb{C}\times\mathbb{R}$ as a braid in n strands connecting $y_i = (z_i, 1)$ to $w_i = (z_i, 0)$ for $i = 1, \ldots n$ (a pure braid). Two such loops are homotopic if and only if the corresponding braids are isotopic. Thus $\pi_1(X) = PB_n$.

Problem 14.1: This is a consequence of the following relations:

$$R_{12}(\Delta\otimes 1)R = R_{12}\Phi_{312}R_{13}\Phi_{132}^{-1}R_{23}\Phi_{123}$$

and

$$\begin{aligned}R_{12}(\Delta\otimes 1)R &= (\Delta^{op}\otimes 1)RR_{12}\\ &= \Phi_{321}R_{23}\Phi_{231}^{-1}R_{13}\Phi_{213}R_{12}.\end{aligned}$$

Alternatively, one can use the picture proof of Proposition 9.3, keeping track of the associators.

Lecture 15

Problem 15.1: By definition, Φ_{KZ}^{321} is the monodromy from 0 to 1 of the system

$$\begin{aligned}\frac{1}{\hbar}\frac{\partial u}{\partial z} &= (\frac{\Omega_{32}}{z}+\frac{\Omega_{21}}{z-1}).u\\ &= (\frac{\Omega_{12}}{z-1}+\frac{\Omega_{23}}{z}).u \qquad (KZ_3'')\end{aligned}$$

We recover the usual (KZ_3') system by the transformation $z\mapsto 1-z$. In particular, the monodromy of (KZ_3'') from 0 to 1 is the monodromy of (KZ_3') from 1 to 0, i.e $\Phi_{KZ}^{321}=\Phi_{KZ}^{-1}$.

Lecture 16

Exercise 16.1: Let $(U\mathfrak{g},\Delta_0,\epsilon_0,1)$ be the classical limit of (H,Δ,ϵ,Φ). By definition of a quasi-Hopf QUE algebra, $\Phi\equiv 1 \pmod h$. If R is a quasitriangular structure then in particular R is invertible. This implies that $R \pmod h \in U(\mathfrak{g})\otimes U(\mathfrak{g})$ is invertible, i.e $R\equiv c \pmod h$, with $c\in\mathbb{C}^*$. The relation $(\epsilon\otimes 1)R=1$ shows that $c=1$.

Problem 16.1: Set $\overline{R}=R(R^{21}R)^{-1/2}$ (notice that square roots make sense as $R\equiv 1 \pmod h$). Then $\overline{R}$ is invertible, $(\epsilon\otimes 1)\overline{R}=(1\otimes\epsilon)\overline{R}=1$, so $\overline{R}\in\Gamma_H$. We have

$$\begin{aligned}\overline{R}^{21}\overline{R} &= R^{21}(RR^{21})^{-1/2}R(R^{21}R)^{-1/2}\\ &= R^{21}(RR^{21})^{-1}R\\ &= 1\end{aligned}$$

(for this reason, $\overline{R}$ is called the unitarization of R). Now let $J\in\Gamma_H$. Then $J_{21}^{-1}RJ$ is symmetric if and only if $J_{21}^{-1}RJ=J^{-1}R_{21}J_{21}$, which can be rewritten as

$$\begin{aligned}(J_{21}^{-1}RJ)^2 &= (J^{-1}R_{21}J_{21})(J_{21}^{-1}RJ)\\ &= J^{-1}R_{21}RJ.\end{aligned}$$

This is equivalent to the condition

$$J_{21}J^{-1} = R(R^{21}R)^{-1/2}.$$

It is easy to check that $J = (\overline{R})^{-1/2}$ satisfies this condition. This gives a canonical way of twisting R into a symmetric R-matrix.

Exercise 16.2: If $J \in \Gamma_A$ then in particular J is invertible and $(\epsilon \otimes 1)J = 1$. Thus $J \equiv 1 \pmod h$.

Exercise 16.3: Let $(H, \Delta, \epsilon, \Phi, R)$ be a quasitriangular quasi-Hopf QUE algebra and let $J \in \Gamma_H$. Then by the (quasi)-version of Problem 9.4, $J_{21}^{-1}RJ$ is a quasitriangular structure for the twist of H by J. If $(H, \Delta, \epsilon, \Phi, R)$ is triangular, (i.e if $RR^{21} = 1$) then

$$(J_{21}^{-1}RJ)(J_{21}^{-1}RJ)^{21} = 1$$

and the twist of H by J is also triangular.

Exercise 16.4: We have

$$\begin{aligned}\frac{1}{h}((J_{21}^{-1}RJ)(J_{21}^{-1}RJ)^{21} - 1) &= J_{21}^{-1}(\frac{RR_{21}-1}{h})J_{21}\\ &\equiv \frac{RR_{21}-1}{h} \pmod h\end{aligned}$$

since $J \equiv 1 \pmod h$. Thus Ω remains unchanged under twisting. In particular, after twisting R into a symmetric form (see Problem 16.1), we obtain the expansion $R = 1 + \frac{h\Omega}{2} + \mathcal{O}(h^2)$.

Problem 16.2: Let $R = 1 + h\frac{\Omega}{2} + \mathcal{O}(h^2)$. Then

$$\begin{aligned}\delta(x) &= [x \otimes 1 + 1 \otimes x, \frac{\Omega}{2}]\\ &= 0\end{aligned}$$

by $\mathfrak{g}$-invariance of Ω. Now let $\Phi \equiv 1 + ch^2 \pmod{h^3}$. Expanding the (quasi-Hopf) quantum Yang-Baxter equation up to order 3 gives

$$\begin{aligned}1 &+ \frac{1}{4}(\Omega_{12}\Omega_{13} + \Omega_{12}\Omega_{23} + \Omega_{13}\Omega_{23}) + c_{312} - c_{132} + c_{123}\\ &= 1 + \frac{1}{4}(\Omega_{13}\Omega_{12} + \Omega_{23}\Omega_{12} + \Omega_{23}\Omega_{13}) + c_{321} - c_{231} + c_{213}.\end{aligned}$$

Thus,

$$\varphi = \frac{\mathrm{Alt}\Phi}{h^2} \pmod{h^3} = \frac{1}{4}CYB(\Omega) = \frac{1}{4}[\Omega_{12}, \Omega_{23}]$$

where the last equality follows from $\mathfrak{g}$-invariance of Ω.

Exercise 16.5: This follows from the fact that Φ is invertible and $(1 \otimes \epsilon \otimes 1)\Phi = 1$.

Problem 16.3: It is clear that the group Hom $(\mathfrak{g}_-, \mathfrak{g}_+)$ acts freely and transitively on the set of all complements V of $\mathfrak{g}_+$ in $\mathfrak{g}$ by

$$\rho \times V \mapsto \begin{pmatrix} 1 & \rho \\ 0 & 1 \end{pmatrix} V.$$

Furthermore, if V is Lagrangian and $v, v' \in V$ then

$$< \rho(v) + v, \rho(v') + v' >=< \rho(v), v' > + < \rho(v'), v > .$$

Thus $\rho.V$ is Lagrangian if and only if ρ is skew-symmetric.

Lecture 17

Problem 17.1: Let us write $\Phi \equiv 1 + c_1 a + c_2 b \pmod{I^2}$. Then, modulo I^2, the first hexagon axiom in $T_3[[h]]$ reads

$$1 + ht_{12,3} = 1 + c_1 ht_{23} + c_2 ht_{31} + ht_{13} - c_1 ht_{13} - c_2 ht_{32} + ht_{23} + c_1 ht_{12} + c_2 ht_{23}.$$

Equating the coefficients of t_{13} gives $c_1 = c_2$. Equating the coefficients of t_{12} gives $c_1 = 0$. Thus $\Phi \equiv 1 \pmod{I^2}$.

Exercise 17.1: Let $(0, f) \in GT_0$ and let $(\mathcal{C}, \beta, \Phi)$ be a braided (quasi-symmetric) monoidal category. By definition, the action of $(0, f)$ on $\mathcal{C}$ transforms β into

$$\beta'_{V_1,V_2} = \beta_{V_1,V_2}(\beta_{V_2,V_1}\beta_{V_1,V_2})^{-\frac{1}{2}}.$$

In particular,

$$\begin{aligned} \beta'_{V_1,V_2}\beta'_{V_2,V_1} &= \beta_{V_1,V_2}(\beta_{V_2,V_1}\beta_{V_1,V_2})^{-\frac{1}{2}}\beta_{V_2,V_1}(\beta_{V_1,V_2}\beta_{V_2,V_1})^{-\frac{1}{2}} \\ &= \beta_{V_1,V_2}(\beta_{V_1,V_2}^{-1}\beta_{V_2,V_1}^{-1})^{+\frac{1}{2}}\beta_{V_2,V_1}(\beta_{V_1,V_2}\beta_{V_2,V_1})^{-\frac{1}{2}} \\ &= 1. \end{aligned}$$

and $(0, f).\mathcal{C}$ is a symmetric monoidal category.

Lecture 19

Exercise 19.1: The counit maps $\epsilon_\pm : M_\pm \to k$ are defined as

$$\epsilon_\pm(u_\mp 1_\pm) = \epsilon(u_\mp), \qquad u_\mp \in U\mathfrak{g}_\mp.$$

In particular, they are morphisms in $\mathcal{M}_\mathfrak{g}$. The relation $(\epsilon_\pm \otimes 1)i_\pm = (1 \otimes \epsilon_\pm)i_\pm = 1$ follows from the usual relation $(\epsilon \otimes 1)\Delta = (1 \otimes \epsilon)\Delta = 1$ by identifying $M_\pm$ with $U\mathfrak{g}_\mp$.

Exercise 19.2: Let us prove the first hexagon relation

$$(\Delta \otimes 1)R = R^{13}R^{23}.$$

The proof of the second hexagon relation $(1 \otimes \Delta)R = R_{13}R_{12}$ is similar. Let $U, V, W \in \mathcal{M}_{\mathfrak{g}}$, $u \in F(U), v \in F(V), w \in F(W)$. By definition, $R(v \otimes w) = \sigma J_{W,V}^{-1} F(\beta) J_{V,W}$. Hence,

$$\begin{aligned}(\Delta \otimes 1)R(v \otimes w \otimes u) &= (J_{V,W}^{-1} \otimes 1)R(J_{V,W} \otimes 1)(v \otimes w \otimes u)\\ &= \sigma_{12,3}(1 \otimes J_{VW}^{-1})J_{U,V\otimes W}^{-1}F(\beta_{V\otimes W,U})J_{V\otimes W,U}(J_{V,W} \otimes 1)(v \otimes w \otimes u).\end{aligned}$$

By functoriality of the braiding, we have $\beta_{V\otimes W,U} = (\beta_{V,U} \otimes 1)(1 \otimes \beta_{W,U})$ and

$$\begin{aligned}J_{U\otimes V,W}^{-1}(F(\beta_{V,U}) \otimes 1)J_{V\otimes U,W} &= F(\beta_{V,U}) \otimes 1\\ J_{V,U\otimes W}^{-1}(1 \otimes F(\beta_{W,U}))J_{V,W\otimes U} &= 1 \otimes F(\beta_{W,U}).\end{aligned}$$

Thus,

$$\begin{aligned}(\Delta \otimes 1)R(v \otimes w \otimes u) &= \sigma_{13}J_{U,V}^{-1}F(\beta_{V,U})J_{V,U}\sigma_{23}J_{U,W}^{-1}F(\beta_{W,U})J_{W,U}\\ &= R^{13}R^{23}.\end{aligned}$$

Exercise 19.3: Let $x, y \in F(M_-)$. Then $i(x) \circ i(y)$ is the composition

$$\begin{aligned}&M_+ \otimes M_- \xrightarrow{i_+\otimes 1} M_+ \otimes M_+ \otimes M_- \xrightarrow{i_+\otimes 1\otimes 1} M_+ \otimes M_+ \otimes M_+ \otimes M_-\\ &\overset{1\otimes 1\otimes y}{M} {}_+ \otimes M_+ \otimes M_- \xrightarrow{1\otimes x} M_+ \otimes M_-.\end{aligned}$$

This is $i(m(x,y))$ where $m(x,y)$ is the composition

$$M_+ \otimes M_- \xrightarrow{i_+\otimes 1} M_+ \otimes M_+ \otimes M_- \xrightarrow{1\otimes y} M_+ \otimes M_- \xrightarrow{x} M_-.$$

Exercise 19.5: If r is unitary, then there exists a triangular Lie bialgebra $(\mathfrak{g}, \overline{r})$ and a Lie algebra map $\pi : \mathfrak{g} \to A$ such that $\pi \otimes \pi)\overline{r} = r$. The quantization $(U_h^{qt}(\mathfrak{g}), \overline{R})$ of $(\mathfrak{g}, \overline{r})$ constructed in theorem 19.3 is also triangular (c.f remark at the end of section 19.3). In particular $R = (\pi \otimes \pi)\overline{R}$ is a triangular (unitary) quantization of r.

Lecture 21

Exercise 21.1: Let $x, y \in F(M_-)$ and $v \in F(V)$. Then $i(x) \circ i(y)(v)$ is represented by

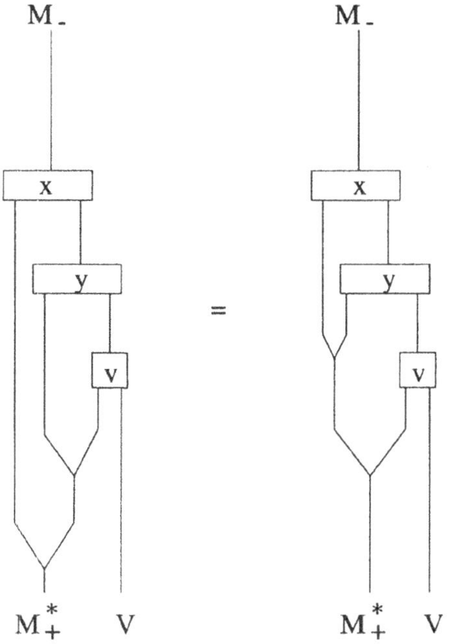

Thus, $i(x) \circ i(y) = i((i_+^* \otimes 1) \circ (1 \otimes y) \circ x)$.

Exercise 21.2: Let $(\mathfrak{g}_+, [\,,\,], \delta)$ be a finite dimensional Lie bialgebra. Let V be a dimodule on $\mathfrak{g}_+$ with action

$$\pi : \mathfrak{g}_+ \otimes V \to V,$$

and coaction

$$\pi^* : V \to \mathfrak{g}_+ \otimes V.$$

Define a map $P : D(\mathfrak{g}_+) \otimes V = (\mathfrak{g}_+ \oplus \mathfrak{g}_+^*) \otimes V \to V$ by

$$\begin{aligned} P(x)(v) &= \pi(x)(v), && x \in \mathfrak{g}_+, \\ P(x)(v) &= (x \otimes 1)(\pi^*(v)), && x \in \mathfrak{g}_+^*. \end{aligned}$$

Then, for $x \in \mathfrak{g}_+, y \in \mathfrak{g}_+^*$, we have

$$\begin{aligned} (P(x)P(y) - P(y)P(x))(v) &= (y \otimes 1)((1 \otimes \pi(x))\pi^*(v) - \pi^*(\pi(x)(v)) \\ &= (y \otimes 1)([x^1, \pi^*(v)] - (1 \otimes \pi)\delta(x)(v)) \\ &= P(ad^*(y)(x) - ad^*(x)(y))(v). \end{aligned}$$

Thus P defines an action of $D(\mathfrak{g}_+)$ over V.

Lecture 22

Exercise 22.1: Let V be a left dimodule over a finite dimensional Hopf algebra A_+ with action $\pi : A_+ \otimes V \to V$ and coaction $\pi^* : V \to A_+ \otimes V$. Define a map

$$P : D(A_+) \otimes V = (A_+ \otimes A_+^*) \otimes V \to V$$

by

$$\begin{aligned} P(a \otimes 1)(v) &= \pi(a_+)(v), \qquad & a \in A_+, \\ P(1 \otimes b)(v) &= (b \otimes 1)\pi^*(v), \qquad & b \in A_+^*, \\ P(a \otimes b)(v) &= P(a)P(b)(v). \end{aligned}$$

It follows from the consistency relation and (12.12) that P defines an action of $D(A_+)$ on V.

Index

Bibliography

[A] Amitsur S., *Commutative linear differential operators*, Pacific Journal of Math. **8** (1958) 1-11.

[Ar] Artin M., *On the solution of analytic equations*, Inv. Math **5** (1968) 277-291.

[BB] Balinsky A.A, Burman Yu.M, *Quadractic Poisson brackets and Drinfeld theory for associative algebras*, Lett. Math. Phys. **38** (1996) no. 1 63-75.

[Be] Beck J., *Braid group action and quantum affine algebras*, Comm. Math. Phys. (1995).

[BD1] Belavin A.A, Drinfeld V.G, *Triangle equation and simple Lie algebras*, Soviet Sci. reviews, Sect C **4** 93-165 (1984).

[BD2] Belavin A.A, Drinfeld V.G, *Solutions of the classical Yang-Baxter equation and simple Lie algebras*, Funktz. Analyz i Ego Prilozh. **16** (1982) n.3 1-29.

[Bel] Belavin A.A, *Dynamical symmetry of integrable quantum systems*, Nucl. Phys. **B180(FS2)** (1981), no.2, 189-200.

[Bo] Borel A. et al, *Algebraic D-modules* Perspectives in Math. Acad. Press (1987).

[Bou] Bourbaki N. *Groupes et algèbres de Lie* Masson (1981).

[BEG] Braverman A., Etingof P., Gaitsgory D., *Quantum integrable systems and differentiable Galois theory*, Transformation Groups **1** (1997), 31-56.

[Br] Broadhurst D.J, *Conjectured enumeration of irreducible multiple Zeta valies, from knot and Feynman diagrams*, hep-th/9612012.

[BrKr] Broadhurst D.J, Kreimer D., *Association of multiple zeta values with positive knots via Feynman diagrams up to 9 loops*, Phys. Lett. B393 (1997) 403-412.

[CP] Chari V., Pressley A., *A guide to quantum groups*, Cambridge Univ. Press (1994).

[Che1] Cherednik I.V, *On properties of factorized S-matrices in terms of elliptic functions*, Yad. Fiz **36** (1982), no.2, 549-557.

[Che2] Cherednik I.V, *Generalized braid groups and local r-matrix systems*, Soviet Math. Doklady **307** (1990), 43-47.

[CG] Chriss N., Ginzburg V., *Representation theory and complex geometry*, Birkhauser (1997)

[CL] Coddington E.A, Levinson N., *Theory of ordinary differential equations* McGraw-Hill, New York, Toronto, London, 1955.

[Del] Deligne P., *Le groupe fondamental de la droite projective moins trois points*, Galois groups on $\mathbb{Q}$., Math. Sci. Res. Inst. Publ. vol **16**, Springer-Verlag (1989) 79-297.

[DM] Deligne P., Milne R., *Tannakian categories*, Lecture Notes in Math. vol **900** Springer-Verlag, 1982, p. 101-228.

[DeLe] De Wilde M., Lecomte P.,*Existence of star-products and of formal deformations of the Poisson Lie algebra of arbitrary sympletic manifolds* Lett. Math. Phys. **7** (1983), no. 6, 487-496.

[DF] Ding J., Frenkel I., *Isomorphism of two realizations of quantum affine algebra* $U_q(\hat{gl}_n)$, Comm. Math. Phys. **156** (1993), 277-300.

[Dr1] Drinfeld V.G, *Quantum groups*, Proceedings ICM (Berkeley 1986) **1** (1987), AMS 798-820.

[Dr2] Drinfeld V.G, *Quasi-Hopf algebras*, Leningrad Math. J. **1** (1990) 1419-1457.

[Dr3] Drinfeld V.G, *On some unsolved problems in quantum groups theory*, Lect. Notes Math. 1510 (1992), 1-8.

[Dr4] Drinfeld V.G, *On quasitriangular quasiHopf algebras and a group closely rlated to* $Gal(\bar{\mathbb{Q}}/\mathbb{Q})$ Leningrad. Math. J. **2** (1991) n. **4** 829-860.

[Dr5] Drinfeld V.G, *A new realization of Yangians and quantum affine algebras*, Sov. Math. Dokl. **36**. (1988), 212-216.

[Dr6] Drinfeld V.G, *On almost-cocommutative Hopf algebras*, Leningrad Math. J. **1** 321-42.

[Dr7] Drinfeld V.G, *On Poisson homogeneous spaces of Poisson-Lie groups*, Theoret. and Math. Phys. **95** (1993), no. 2, 524–525.

[E] Etingof P., *Representations of affine Lie algebras, elliptic r-matrix systems and special functions* Comm. Math. Phys. **159** 471-502 (1994).

[EFK] Etingof P.,Frenkel I., Kirillov A. Jr *Lectures on the KZ equations*, Mathematical Surveys and Monographs, **58**. American Mathematical Society, Providence, RI, (1998).

[EK1] Etingof P., Kazhdan D. *Quantization of Lie bialgebras, I*, Selecta Math. **2** (1996) n. **1** 1-41.

[EK2] Etingof P., Kazhdan D. *Quantization of Poisson algebraic groups and Poisson homogeneous spaces*, Symétries quantiques (Les Houches, 1995), 935-946, North-Holland, Amsterdam, (1998).

[EK3] Etingof P., Kazhdan D. *Quantization of Lie bialgebras, II*, Selecta Math. (N.S.) **4** (1998), no. 2, 213-231.

[EK4] Etingof P., Kazhdan D. *Quantization of Lie bialgebras, III*, Selecta Math. (N.S.) **4** (1998), no. 2, 233–269.

[FRT] Fadeev L., Reshetikhin N., Takthadjan L. *Quantization of Lie groups and Lie algebras* Algebraic Analysis Vol **1** 129-139 Acad. Press (1988).

[Fed] Fedosov, B V, *A simple geometrical construction of deformation quantization* J. Differential Geom. **40** (1994), no. 2, 213–238.

[GK] Gabber, O., Kac, V.,*On defining relations of certain infinite dimensional Lie algebras* Bull. AMS **5**, 1981, p. 185-189.

[G] Gerstenhaber M., *On the deformation of rings and algebras*, Ann. of Math. (2) 79 1964 59–103.

[GG] Goto M., Grosshans F., *Semisimple Lie algebras*, Marcel Dekker, INC, New York and Basel, 1978.

[GuSt] Guillemin, V., and Sternberg, S., Symplectic techniques in physics, Cambridge University Press, Cambridge, 1990.

[Gro] Grothendieck A. *Esquisse d'un programme*, Université de Montpellier (1984).

[J] Jantzen J., *Lectures on quantum groups*, Grad. studies in Math. Vol **6**, AMS (1996).

[Jos] Joseph, A. *Quantum groups and their primitive ideals*, Springer-Verlag **29** (1995).

[Kac] Kac V.G, *Infinite dimensional Lie algebras*, Cambridge Univ. Press (1990).

[Kass] Kassel C. *Quantum groups*, Springer-Verlag GTM 155 (1994).

[Ko] Kontsevich M., *Deformation quantization of Poisson manifolds*, q-alg/9709040.

[KS] Kupershmidt B., Stoyanov, *Classification of all Poisson-Lie structures on an infinite-dimensional jet group* Lett. Math. Phys. **37** (1996), no. 1, 1–9.

[La] Lawvere, F-W *Functorial semantics of algebraic theories*, Proc. Nat. Acad. Sci. USA (**50**) (1963), 869-872.

[LS1] Levendorskii S, Soibelman Y., *Some applications of the quantum Weyl groups* J. Geom. Phys. **7** (1990), no. 2, 241–254.

[LS2] Levendorskii S, Soibelman Y., *The quantum Weyl group and a multiplicative formula for the R-matrix of a simple Lie algebra* (Russian) Funktsional. Anal. i Prilozhen **25** (1991), no. 2, 73–76; translation in Funct. Anal. Appl. **25** (1991), no. 2, 143–145.

[Lu] Lusztig G., *Introduction to quantum groups*, Progress in Math. Birkhauser Boston (1993).

[Mac] MacLane S., *Categories for the working mathematician*, Springer-Verlag 1971.

[MOY] Maeda Y., Omori H., Yoshioka A., *Weyl manifolds and deformation quantization* Adv. Math. **85** (1991), no. 2, 224–255.

[Maj] Majid S, *Foundations of quantum group theory* Cambridge University Press (1995)

[ML] Makar-Limanov, L. G. Commutativity of certain centralizers in the rings $R_{n,k}$. (Russian) Funkcional. Anal. i Priložen. 4 (1970), no. 4, 78.

[Ma] Mathieu O., *Homologies associated with Poisson structures* Deformation theory and symplectic geometry (Ascona, 1996), 177–199, Math. Phys. Stud. **20**, Kluwer Acad. Publ., Dordrecht, (1997).

[Mum] Mumford D., *Abelian varieties*, Tat Studies in Math., Oxford Univ. Press, 2nd ed. (1975).

[R] Radford D., *The order of the antipode of a finite dimensional Hopf algebra is finite* Amer. J. Math. **98** (1976), no. 2, 333–355.

[RS] Reshetikhin, N., Semenov-Tian-Shansky, M., *Central extensions of quantum current groups*. Lett. Math. Phys. **19** (1990), no. 2, 133-142.

[ShSt] Shnider S., Sternberg S. *Quantum groups: from coalgebras to Drinfeld algebras*, International Press, 1993.

[Sem] Semenov-Tian-Shanky M., *What is a classical r-matrix ?*, Funct. Anal. Appl. **17** (1983) 259-72.

[So] Soibelman Y., *Quantum Weyl groups and some of its applications*, Rend. Circ. Mat. Palermo. Suppl **20**(2) (1991) 233-5.

[Sto] Stolin *On rational solutions of the Classical Yang-Baxter equation for* $sl(n)$, Math. Scand. **69** 57-80.

[Var] Varchenko A., *Multidimensional hypergeometric functions and representation theory of Lie algebras and quantum groups* Advanced Series in Mathematical Physics, 21. World Scientific Publishing Co., Inc., River Edge, NJ, 1995.

Made in the USA
Middletown, DE
05 November 2026

36599691R00144